Lecture Notes in Mathematics

Edited by A. Dold and B. Eckmann

839

Cabal Seminar 77 – 79

Proceedings, Caltech-UCLA Logic Seminar 1977 – 79

Edited by
A. S. Kechris, D. A. Martin, and Y. N. Moschovakis

Springer-Verlag
Berlin Heidelberg New York 1981

Editors

Alexander S. Kechris
Department of Mathematics
California Institute of Technology
Pasadena, CA 91125/USA

Donald A. Martin
Yiannis N. Moschovakis
Department of Mathematics
University of California
Los Angeles, CA 90024/USA

AMS Subject Classifications (1980): 03DXX, 03EXX, 04-XX

ISBN 3-540-10288-4 Springer-Verlag Berlin Heidelberg New York
ISBN 0-387-10288-4 Springer-Verlag New York Heidelberg Berlin

Printing and binding: Beltz Offsetdruck, Hemsbach/Bergstr.
2141/3140-543210

ΔΙΣ ΕΞΑΜΑΡΤΕΙΝ ΤΑΥΤΟΝ ΟΥΚ ΑΝΔΡΟΣ ΣΟΦΟΥ

- ΜΕΝΑΝΔΡΟΣ

INTRODUCTION

This is the second volume of the proceedings of the Caltech-UCLA Logic Seminar, based essentially on material which was presented and discussed in the period 1977-1979.

Los Angeles
July 1980

Alexander S. Kechris
Donald A. Martin
Yiannis N. Moschovakis

TABLE OF CONTENTS

1. CAPACITIES AND ANALYTIC SETS, C. Dellacherie 1

2. HOMOGENEOUS TREES AND PROJECTIVE SCALES, A. S. Kechris 33

3. THE AXIOM OF DETERMINACY, STRONG PARTITION PROPERTIES AND NON-SINGULAR MEASURES, A. S. Kechris, E. M. Kleinberg, Y. N. Moschovakis and W. H. Woodin 75

4. THE AXIOM OF DETERMINACY AND THE PREWELLORDERING PROPERTY, A. S. Kechris, R. M. Solovay and J. R. Steel 101

5. SOUSLIN CARDINALS, κ-SOUSLIN SETS AND THE SCALE PROPERTY IN THE HYPERPROJECTIVE HIERARCHY, A. S. Kechris 127

6. CLOSURE PROPERTIES OF POINTCLASSES, J. R. Steel 147

7. A NOTE ON WADGE DEGREES, A. S. Kechris 165

8. ORDINAL GAMES AND PLAYFUL MODELS, Y. N. Moschovakis 169

9. MEASURABLE CARDINALS IN PLAYFUL MODELS, H. S. Becker and Y. N. Moschovakis 203

10. Π^1_2 MONOTONE INDUCTIVE DEFINITIONS, D. A. Martin 215

11. TREES AND DEGREES, P. Odifreddi 235

APPENDIX. PROGRESS REPORT ON THE VICTORIA DELFINO PROBLEMS 273

CAPACITIES AND ANALYTIC SETS

Claude Dellacherie
CNRS, Departément de Mathématiques
Universite de Rouen
France

Introduction

In these notes I have tried to convince the logician reader that the notion of capacity, far from being merely a "refinement for analyst" of a measure (several examples more and more deviating from the expected behavior of a measure are given in Chapter I), lies in the heart of the theory of analytic sets. Therefore I chose in Chapter II to define the analytic sets (in a more general setting than the usual Polish frame) with the help of capacities, as I did in Dellacherie [1978]: that allows, for example, to get the separation theorem as a natural consequence of the separation of two disjoint compact sets by open sets. On the way is defined a large class of monotone set operations closely related to capacities, stable under composition and preserving the property to be analytic. In Chapter III, going back into the Polish frame, I consider under the name of caliber a natural and useful extension of the notion of capacity (there is a kind of analogy between the couple compact/analytic and the couple capacity/caliber) and under the name of analytic operation the related notion of monotone set operation. Finally, in Chapter IV, I give an extension to capacities of Suslin's theorem on uncountable analytic sets, improving a result of Dellacherie [1972]. Applications to Hausdorff measure theory are succinctly given in almost all chapters. For a more comprehensive treatment (and also for substantial bibliographies) the reader may consult my already quoted former works.

I am very happy to thank here Martin and Moschovakis for their hospitality (I hope they will not forget I am still the best in ping-pong). It is also a great pleasure, and a duty, to say my indebtedness to Miss Sarbadhikari who has written carefully all of these notes from a confused version in frenglish jargon.

Chapter 1

Introduction to the notion of capacity

§1. Definition and basic properties. We use $\mathbb{R}_+$ and $\overline{\mathbb{R}}_+$ to mean the nonnegative real numbers and the extended nonnegative real numbers, respectively. For any set E, $\mathcal{P}(E)$ denotes the power set of E. $\mathcal{K}$, $\mathcal{K}_\sigma$ denotes the class of compact and σ-compact subsets of Hausdorff spaces respectively.

Definition. A capacity I on a Hausdorff space E is a function on $\mathcal{P}(E)$ into $\overline{\mathbb{R}}_+$ such that

(1) $A \subseteq B$ implies $I(A) \leq I(B)$ i.e. I is non-decreasing

(2) If $A_n \uparrow A$ then $I(A_n) \uparrow I(A)$ i.e. if $A_1 \subseteq A_2 \subseteq \cdots$ and $\bigcup_n A_n = A$, $I(A_1) \leq I(A_2) \leq \cdots$ and $\sup_n I(A_n) = I(A)$.

We express this by "I is going up."

(3) For compact K, $I(K)$ is finite and if $I(K) < t$, then there exists an open $U \supseteq K$ such that $I(U) < t$, i.e. $I(K) = \inf_{\substack{U \supseteq K \\ U \text{ open}}} I(U)$.

We express this by saying "I is right continuous over the compacts."

Remark. If E is compact metrisable, (3) can be replaced by

(3') For compact K, $I(K)$ is finite and if $K_1 \supseteq K_2 \supseteq \cdots$ are compact sets with $\bigcap_n K_n = K$, then $I(K_1) \geq I(K_2) \geq \cdots$ and $\inf_n I(K_n) = I(K)$, i.e. I is going down on compacts. In symbols, $K_n \downarrow K \Rightarrow I(K_n) \downarrow I(K)$.

In general, however, (1), (2), (3) $\Rightarrow$ (1), (2), (3') but not conversely. For example, let $E = N^N$, with the product of discrete topologies, and let $I(A) = 0$ if A is contained in some $\mathcal{K}_\sigma$-subset of E and $I(A) = 1$ otherwise. I satisfies (1), (2), (3') but not (3).

In the rest of this chapter, we take E, F to be Polish spaces.

Examples of capacities

1.1 Let m be a nonnegative, σ-additive measure on the Borel σ-field $\mathcal{B}(F)$ of a Polish space F. Extend m to an exterior measure on $\mathcal{P}(F)$ so that for any $A \subseteq F$, $m(A) = \inf\{m(B) : B \supseteq A, B \in \mathcal{B}(F)\}$. Then m is a capacity.

1.2 Suppose J is a capacity on F and f a continuous map from E into F. Define I on $\mathcal{P}(E)$ by $I(A) = J(f(A))$. Then I is a capacity.

In particular, we can take J to be an exterior measure on $\mathcal{P}(F)$. The induced capacity I on $\mathcal{P}(E)$ is not, in general, an exterior measure. However, if f is injective, I is also an exterior measure.

Definition. Let I be a capacity on E. A subset A of E is called I-capacitable if $I(A) = \sup\{I(K) : K \subseteq A, K \in \mathcal{K}\}$. A is called universally capacitable if it is I-capacitable for all capacities I on E.

Remark. Using example 1.2, it is easy to see that if $A \subseteq E$ is universally capacitable and $f : E \to F$ a continuous map, then $f(A)$ is universally capacitable. As a matter of fact, this is almost the only known stability property of the class of universally capacitable sets.

We now prove the main theorem of this chapter. It is a somewhat stronger version of Choquet's capacitability theorem. This version is due to Sion.

Theorem 1.1 (Choquet [1955], [1959]; Sion [1963]). Any Borel subset of a Polish space is universally capacitable. More generally, any analytic subset of a Polish space (i.e. a continuous image of a Borel subset of a Polish space) is universally capacitable.

Proof. Note that the analytic subsets of Polish spaces are just the continuous images of N^N with the product of discrete topologies (see Kuratowski [1958], §35, I). Hence in view of the above remark, it is enough to prove that N^N is universally capacitable.

Let I be a capacity on N^N with $I(N^N) > t$. We will show that there exists a compact set $K \subseteq N^N$ such that $I(K) \geq t$.

Put $A_n = \{1,2,\ldots,n\} \times N^N$. Then $A_n \uparrow N^N$. Hence for some n_1, $I(A_{n_1}) > t$. Repeating this argument, we can find natural numbers $n_1, n_2, \ldots, n_k, \ldots$ such that $I(\{1,\ldots,n_1\} \times \cdots \times \{1,\ldots,n_k\} \times N^N) > t$ for all k. Put $K = \prod_k \{1,\ldots,n_k\}$. As I is right continuous on compacts, $I(K) \geq t$.

Remark 1. We shall see later that there exist coanalytic sets (i.e. complements of analytic sets in Polish spaces) which are not universally capacitable. On the other hand, assuming the continuum hypothesis, it is possible to construct a nonanalytic (even nonprojective) universally capacitable set. This construction is due to Martin and will be given in an appendix to this chapter.

Remark 2. The same proof shows that any subset of a Hausdorff space which is a continuous image of N^N (i.e. any Souslin space in the sense of Bourbaki [1975]) is universally capacitable. In the next chapter, we will prove the following more general result: Any $\mathcal{K}$-analytic set is universally capacitable.

Definition. Two capacities on a set E are called equivalent if they agree on compact sets (equivalently on universally capacitable sets or on Borel sets or on analytic sets).

The next theorem implies the first principle of separation for analytic sets, as we shall see later.

Theorem 1.2. Let I be a capacity on E. There exists a greatest capacity J on E equivalent to I. J is given by $J(A) = \inf\{I(B) : B \supseteq A, B \text{ Borel}\}$.

Proof. Since I and J agree on Borel sets, it is enough to check that J is a capacity.

Clearly, J satisfies (1) and (3). To verify (2), let $A_n \uparrow A$ and let $B_n \supseteq A_n$ be Borel sets such that $I(B_n) = J(A_n)$. Without loss of generality, we can take $B_n = \bigcap_{m \geq n} B_m$ so that $B_n \uparrow B$ (say). Then $I(B) \geq J(A) \geq J(A_n) = I(B_n)$ for all n. But $I(B_n) \uparrow I(B)$, hence $J(A_n) \uparrow J(A)$.

Corollary. If A is a universally capacitable subset of E, then $\sup_{\substack{K \in \mathcal{K} \\ K \subseteq A}} I(K) = I(A) = \inf_{\substack{B \supseteq A \\ B \text{ Borel}}} I(B)$. In particular, there exist Borel sets B_1 and B_2 such that $B_1 \subseteq A \subseteq B_2$ and $I(B_1) = I(A) = I(B_2)$.

Note however that $I(B_2 - B_1) \neq 0$ in general.

§2. Further examples of capacities. In this section, we give several examples of capacities together with some of their properties. E, F are taken to be compact metrisable spaces.

Example 2.1. Let m be an exterior measure on E. Define I on $\mathcal{P}(E \times F)$ by $I(A) = m[\Pi_E(A)]$, where Π_E denotes projection to E. I is a capacity. The I-capacitable sets are characterized by the following:

Theorem 2.1. A is I-capacitable iff for all $\varepsilon > 0$, there exist a Borel set $B_\varepsilon \subseteq \Pi_E(A)$ and a Borel map $f_\varepsilon : B_\varepsilon \to F$ such that

(1) $I(A) \leq m(B_\varepsilon) + \varepsilon$

(2) graph $f_\varepsilon \subseteq A$.

This theorem may be proved using the capacitability theorem i.e. Theorem 1.1.

Remark. This type of capacity is often met in the theory of stochastic processes.

Example 2.2. Let m be as above and $L \subseteq E \times F$ be compact. We define a capacity I_L on F by

$$I_L(A) = m[\Pi_E(L \cap (E \times A))] .$$

Concerning this kind of capacity, we have the following deep result of Mokobodzki [1978].

Theorem 2.2.

(i) There exists a measure λ on F satisfying "$\lambda(K) = 0 \Rightarrow I_L(K) = 0$ for all compact $K \subseteq F$" iff for m-almost all $x \in E$, $L(x) = L \cap (\{x\} \times F)$ is at most countable.

(ii) There exists a measure λ on F satisfying for all $\varepsilon > 0$, there is a $\delta > 0$ such that $\lambda(K) < \delta \Rightarrow I_L(K) < \varepsilon$ for any compact $K \subseteq F$" iff for m-almost all $x \in E$, $L(x)$ is finite.

Definition. A capacity I on F is called strongly subadditive or alternating of order two if for all compact $K_1, K_2 \subseteq F$,

$$I(K_1 \cup K_2) + I(K_1 \cap K_2) \leq I(K_1) + I(K_2) \qquad (*)$$

Such a capacity is interesting because it can be constructed from a function, satisfying suitable conditions, defined on the compact sets as the following theorem shows.

Theorem 2.3 (Choquet [1955]). Let J be a function on the compact subsets of F into the nonnegative real numbers satisfying

(1) For every compact $K \subseteq F$ and every $t > J(K)$, there exists an open subset U of F, $U \supseteq K$ such that if $L \subseteq U$ is compact, $J(L) < t$.

(2) $J(K_1 \cup K_2) + J(K_1 \cap K_2) \leq J(K_1) + J(K_2)$.

Define I as follows:

If U is open, let $I(U) = \sup\{I(K) : K \subseteq U, K \in \mathcal{K}\}$. For any $A \subseteq F$, put $I(A) = \inf\{I(U) : U \supseteq A, U \text{ open}\}$.

Then I is a strongly subadditive capacity agreeing with J on the compact sets. (Compare with the Caratheodory extension theorem in measure theory). Moreover, it is clear that any strongly subadditive capacity is equivalent to such an I. In particular, for such an I and a universally capacitable set A, we have $\sup_{\substack{K \subseteq A \\ K \in \mathcal{K}}} I(K) = I(A) = \inf_{\substack{U \supseteq A \\ U \text{ open}}} I(U)$.

Remark. A capacity alternating of order two is a "weak" version of a capacity alternating of order infinity. The latter is characterized by a series of inequalities among which (*) occurs. A typical example of such a capacity is I_L in Example 2.2. As a matter of fact, Choquet proved that any capacity alternating of order infinity is equivalent to I_L for some L, E

and m. These capacities occur in potential theory. In particular, the Newtonian capacity is one such.

Example 2.3. Let $\mathfrak{M}^1(F)$ be the set of all probability measures on F with the weakest topology making the maps $m \to m(K)$ upper semicontinuous for all compact $K \subseteq F$. It can be proved that $\mathfrak{M}^1(F)$ is compact and metrizable.

Let $L \subseteq \mathfrak{M}^1(F)$ be compact. For $A \subseteq F$, define $I_L(A) = \sup_{m \in L} m(A)$ where, as usual, $m(A)$ stands for the exterior measure of A induced by m. I_L is a capacity on F.

Using the Hahn-Banach theorem, it is possible to prove that any strongly subadditive capacity on F is equivalent to some I_L. The converse is, however, not true as can be seen from the next theorem.

Theorem 2.4 (Preiss [1973]). If $B \subseteq F$ is a Borel set which is not in $\mathcal{K}_\sigma$, then there exists a compact subset L of $\mathfrak{M}^1(F)$ such that

(1) $I_L(B) = 0$

(2) $\inf_{\substack{U \supseteq B \\ U \text{ open}}} I_L(U) > 0$.

Example 2.4. Let (E,d) be a compact metric space and $h : \mathbb{R}_+ \to \mathbb{R}_+$ a continuous nondecreasing function with $h(t) > 0$ if $t > 0$. For any $\varepsilon > 0$, we define Λ^h_ε as follows:

$$\Lambda^h_\varepsilon(\varphi) = 0 .$$

For $A \neq \phi$,

$$\Lambda^h_\varepsilon(A) = \inf \left\{ \sum_{n \geq 1} h(\delta(F_n)) : \{F_n\}_{n \geq 1} \text{ is a countable cover of } A \text{ by closed sets with } \delta(F_n) \leq \varepsilon \text{ for all } n \right\},$$

where $\delta(F_n)$ stands for the diameter of F_n.

If $\varepsilon \downarrow 0$, $\Lambda^h_\varepsilon(A) \uparrow \Lambda^h(A)$ (say). Λ^h is called the h-Hausdorff measure on E. Λ^h, restricted to the Borel subsets of E is certainly a nonnegative, σ-additive measure but it is not, in general, bounded or even σ-finite. For example, let $h(t) \equiv 1$. Then

$$\Lambda^h(A) = \text{cardinality of } A \text{ if } A \text{ is finite}$$
$$= \infty \text{ otherwise} .$$

So, in general Λ^h, Λ^h_ε are not capacities although they are pointwise limits of increasing sequences of capacities.

However, if we take $\varepsilon \geq \delta(E)$, Λ^h_ε, also denoted by Λ^h_∞, is a capacity. Note that taking $\varepsilon \geq \delta(E)$ is the same as removing all restrictions on $\delta(F_n)$ in the definition of Λ^h_ε.

Theorem 2.5 (Davies and Rogers [1969]). There exists a triplet (E,d,h) as in the example such that for any measure m on E, there is a partition of E into two Borel sets B_1, B_2 with $m(B_1) = 0$ and $\Lambda^h_\infty(B_2) = \Lambda^h(B_2) = 0$.

Example 2.5. Let $\mathcal{K}(E)$ be the set of compact subsets of E equipped with the topology induced by the Hausdorff metric. Note that this is the weakest topology such that the restriction of any capacity on E to compact sets, considered as a function on $\mathcal{K}(E)$, is upper semicontinuous.

Let L be a compact subset of $\mathcal{K}(E)$ such that $K \in L \Rightarrow K$ is finite. Define I by

$$I(A) = 1 \text{ if there is some } K \in L \text{ with } K \subseteq A$$
$$= 0 \text{ otherwise .}$$

It is easy to check that I is a capacity. If $K \in L \Rightarrow$ cardinality of $K = 1$, then I is a capacity alternating of order infinity. In general, however, it is not even strongly subadditive. For example, suppose $\{a_1,a_2\} \in L$, $\{a_1\} \notin L$, $\{a_2\} \notin L$, $a_1 \neq a_2$. Then $I(\{a_1\}) = 0$, $I(\{a_2\}) = 0$ but $I(\{a_1\} \cup \{a_2\}) = 1$.

On the other hand, any capacity on E with values in $\{0,1\}$ is the limit of an increasing sequence of such capacities.

§3. Theorem of Separation.

We now prove the theorem of separation for universally capacitable sets, using the capacitability theorem.

Theorem 3.1. Let E be a Polish space and A_1, A_2 two disjoint subsets of E such that $A_1 \times A_2$ is universally capacitable in $E \times E$. There exist Borel subsets B_1, B_2 of E such that $B_1 \supseteq A_1$, $B_2 \supseteq A_2$ and $B_1 \cap B_2 = \emptyset$.

Proof. Without loss of generality, take $A_1, A_2 \neq \emptyset$. For $A \subseteq E \times E$, define $\boxed{A} = \Pi_1(A) \times \Pi_2(A)$, Π_1, Π_2 denoting projections to the first and second coordinate, respectively. $\boxed{A}$ is the smallest rectangle containing A.

Put

$$I(A) = 1 \text{ if } \boxed{A} \text{ meets the diagonal}$$
$$= 0 \text{ otherwise .}$$

Note that I is a capacity and $I(A_1 \times A_2) = 0$. Thus there is a Borel set $C_1 \supseteq A_1 \times A_2$ such that $I(C_1) = 0$. Put $C_2 = \boxed{C_1}$. Then C_2 is analytic, hence universally capacitable, and $I(C_2) = 0$. Repeating the process, we can find C_n, $n \geq 1$, such that $A_1 \times A_2 \subseteq C_1 \subseteq C_2 \subseteq \cdots$ and, for all n, C_{2n-1} is Borel and C_{2n} is an analytic rectangle of null capacity. Hence $C = \bigcup_n C_n$ is a Borel rectangle of null capacity. Put finally, $B_1 = \Pi_1(C)$ and $B_2 = \Pi_2(C)$.

Remark. It is known that there exist a pair (A_1, A_2) of coanalytic sets which are disjoint but cannot be separated by Borel sets. For such a pair, $A_1 \times A_2$ is a coanalytic set which is not universally capacitable. On the other hand Shochat [1972] and Busch [1973] proved assuming (PD) that any projective set is capacitable for any capacity alternating of order infinity.

Appendix

We denote by E a fixed perfect Polish space. We give first a lemma (due to Martin) to show the inability of the notion of capacity to grasp the notion of "topological thickness."

Lemma. Let I be a capacity on E and A an I-capacitable subset of E. Then there exist a meager K_σ set $L \subseteq A$ such that $I(A) = I(L)$.

Proof. Clearly it is sufficient to prove that if A is compact and $I(A) > t$, then there is a nowhere dense compact set $K \subseteq A$ such that $I(K) \geq t$. Note that for this it is enough to prove:

(*) If A is compact, B open and $I(A) > t$, then there is an open set $B' \subseteq B$ such that $I(A - B') > t$.

To see this, assume (*) and let $B_1, B_2, \ldots$ be an open base for E. By induction on n, get open sets $B_1', B_2', \ldots$ such that for all n, $B_n' \subseteq B_n$ and $I(A - \bigcup_{i=1}^n B_i') > t$. Put $K = A - \bigcup_n B_n'$. K is clearly nowhere dense and compact. Since I is going down on compacts, $I(K) \geq t$.

Now we prove (*). Let A be compact, B open and $I(A) > t$. Let $x \in B$ and $B \supseteq U_1 \supseteq U_2 \supseteq \cdots$ where each U_i is open and $\bigcap_i U_i = \{x\}$. Now as $\bigcup_i (A - (U_i - \{x\})) = A$, $I(A - (U_i - \{x\})) > t$ for some i. Put $B' = U_i - \{x\}$ for such an i.

Theorem. Assuming (CH), there exists a nonprojective, universally capacitable subset of E.

Proof. Without loss of generality, take E to be the Cantor set. Restrict the capacities on E to the set of universally capacitable subsets of E. Then we get c many capacities, say $\{I_i : i < c\}$. Enumerate the projective subsets of E as $\{P_i : i < c\}$.

We now construct, by transfinite induction, two families $\{A_i : i < c\}$, $\{B_i : i < c\}$ of subsets of E such that

(a) Each A_i is a meager K_σ set, each G_i is a comeager G_δ set.

(b) $\{A_i : i < c\}$ is an increasing family and $\{B_i : i < c\}$ is a decreasing family.

(c) For each i, $A_i \subseteq B_i$ and $I_i(A_i) = I_i(B_i)$.

(d) For each i, either $A_i - P_i \neq \emptyset$ or $P_i - B_i \neq \emptyset$.

Then any subset H of E such that $\bigcup_i A_i \subseteq H \subseteq \bigcup_i B_i$ will be universally capacitable and not equal to any P_i.

Suppose A_i, B_i have been constructed for $i < j < c$. Let L_j be a meager K_σ set contained in $B'_j = \bigcap_{i<j} B_i$ such that $I_j(L_j) = I_j(B'_j)$. This is possible since by (CH), B'_j is a countable intersection of G_δ sets and is hence a G_δ and therefore a universally capacitable set. Put $A'_j = L_j \cup \bigcup_{i<j} A_i$. By (CH) A'_j is a meager K_σ set.

If $A'_j - P_j \neq \emptyset$ or $P_j - B'_j \neq \emptyset$, take $A_j = A'_j$, $B_j = B'_j$. If $A'_j \subseteq P_j \subseteq B'_j$, then either $B'_j - P_j \neq \emptyset$ or $P_j = B'_j \neq A'_j$ (since B'_j is comeager and A'_j is meager). In the first case, pick a point $x \in (B'_j - P_j)$ and put $A_j = A'_j \cup \{x\}$, $B_j = B'_j$. In the second case pick some $x \in (P_j - A'_j)$ and put $A_j = A'_j$ and $B_j = B'_j - \{x\}$.

Chapter 2

Multicapacities, capacitary operations and $\mathcal{K}$-analytic sets

In this chapter E, F, with or without suffices, denote nonempty Hausdorff spaces.

§1. Multicapacities.

Definition. If $\{E_n\}_{n\geq 1}$ is a sequence (finite or countably infinite) of spaces, a multicapacity on ΠE_n (i.e. $E_1 \times E_2 \times \cdots$) is a function I on $\Pi \mathcal{P}(E_n)$ into $\overline{\mathbb{R}}_+$ which is

(1) globally increasing, i.e. $A_n \subseteq B_n$ for all n implies $I(A_1,A_2,\dots) \leq I(B_1,B_2,\dots)$,

(2) separately going up, i.e. if for some n, $A_n^m \uparrow A_n$, then $I(A_1,A_2,\dots,A_{n-1},A_n^m,A_{n+1},\dots) \uparrow I(A_1,A_2,\dots,A_{n-1},A_n,A_{n+1},\dots)$,

(3) globally right continuous over compact sets, i.e. if $K_1,K_2,\dots$ are compact, then $I(K_1,K_2,\dots)$ is finite and if $I(K_1,K_2,\dots) < t$, then there exist open sets $U_1,U_2,\dots,K_n \subseteq U_n \subseteq E_n$ and $U_n = E_n$ for all but finitely many n, such that $I(U_1,U_2,\dots) < t$.

Note that in (3), for any open set V in ΠE_n which contains ΠK_n, there must exist a ΠU_n such that $\Pi K_n \subseteq \Pi U_n \subseteq V$.

Example 1.1. Any capacity is a multicapacity with one argument.

Example 1.2. Consider the capacities I_L in Examples 2.2 and 2.3 of Chapter 1. Remove the restriction L is compact and let $I(L,A) = I_L(A)$. The functions I thus obtained are multicapacities with two arguments.

Example 1.3. Let $E_n = E$ for all n. Let

$$I(A_1,A_2,\dots) = 0 \text{ if } \bigcap_n A_n = \emptyset$$
$$= 1 \text{ otherwise .}$$

I is a multicapacity. (The number of arguments may be taken to be finite or countably infinite).

Definition. If I is a multicapacity on ΠE_n, a sequence $\{A_n\}_{n\geq 1}$, $A_n \subseteq E_n$, is I-capacitable if $I(A_1,A_2,\dots) = \sup\{I(K_1,K_2,\dots) : K_n \subseteq A_n, K_n \in \mathcal{K}\}$. $\{A_n\}_{n\geq 1}$ is called universally capacitable if it is I-capacitable for any multicapacity I having $\{A_n\}$ as a sequence of arguments.

Theorem 1.1. Any sequence $\{A_n\}_{n\geq 1}$ of $\mathcal{K}_\sigma$-sets is universally capacitable.

Proof. For each n, let $\{K_n^m\}_{m\geq 1}$ be a sequence of compact sets with $K_n^m \uparrow A_n$. Let I be any multicapacity with $I(A_1,A_2,\dots) > t$. By (2) of the definition of multicapacity, we can find $m_1,m_2,\dots$ such that for all n, $I(K_1^{m_1},\dots,K_n^{m_n},A_{n+1},A_{n+2},\dots) > t$. Hence by (1) $I(K_1^{m_1},\dots,K_n^{m_n},E_{n+1},E_{n+2},\dots) > t$. Finally by (3), $I(K_1^{m_1},\dots,K_n^{m_n},K_{n+1}^{m_{n+1}},\dots) \geq t$.

§2. Capacitary operations. In this section, we are going to consider what is a "good change of variables" in a multicapacity.

Definition. If $\{E_n\}$ is a sequence (finite or infinite) of spaces and F is a space, a capacitary operation on $\Pi\, E_n$ with values in F is a map $J : \Pi\, \mathcal{P}(E_n) \to \mathcal{P}(F)$ which is

(1) globally increasing (with the obvious meaning)

(2) separately going up (with the obvious meaning)

(3) globally right continuous on compacts, i.e.

  (a) if $K_1,K_2,\dots$ are compact, then so is $J(K_1,K_2,\dots)$ and

  (b) if $V \subseteq F$ is open and $J(K_1,K_2,\dots) \subseteq V$, then there exist open sets $U_1,U_2,\dots$ in $E_1,E_2,\dots$ such that $K_n \subseteq U_n$, $U_n = E_n$ for all but finitely many n and $J(U_1,U_2,\dots) \subseteq V$.

Note that for a fixed $y \in F$, if we define

$$J_y(A_1,A_2,\dots) = 1 \quad \text{if } y \in J(A_1,A_2,\dots)$$
$$= 0 \quad \text{otherwise}$$

then J_y is a multicapacity on $\Pi\, E_n$ with values in $\{0,1\}$.

The next theorem, although easy, is very important.

Theorem 2.1 (Composition). If J^i, $i = 1,2,\dots$, are capacitary operations on $\Pi_j\, E_j^i$ with values in E^i and if I is a multicapacity (or capacitary operation) on $\Pi_i\, E^i$, then the composition $I(J^1(A_1^1,A_2^1,\dots),J^2(A_1^2,A_2^2,\dots),\dots)$ is a multicapacity (or capacitary operation) on $\Pi_{(i,j)}\, E_j^i$.

Example 2.1. f is a continuous map from $\Pi\, E_n$ into F. For $A_n \subseteq E_n$, $n = 1,2,\dots$ put $J(A_1,A_2,\dots) = f(\Pi\, A_n)$. Then J is a capacitary operation.

Example 2.2. If $E_n = E$ for all n, then $J(A_1,A_2,\dots) = \bigcap_n A_n$ is a capacitary operation.

Example 2.3. If $E_n = E$ for all n, $\{A_n\} \to \bigcup_n A_n$ is not, in general, a capacitary operation. However, add an argument $A_0 \subseteq N = \{1,2,\dots\}$ and put $J(A_0,A_1,A_2,\dots) = \bigcup_{n\in A_0} A_n$. Then J is a capacitary operation and $\bigcup_{n\geq 1} A_n = J(N,A_1,A_2,\dots)$.

Example 2.4. Let $A_1 \subseteq E \times F$, $A_2 \subseteq E$, $A_3 \subseteq F$ and put $J(A_1, A_2, A_3, \ldots) = \Pi_F[A_1 \cap (A_2 \times A_3)]$. J is a capacitary operation by the composition theorem.

Note that if A_1 is the graph of a function $f : E \to F$ and $A_3 = F$, then $J(A_1, A_2, A_3) = f(A_2)$. If A_1 is the graph of a function $f : F \to E$ and $A_3 = F$, then $J(A_1, A_2, A_3) = f^{-1}(A_2)$.

Example 2.5. Let $E = N^N$ and $f : N^N \to \mathcal{K}(F)$ (where $\mathcal{K}(F)$ is the family of compact subsets of F) be upper semicontinuous i.e. if $f(\sigma) \subseteq V \subseteq F$ for some $\sigma \in N^N$ and open V, then there is an open subset U of N^N such that $\sigma \in U$ and $\tau \in U \Rightarrow f(\tau) \subseteq V$ for all τ. Now for any $A \subseteq N^N$, put $J(A) = \bigcup_{\sigma \in A} f(\sigma)$. Then J is a capacitary operation.

Example 2.6. In this example we shall see that the Souslin operation, although it is not a capacitary operation, can be obtained from a capacitary operation.

Let $E_s = E$ for any finite sequence s of positive integers. If $A_s \subseteq E$, $\mathcal{A}(\ldots A_s \ldots)$ is defined to be $\bigcup_{\sigma \in N^N} \bigcap_{s \prec \sigma} A_s$ where $s \prec \sigma$ means "s is an initial segment of σ." Add an argument $A_0 \subseteq N^N$ and put $J(A_0, \ldots, A_s, \ldots) = \bigcup_{\sigma \in A_0} \bigcap_{s \prec \sigma} A_s$. Using the composition theorem, it is easy to prove that J is a capacitary operation. Note that $\mathcal{A}(\ldots A_s \ldots) = J(N^N, \ldots, A_s, \ldots)$.

§3. $\mathcal{K}$-analytic sets. Consider a capacitary operation J. We know that if $A_1, A_2, \ldots$ are all compact, then so is $J(A_1, A_2, \ldots)$. However, since J is only separately (and not globally) going up, we do not know what kind of a set $J(A_1, A_2, \ldots)$ is when the A_n's are σ-compact (and there are infinitely many arguments). We now introduce a definition for such sets.

Definition. A subset A of a space F is called $\mathcal{K}$-analytic if there is a sequence of spaces $\{E_n\}_{n \geq 1}$, a capacitary operation J on ΠE_n with values in $\mathcal{P}(F)$ and $\mathcal{K}_\sigma$-sets $H_n \subseteq E_n$ such that $J(H_1, H_2, \ldots) = A$.

Note that in the above definition, if we take $E_n = H_n$ for all n and restrict J to ΠH_n, then J takes values in $\mathcal{P}(A)$. Thus the property of being $\mathcal{K}$-analytic is intrinsic.

Theorem 3.1 (Invariance). If J is a capacitary operation and $\{A_n\}$ a sequence of $\mathcal{K}$-analytic arguments for J, then $J(A_1, A_2, \ldots)$ is $\mathcal{K}$-analytic.

Proof. This is immediate from the composition theorem.

From this theorem and our examples we get the stability properties of the class of $\mathcal{K}$-analytic sets.

Corollary 1. The class of $\mathcal{K}$-analytic sets contains all $\mathcal{K}_\sigma$ sets and is closed under countable unions, countable intersections and Souslin operations.

Proof. It is enough to note that in Example 2.3, N is a $\mathcal{K}_\sigma$ set and in Example 2.6, N^N is a $\mathcal{K}_{\sigma\delta}$ (and hence $\mathcal{K}$-analytic) set in the compact metrisable space $\overline{N}^N$, $\overline{N}$ being the natural one-point compactification of N.

Corollary 2. Any closed subset of a $\mathcal{K}$-analytic space A is $\mathcal{K}$-analytic.

Proof. Suppose $A = J(H_1,H_2,\ldots)$ where $H_1,H_2,\ldots$ are $\mathcal{K}_\sigma$ sets and J a capacitary operation. If X is a closed subset of A, define for $A_i \subseteq H_i$, $i = 1,2,\ldots$, $J_X(A_1,A_2,\ldots) = J(A_1,A_2,\ldots) \cap X$. Clearly, J_X is a capacitary operation into A and $X = J_X(H_1,H_2,\ldots)$.

Corollary 3.

(a) Any Borel subset of a Polish space is $\mathcal{K}$-analytic.

(b) If E, F are Polish and $f : E \to F$ is a Borel map, then the direct image of $\mathcal{K}$-analytic subsets of E and the inverse image of $\mathcal{K}$-analytic subsets of F are $\mathcal{K}$-analytic.

Proof. (a) By Corollary 1, any Borel subset of a compact metrisable space is $\mathcal{K}$-analytic. As any Polish space can be imbedded as a G_δ subset in a compact metrisable space, the result follows.

(b) Use Example 2.4 and the fact that graph f is Borel in $E \times F$.

Actually, a subset of a Polish space is $\mathcal{K}$-analytic iff it is "classical" analytic i.e. a continuous image of N^N. This is an easy consequence of the following:

Theorem 3.2. A subset A of F is $\mathcal{K}$-analytic iff there is an upper semicontinuous map $f : N^N \to \mathcal{K}(F)$ (the compact subsets of F) such that $A = \bigcup_{\sigma\in N^N} f(\sigma)$ i.e. iff it is "$\mathcal{K}$-analytic" in the sense of Frolik.

Remark. By a result of Jayne [1976], this implies that our $\mathcal{K}$-analytic sets are precisely the "$\mathcal{K}$-analytic" sets defined by Choquet, i.e. continuous images of $\mathcal{K}_{\sigma\delta}$ subsets of compact spaces.

Proof of the Theorem. The "if" part follows by Example 2.5. For the "only if" part, let $A = J(H_1,H_2,\ldots)$ where J is a capacitary operation and $H_1,H_2,\ldots$ are $\mathcal{K}_\sigma$ sets. For each n, let $K_n^m \uparrow H_n$, where $K_n^1,K_n^2,\ldots$ are compact. By Theorem 1.1, for $y \in F$, we have $J_y(H_1,H_2,\ldots) =$

$\sup_y J_y(K_1^{m_1}, K_2^{m_2}, \ldots)$. Hence $J(H_1, H_2, \ldots) = \bigcup_{\sigma \in N^N} J(K_1^{\sigma_1}, K_2^{\sigma_2}, \ldots)$ where $\sigma = (\sigma_1, \sigma_2, \ldots)$. Let $f(\sigma) = J(K_1^{\sigma_1}, K_2^{\sigma_2}, \ldots)$. As J is right continuous on the compacts, f is upper semicontinuous.

This completes the proof.

Remark. Actually, in our definition of a $\mathcal{K}$-analytic set A, it is always possible to take $E_n = H_n = N$ for each n so that our definition really reduces to Frolik's definition. To see this, we argue as follows.

Let $A = J(H_1, H_2, \ldots)$ where J is a capacitary operation and $H_1, H_2, \ldots$ are $\mathcal{K}_\sigma$ sets. Note that $H_n = J_n(N, K_1^n, K_2^n, \ldots)$ where J_n is a capacitary operation and $K_1^n, K_2^n, \ldots$ are compact sets. Now define $I_n(H) = J_n(H, K_1^n, K_2^n, \ldots)$, $H \subseteq N$. It is easy to see that I_n is a capacitary operation. Put $I(A_1, A_2, \ldots) = J(I_1(A_1), I_2(A_2), \ldots)$. Note $A = I(N, N, \ldots)$ and I is a capacitary operation.

§4. Approximation properties of $\mathcal{K}$-analytic sets.

Theorem 4.1 (Capacitability). If I is a multicapacity or a capacitary operation and $\{A_n\}$ is a sequence of $\mathcal{K}$-analytic arguments for I, then $I(A_1, A_2, \ldots) = \sup\{I(K_1, K_2, \ldots) : K_n \subseteq A_n,\ K_n \in \mathcal{K}\}$.

Proof. Case 1. I is a multicapacity.

For each n, let $H_1^n, H_2^n, \ldots$ be $\mathcal{K}_\sigma$ sets and J^n a capacitary operation such that $A_n = J^n(H_1^n, H_2^n, \ldots)$. Let $J(M_i^n : i \geq 1,\ n \geq 1) = I(J^1(M_1^1, M_2^1, \ldots), J^2(M_1^2, M_2^2, \ldots), \ldots)$ for any arguments M_i^n. Then by the composition theorem J is a multicapacity. Note that $I(A_1, A_2, \ldots) = J(H_i^n : i \geq 1,\ n \geq 1)$.

Now suppose $I(A_1, A_2, \ldots) > t$. By Theorem 1.1, there exist compact sets $K_i^n \subseteq H_i^n$ such that $J(K_i^n,\ i \geq 1,\ n \geq 1) \geq t$. Let $K_n = J^n(K_1^n, K_2^n, \ldots)$. Then K_n is compact, $K_n \subseteq J^n(H_1^n, H_2^n, \ldots) = A_n$ and $I(K_1, K_2, \ldots) = J(K_i^n,\ i \geq 1,\ n \geq 1) \geq t$.

Case 2. I is a capacitary operation.

Note that for each y, I_y is a multicapacity and use Case 1.

Definition. A subset B of a space F is called $\mathcal{G}$-Borel if it belongs to the smallest class of subsets containing open sets and closed under countable unions and countable intersections.

Note that a $\mathcal{G}$-Borel set is always Borel. If F is Polish, the converse is true. However, in general a Borel set, or even a compact set, need not be $\mathcal{G}$-Borel. Again, in a Polish space, every $\mathcal{K}$-analytic (equivalently "classical" analytic) is $\mathcal{G}$-Borel but in general a $\mathcal{G}$-Borel set need not be $\mathcal{K}$-analytic.

Theorem 4.2 (Borel approximation). If I is a multicapacity or a capacitary operation on ΠE_n and if $\{A_n\}$ is a sequence of $\mathcal{K}$-analytic arguments for I, then $I(A_1,A_2,\ldots) = \inf\{I(B_1,B_2,\ldots) : B_n \supseteq A_n$ and B_n is $\mathcal{G}$-Borel in $E_n\}$.

Proof. As in Theorem 4.1, it is enough to consider the case when I is a multicapacity. Define J on $\Pi \mathcal{P}(E_n)$ by $J(H_1,H_2,\ldots) = \inf\{I(B_1,B_2,\ldots) : B_n \supseteq H_n$, B_n is $\mathcal{G}$-Borel in $E_n\}$.

Since I is right continuous on compacts, J and I agree if $H_1,H_2,\ldots$ are compact. Hence to show that they agree when $H_1,H_2,\ldots$ are $\mathcal{K}$-analytic, it is enough to show that J is a multicapacity (capacitability theorem).

Clearly, J satisfies (1) and (3). We now show J satisfies (2).

Suppose $H_1^m \uparrow H_1$. We can choose $\mathcal{G}$-Borel sets B_n^m such that $B_1^m \supseteq H_1^m$, $B_n^m \supseteq H_n$ for $n \geq 2$ and $J(H_1^m,H_2,\ldots) = I(B_1^m,B_2^m,\ldots)$. Let $B_n = \bigcap_m B_n^m$ for $n \geq 2$. Then $J(H_1^m,H_2,\ldots) = I(B_1^m,B_2,\ldots)$. Replacing B_1^m by $\bigcap_{p \geq m} B_1^p$ if necessary, we suppose $\{B_1^m\}$ to be an increasing sequence, say $B_1^m \uparrow B_1$. Now $J(H_1,H_2,\ldots) \geq J(H_1^m,H_2,\ldots) = I(B_1^m,B_2,\ldots)$ and $I(B_1^m,B_2,\ldots) \uparrow I(B_1,B_2,\ldots) \geq J(H_1,H_2,\ldots)$. Thus $J(H_1^m,H_2,\ldots) \uparrow J(H_1,H_2,\ldots)$.

As corollaries, we get several separation theorems.

Corollary 1. If I is a multicapacity (capacitary operation) on ΠE_n and $I(A_1,A_2,\ldots) = o(\emptyset)$ where $A_1,A_2,\ldots$ are $\mathcal{K}$-analytic, then there exist $\mathcal{G}$-Borel subsets B_n of E_n, $n = 1,2,\ldots$ such that $B_n \supseteq A_n$ and $I(B_1,B_2,\ldots) = o(\emptyset)$.

Proof. By the theorem, the result is clearly true when I is a multicapacity. Let I be a capacitary operation with values in F. Compose I with the capacity J on F such that

$$\begin{aligned} J(H) &= 0 \text{ if } H = \emptyset \\ &= 1 \text{ otherwise .} \end{aligned}$$

Get $B_n \supseteq A_n$, $\mathcal{G}$-Borel in E_n, such that $J(I(B_1,B_2,\ldots)) = 0$. Then $I(B_1,B_2,\ldots) = \emptyset$.

Corollary 2.

(a) (Extension of Novikov's separation theorem). If $A_1,A_2,\ldots$ are $\mathcal{K}$-analytic subsets of E such that $\bigcap_n A_n = \emptyset$, then there exist $\mathcal{G}$-Borel subsets $B_1,B_2,\ldots$ of E such that $B_n \supseteq A_n$ and $\bigcap_n B_n = \emptyset$.

(b) (Extension of Liapunov's separation theorem). Suppose for each finite sequence s of natural numbers, A_s is a $\mathcal{K}$-analytic subset of E such that $\bigcup_{\sigma \in N^N} \bigcap_{s<\sigma} A_s = \emptyset$. Then there exist $\mathcal{G}$-Borel subsets B_s of E, $B_s \supseteq A_s$, such that $\bigcup_{\sigma \in N^N} \bigcap_{s<\sigma} B_s = \emptyset$.

Proof. These are particular cases of Corollary 1.

Definition. A subset H of E is $\mathcal{F}$-Souslin if $H = \bigcup_{\sigma \in N^N} \bigcap_{s<\sigma} F_s$, where the F_s's are closed in E. H is called $\mathcal{K}$-Souslin if the F_s's are compact.

It follows from Example 2.6 that every $\mathcal{K}$-Souslin set is $\mathcal{K}$-analytic and easily, from Frolik's definition, that every $\mathcal{K}$-analytic set is $\mathcal{F}$-Souslin. Clearly, if E is compact or Polish every $\mathcal{F}$-Souslin set is $\mathcal{K}$-analytic but this is not necessarily true for general E. Nevertheless, we do have the following stronger version of Corollary 1.

Corollary 3. Suppose I is a capacitary operation on ΠE_n with values in F, for $n \geq 1$, $A_n \subseteq E_n$ is $\mathcal{K}$-analytic and $A' \subseteq F$ is $\mathcal{F}$-Souslin. If $I(A_1,A_2,\dots) \cap A' = \emptyset$, then there exist $\mathcal{G}$-Borel subsets B_n of E_n such that $B_n \supseteq A_n$ and $I(B_1,B_2,\dots) \cap A' = \emptyset$.

In particular if $A \subseteq F$ is $\mathcal{K}$-analytic and $A \cap A' = \emptyset$, then there exist a $\mathcal{G}$-Borel subset B of F such that $B \supseteq A$ and $B \cap A' = \emptyset$.

Proof. The last statement clearly follows from the first. We now prove the first statement. First note that if J is a capacitary operation with values in F and H is a closed subset of F, then $(X_1,X_2,\dots) \to J(X_1,X_2,\dots) \cap H$ is a capacitary operation.

Write A' as $\bigcup_{\sigma \in N^N} \bigcap_{s<\sigma} H_s$, H_s being closed in F.

If $X_0 \subseteq N^N$ and for each sequence s of natural numbers, $J_s(X_1^s,X_2^s,\dots)$ is a capacitary operation, then by the composition theorem $(X_0,X_n^s : n \geq 1,\ s$ is a finite sequence of natural numbers$) \to \bigcup_{\sigma \in X_0} \bigcap_{s<\sigma} [J_s(X_1^s,X_2^s,\dots) \cap H_s]$ is a capacitary operation. Now for all s, take $J_s = I$ and $X_n^s = A_n$. Applying Corollary 1, we get $\mathcal{G}$-Borel sets $B_n^s \supseteq A_n$ such that $\bigcup_{\sigma \in N^N} \bigcap_{s<\sigma} [I(B_1^s,B_2^s,\dots) \cap H_s] = \emptyset$. Let $B_n = \bigcap_s B_n^s$.

Corollary 4. Suppose I is a capacitary operation on ΠE_n with values in F and let $A_n \subseteq E_n$, $n = 1,2,\dots$ be $\mathcal{K}$-analytic. For every $\mathcal{G}$-Borel subset B of F such that $I(A_1,A_2,\dots) \subseteq B$, there exist $\mathcal{G}$-Borel subsets B_n of E_n such that $A_n \subseteq B_n$ and $I(B_1,B_2,\dots) \subseteq B$.

Proof. $F - B$ belongs to the smallest family of subsets of F containing closed sets and closed under countable unions and intersections. Hence $F - B$

is $\mathfrak{F}$-Souslin. Now we use Corollary 3 with $A' = F - B$.

§5. An application of the separation theorem. Let E be a compact metrisable space with $\text{card } E \geq 2$ and $\Omega \subseteq E^{\mathbb{R}_+}$ the space of all right continuous maps from $\mathbb{R}_+$ into E. Since the elements of Ω are completely determined by their values on the rationals, we can suppose $\Omega \subseteq E^{\mathbb{Q}_+}$ and give it the relative topology. It can be shown that Ω is coanalytic but not Borel in $E^{\mathbb{Q}_+}$.

Let X_t, $t \in \mathbb{R}_+$, be defined on Ω into E by $X_t(\omega) = \omega(t)$. Let $\mathcal{B}$ be the smallest σ-field on Ω making $\{X_t : t \in \mathbb{R}_+\}$ measurable. Then $\mathcal{B}$ can be shown to equal the Borel σ-field on Ω.

Proposition A.

(1) The map $X : \mathbb{R}_+ \times \Omega \to E$ defined by $X(t,\omega) = \omega(t)$ is measurable.

(2) The map $\theta : \mathbb{R}_+ \times \Omega \to \Omega$ given by $\theta(t,\omega) = \theta_t(\omega)$ is measurable where $\theta_t(\omega)(s) = \omega(t + s)$.

Theorem 5.1. Let $\{m_x : x \in E\}$ be a family of probability measures on Ω such that for all $B \in \mathcal{B}$, the map $x \to m_x(B)$ is Borel measurable. There exists a subset Ω_0 of Ω such that

(1) $\theta_t(\Omega_0) \subseteq \Omega_0$ for all $t \in \mathbb{R}_+$

(2) $m_x(\Omega - \Omega_0) = 0$ for all $x \in E$

(3) Ω_0 is a Borel subset of $E^{\mathbb{Q}_+}$.

Proof. Let $\mathfrak{M}^1(E^{\mathbb{Q}_+})$ be as described in Chapter 1, Example 2.3. Define I on $\mathcal{P}(\mathfrak{M}^1(E^{\mathbb{Q}_+})) \times \mathcal{P}(E^{\mathbb{Q}_+})$ by $I(H_1,H_2) = \sup_{m \in H_1} m(H_2)$. Note that I is a multicapacity. Let $A = \{m_x : x \in E\} \subseteq \mathfrak{M}^1(E^{\mathbb{Q}_+})$, where we extend m_x to $E^{\mathbb{Q}_+}$ by taking $m_x(E^{\mathbb{Q}_+} - \Omega) = 0$. The map $x \to m_x$ is Borel. Hence A is analytic. $I(A, E^{\mathbb{Q}_+} - \Omega) = \sup_{x \in E} m_x(E^{\mathbb{Q}_+} - \Omega) = 0$. Find Borel sets B_1, B_2 such that $A \subseteq B_1$, $E^{\mathbb{Q}_+} - \Omega \subseteq B_2$, and $I(B_1,B_2) = 0$. Put $\Omega_1 = E^{\mathbb{Q}_+} - B_2$. Then $\Omega_1 \subseteq \Omega$ and Ω_1 is a Borel set satisfying $m_x(\Omega - \Omega_1) = 0$ for all x. Let Ω_2 be the smallest set $\supseteq \Omega_1$ such that $\theta_t(\Omega_2) \subseteq \Omega_2$ for all $t \in \mathbb{R}_+$. Clearly Ω_2 is analytic and $\Omega_2 \subseteq \Omega$. Let Ω_3 be a Borel subset of $E^{\mathbb{Q}_+}$ such that $\Omega_2 \subseteq \Omega_3 \subseteq \Omega$. Clearly $m_x(\Omega - \Omega_3) = 0$ for all x. In general, we can get for all $n \geq 1$, $\Omega_n \subseteq \Omega$ such that $\Omega_n \subseteq \Omega_{n+1}$, Ω_n is Borel if n is odd, $\theta_t(\Omega_n) \subseteq \Omega_n$ for all t if n is even and $m_x(\Omega - \Omega_n) = 0$ for all n. Put $\Omega_0 = \bigcup_n \Omega_n$.

Chapter 3
Multicalibers and analytic operations

In this chapter E, F (with or without indices) are taken to be Polish spaces. Thus $\mathcal{K}$-analytic sets are just the "classical" analytic sets. Here we consider operations obtained from multicapacities (or capacitary operations) by

(a) identifying some arguments and

(b) replacing some other arguments by fixed analytic sets (called analytic parameters).

§1. Calibers. Suppose I is a multicapacity on $\Pi(E_n : n \in M)$, where M is a countable set. Let $\emptyset \neq M_1 \subseteq M$ and suppose E_n = a fixed space E for $n \in M_1$. For each $n \in M - M_1$, let $A_n \subseteq E_n$ be analytic.

We say $(I, M_1, \{A_n : n \in M - M_1\})$ is an analytic representation of some map $J : \mathcal{P}(E) \to \overline{\mathbb{R}}_+$ if

(a) J is nondecreasing and

(b) for every analytic $X \subseteq E$, $J(X) = I(X_1, X_2, \dots)$ where

$$X_n = X \text{ for } n \in M_1$$
$$= A_n \text{ if } n \notin M_1 .$$

Remark. It is possible to take $M = M_1$.

Definition. A caliber J on E is a nondecreasing map $J : \mathcal{P}(E) \to \overline{\mathbb{R}}_+$ which has an analytic representation.

Example 1.1. Let E be compact metrisable. Define J on $\mathcal{P}(E)$ by

$$J(X) = 0 \text{ if } X \text{ is countable}$$
$$= 1 \text{ otherwise .}$$

J is a caliber.

Proof. Let $E_1 = \mathfrak{m}_1(E)$ be the space of probability measures on E with the natural topology (see Chapter 1, Example 2.3), $E_2 = E$. Define I on $\mathcal{P}(E_1) \times \mathcal{P}(E_2)$ by $I(X_1, X_2) = \sup_{m \in X_1} m(X_2)$, where $m(X_2)$ is the outer measure of X_2 induced by m. Then I is a multicapacity. Let $D \subseteq \mathfrak{m}^1(E)$ be the set of diffuse probabilities (i.e. probabilities taking the value zero on singletons). $(I, \{2\}, D)$ is an analytic representation of J. To see this, it is enough to note that if X_2 is uncountable analytic, there is a diffuse probability measure m with $m(X_2) = 1$. This is true because the Cantor set carries such a measure, namely the product measure of the measures μ on $\{0,1\}$ with $\mu(\{0\}) = \mu(\{1\}) = \frac{1}{2}$, and X_2 contains a homeomorph of the Cantor set.

Note that $J(X) = I(D,X)$ need not hold when X is not analytic.

Example 1.2. Let $\{I_n\}$ be a sequence of capacities on E. Then $J_1(X) = \sup_n I_n(X)$ and $J_2(X) = \inf_n I_n(X)$ are calibers.

Proof is similar to Examples 2.2 and 2.3 in Chapter 2.

Example 1.3. More generally, suppose for each finite sequence of integers s, I_s is a capacity on E. For $X \subseteq E$, put $J(X) = \sup_{\sigma \in N^N} \inf_{s<\sigma} I_s(X)$. Then J is a caliber.

Remark. It is not known if every caliber agrees on analytic arguments with a J of this form. However, it is possible to prove the following weaker fact:

Call I a primitive caliber if it has an analytic representation with no parameters (i.e. $M = M_1$). Every caliber on E agrees on analytic arguments with some J given by $J(X) = \sup_{\sigma \in N^N} \inf_{s<\sigma} I_s(X)$ where I_s is a primitive caliber on E for every finite sequence of integers s.

Following Cenzer and Mauldin, call a class $\mathcal{C}$ of subsets of a space E a Π_1^1-monotone class iff there is an analytic subset T of E^N such that $X \in \mathcal{C} \leftrightarrow X^N \cap T = \emptyset$. Examples of Π_1^1-monotone classes are

(a) $\mathcal{C} = \{X : \text{card } X \leq 1\}$

(b) $\mathcal{C} = \{X : X$ is relatively compact, i.e. $\bar{X}$ is compact$\}$

(c) $\mathcal{C} = \{X : X$ is nowhere dense$\}$.

Example 1.4. Let $\mathcal{C}$ be a Π_1^1-monotone class. Let

$$J(X) = 0 \quad \text{if } X \in \mathcal{C}$$
$$= 1 \quad \text{otherwise} .$$

J is a caliber.

Proof. Let I be the multicapacity on $E^N \times E \times E \times \cdots$ defined by

$$I(H, X_1, X_2, \ldots) = 0 \quad \text{if } H \cap \Pi X_n = \emptyset$$
$$= 1 \quad \text{otherwise} .$$

$J(X) = I(H, X_1, X_2, \ldots)$ where $H = T$, the analytic set in the definition of $\mathcal{C}$, and $X_1 = X_2 = \cdots = X$.

Example 1.5. In this example, we take E to be compact metrisable and $\mathcal{K}(E)$ to be the set of closed subsets of E, endowed with the natural topology. Thus $\mathcal{K}(E)$ is compact metrisable. If J is a caliber on E and $\overline{\mathcal{C}} = \{K \in \mathcal{K}(E) : J(K) = 0\}$, then it is easy to show that $\overline{\mathcal{C}}$ is coanalytic in $\mathcal{K}(E)$. Trivially, $\overline{\mathcal{C}}$ is also hereditary, i.e. if $K_1 \subseteq K_2$ are in $\mathcal{K}(E)$ and $K_2 \in \overline{\mathcal{C}}$, then $K_1 \in \overline{\mathcal{C}}$.

Conversely, let $\overline{\mathcal{C}}$ be any coanalytic hereditary subset of $\mathcal{K}(E)$. For $X \in \mathcal{P}(E)$, define

$$J(X) = 0 \text{ if } \overline{X} \in \overline{\mathcal{C}}$$
$$= 1 \text{ otherwise .}$$

Then J is a caliber.

Proof. $\mathcal{C} = \{X : J(X) = 0\}$ is a Π_1^1-monotone class. To see this, take $T \subseteq E^N$ to be all sequences $\{x_n\}$ such that the closure of $\{x_1, x_2, \ldots\} \notin \overline{\mathcal{C}}$. Then T is analytic and $X \in \mathcal{C} \leftrightarrow X^N \cap T = \emptyset$.

§2. Multicalibers and analytic operations. We now generalize the notion of analytic representation for a caliber.

Let I be a multicapacity (or capacitary operation with values in F) on $\Pi_{n\in M} E_n$, where M is countable. Let $\{M_k\}$ be a sequence (finite or infinite) of nonempty, pairwise disjoint subsets of M. Suppose

(a) for $n \in M_k$, $E_n = E^k$, a fixed space depending only on k and

(b) $A_n \subseteq E_n$ are analytic for $n \in M - \cup_k M_k$ (note that this set may be empty).

We say $(I, \{M_k\}, \{A_n\}_{n\in M-\cup_k M_k})$ is an analytic representation of some map J from $\Pi_k \mathcal{P}(E_k)$ into $\overline{R}_+$ (or $\mathcal{P}(F)$) if J is globally nondecreasing and for $X_1, X_2, \ldots$ analytic subsets of $E^1, E^2, \ldots$ respectively, $J(X_1, X_2, \ldots) = I(Y_1, Y_2, \ldots)$ where

$$Y_n = X_k \text{ if } n \in M_k$$
$$= A_n \text{ if } n \in M - \bigcup_k M_k .$$

Definition. A multicaliber on ΠE^k (or an analytic operation on ΠE^k with values in F) is a nondecreasing map J from $\Pi \mathcal{P}(E^k)$ into $\overline{R}_+$ (or $\mathcal{P}(F)$) which admits an analytic representation.

Example 2.1. The Souslin operation is an analytic operation (see Chapter 2). It is obtained by taking one parameter = N^N in a capacitary operation.

Example 2.2. $X \to$ closure of X is an analytic operation.

Proof. Let, for $n \geq 1$, $X_n, Y_n \subseteq E$ and $Z_n \subseteq E \times E$. Define $I(\{X_n, Y_n, Z_n : n \geq 1\}) = \bigcap_n \Pi [(X_n \times Y_n) \cap Z_n]$ where Π denotes projection to the second coordinate. By the composition theorem, I is a capacitary operation. Put $J(X) = I(\{X_n, Y_n, Z_n : n \geq 1\})$ where for $n \geq 1$,

(a) $X_n = X$

(b) $Y_n = E$

(c) $Z_n = \{(x,y) : \text{distance}(x,y) < \frac{1}{n}$ in the metric of $E\}$.

Clearly $J(X)$ is an analytic operation and $J(X) = \overline{X}$.

Remark. $X \to$ interior of X is not an analytic operation but $X \to$ interior of $\overline{X}$ is one.

Example 2.3. Let R be an analytic equivalence relation on E. Then $J(A) = \{x : (\exists y \in A) xRy\}$ is an analytic operation.

In general, multicalibers and analytic operations are not "separately going up" or "right continuous over compact sets." However they have many of the regularity properties of multicapacities and capacitary operations. For example we have

Theorem 2.1 (Theorem of Invariance). An analytic operation applied to analytic arguments results in an analytic set. Moreover, any analytic set is the image of N under some analytic operation with one argument.

Theorem 2.2 (Theorem of Composition). The composition of multicalibers and analytic operations yields multicalibers or analytic operations. Identification of arguments in a multicaliber (or an analytic operation) yields a multicaliber (or an analytic operation).

Theorem 2.3 (Theorem of Capacitability). Here we do not have approximation from below by compact sets. However, we do have approximation from below by $\mathcal{K}_\sigma$ sets, i.e. if I is a multicaliber (or an analytic operation) and $X_1, X_2, \ldots$ is a sequence of analytic arguments, then $I(X_1, X_2, \ldots) = \sup\{I(K_1, K_2, \ldots) : K_n \subseteq X_n, K_n$ are $\mathcal{K}_\sigma$ sets$\}$.

Proof. We prove the theorem for a caliber J. Proof in the other cases are similar. Let I be a multicapacity and $A_1, A_2, \ldots$ analytic sets such that for any analytic argument X, $J(X) = I(A_1, A_2, \ldots, X_1, X_2, \ldots)$ where $X = X_1 = X_2 = \cdots$. Given $\varepsilon > 0$, we can find by the capacitability theorem for multicapacities, compact sets $K_1, K_2, \ldots \subseteq X$ such that $J(X) < I(A_1, A_2, \ldots, K_1, K_2, \ldots) + \varepsilon$. Let $K = \bigcup_n K_n$. Clearly $K \subseteq X$, K is a $\mathcal{K}_\sigma$ set and $J(X) < J(K) + \varepsilon$.

Theorem 2.4. If J is a multicaliber (or an analytic operation) and $A_1, A_2, \ldots$ a sequence of analytic arguments for J, then $J(A_1, A_2, \ldots) = \inf\{J(B_1, B_2, \ldots) : A_n \subseteq B_n,\ B_n$ Borel for $n \geq 1\}$.

Theorem 2.5 (Theorem of Separation). If J is a multicaliber (or an analytic operation) and $A_1, A_2, \ldots$ is a sequence of analytic arguments for J with $J(A_1, A_2, \ldots) = 0$ (or $\emptyset$), then there exist Borel sets $B_1, B_2, \ldots$ with $A_n \subseteq B_n$ for all n and $J(B_1, B_2, \ldots) = 0$ (or $\emptyset$).

§3. An application of the separation theorem.

Definition. We call a nonnegative function f defined on a Polish space an analytic function if for every $t \in R_+$, $\{x : f(x) > t\}$ is analytic (equivalently for every $t \in R_+$, $\{x : f(x) \geq t\}$ is analytic).

The point (2) of the next theorem is an extension of a recent result of Cenzer and Mauldin on Π^1_1-monotone classes.

Theorem 3.1. Let J be a caliber on E with $J(\emptyset) = 0$ and $A \subseteq E \times F$ an analytic set. Then

(1) The function on F defined by $y \to J[A(y)]$, where $A(y)$ is the section of A with respect to $y \in F$, is analytic.

(2) If $J[A(y)] = 0$ for each $y \in F$, then there exists a Borel set B in $E \times F$ such that $A \subseteq B$ and $J[B(y)] = 0$ for each $y \in F$.

Proof. We can suppose E and F to be compact metrisable since otherwise, we can imbed them as G_δ subsets of compact metrisable spaces $\tilde{E}$, $\tilde{F}$ respectively and extend J to $\tilde{E}$ by $\tilde{J}(X) = J(X \cap E)$. $\tilde{J}$ is a caliber on $\tilde{E}$. This simplifies the proof to some extent.

Fix $t \in R_+$. We shall prove that the operation J_t on $E \times F$ with values in $\mathcal{P}(F)$ given by

$$J_t(H) = \{y : J[H(y)] > t\}$$

is analytic.

Now (1) follows from the invariance theorem and (2) from the separation theorem.

Define an operation Γ on $(E \times F) \times \bar{R}_+$ into $F \times \bar{R}_+$ as follows:

For $X \subseteq \bar{R}_+$, let $i(X)$ denote the smallest interval containing $\{0\}$ and X. Let $H \subseteq E \times F$. Put $\Gamma(H,X) = \{(y,v) \in F \times \bar{R}_+ : \exists u \in i(X)(v \leq J[H(y)] + u)\}$, in other words, $\Gamma(H,X)$ is obtained by adding the interval $i(X)$ above the graph of $y \to J(H,y)$. It is not difficult to see that Γ is a capacitary operation. (Note that if H is compact, then $y \to J[H(y)]$ is an upper semicontinuous function as J is right continuous on compact sets and hence $\{(y,v) : J[H(y)] \geq v\}$ is compact).

Finally we have $J_t(H) = \bigcap_n \Pi_F[\Gamma(H,\emptyset) \cap F \times [t + \frac{1}{n}, \infty]]$ so that J_t is an analytic operation.

When J is a caliber, we look at its analytic representation and work with the corresponding multicapacity.

Remarks.

(a) In part (2), we can replace "$J[A(y)] = 0$" by "$J[A(y)] \leq t$" and "$J[B(y)] = 0$" by "$J[B(y)] \leq t$" for any t.

To prove this, replace the caliber J by the caliber J' where

$$J'(X) = J(X) - t \quad \text{if} \quad J(X) \geq t$$
$$= 0 \quad \text{otherwise} .$$

(b) If the caliber J satisfies $J(X) = J(\overline{X})$ then in part (2) we can obtain a B with its sections closed. Note that this is not trivial!

(c) In his recent work, Louveau [1979] proved another extension of the Cenzer-Mauldin result which is "disjoint" from ours but nevertheless deeper. In any case, it is more interesting to Descriptive Set Theorists.

We finish with an application of Theorem 3.1 to the theory of Hausdorff measure (see Chapter 1).

Suppose m is a (finite) measure on E and B is a Borel subset of $E \times F$. Then by Fubini's theorem $y \to m[B(y)]$ is a Borel function on F.

Now assume E is a compact metric space and replace m by some Hausdorff measure Λ^h. It is not generally true that for any Borel B, $y \to \Lambda^h[B(y)]$ is Borel. For example, let Λ^h be the counting measure. Then, $\{y : \Lambda^h[B(y)] > 0\}$ is the projection of B into F. Nevertheless, since Λ^h is the limit of an increasing sequence of capacities and hence is a caliber, we get

Corollary. Suppose E is a compact metric space and Λ^h a Hausdorff measure on E. If A is an analytic subset of a product space $E \times F$, then the map $y \to \Lambda^h[A(y)]$ is an analytic function on F.

Chapter 4

Thick and thin sets with respect to a capacity

In this chapter, we take E to be a compact metrisable space.

§1. Definitions and examples. Suppose I is a capacity on E such that

(a) $I(\emptyset) = 0$

(b) for all $A \in \mathcal{P}(E)$, $I(A) = \inf\{I(B) : B \supseteq A$ and B is Borel$\}$

(c) $I(A_1) = 0$ and $I(A_2) = 0$ implies $I(A_1 \cup A_2) = 0$.

Condition (a) is imposed to avoid trivialities. Condition (b) is also not very important since we generally work with analytic A for which it is anyway true. Condition (c), together with the fact that I is a capacity, implies that the class $\mathfrak{n}$ of subsets of E of null capacity is a σ-ideal i.e. it is closed under taking of subsets and countable unions. We call a member of $\mathfrak{n}$ a null set.

Suppose we have fixed I.

Definition. An analytic subset A of E is called thin if for any family $\{A_\gamma : \gamma \in \Gamma\}$ of disjoint analytic subsets of A, A_γ is a null set for all but countably many $\gamma \in \Gamma$. A subset of E is called thin if it is contained in some thin analytic set. Call a subset thick if it is not thin.

Let

$$\tilde{J}(A) = 0 \text{ if } A \text{ is thin}$$
$$= 1 \text{ otherwise .}$$

The main result of this chapter is that $\tilde{J}$ is a caliber which is "going up."

We begin with an easy proposition which implies that $\tilde{J}$ is "going up."

Proposition 1.1. The class $\mathfrak{m}$ of the thin sets is a σ-ideal containing $\mathfrak{n}$.

Proof. The only nontrivial part is the proof of the fact that the countable union of thin analytic sets is thin. Let $\{A_n : n \geq 1\}$ be thin analytic sets and $A = \bigcup_n A_n$. Suppose $\{A_\gamma : \gamma \in \Gamma\}$ is an uncountable family of disjoint analytic subsets of A with $I(A_\gamma) > 0$ for all $\gamma \in \Gamma$. Since $\mathfrak{n}$ is a σ-ideal, for each γ there is an integer $n(\gamma) \geq 1$ such that $I(A_{n(\gamma)} \cap A_\gamma) > 0$. As Γ is uncountable, there is some n_0 such that $n(\gamma) = n_0$ for uncountably many values of γ. Thus A_{n_0} is thick. Contradiction!

In the next section, we show that any σ-ideal ζ such that $\mathfrak{n} \subseteq \zeta \subseteq \mathfrak{m}$ is completely determined by its compact elements. Such an ζ is called a

<u>σ-ideal of thin sets.</u> For the time being we assume this result. We now give some examples.

<u>Example 1.1.</u> If $I(\emptyset) = 0$ and $I(A) = 1$ for $A \neq \emptyset$, then $\mathfrak{n} = \{\emptyset\}$ and $\mathfrak{m}$ is the class of countable sets.

<u>Example 1.2.</u> Let I be a capacity, alternating of order infinity. It can be deduced from Mokobodzki's theorem (cf. Chapter 1) that I has a unique decomposition of the form $I = I_1 + I_2$ where I_1, I_2 are alternating of order infinity, every I_1 thin set is I_1-null and E is I_2 thin. In the case of the classical potential theory with I as the Newtonian capacity, we have $I_1 = I$ and $I_2 = 0$.

<u>Example 1.3.</u> Suppose I is the capacity Λ^h in Chapter 1, Example 2.4. It is easy to check that $\Lambda^h_\infty(A) = 0$ iff $\Lambda^h(A) = 0$. A set A is said to be σ-finite (for Λ^h) if there exists a sequence $\{A_n : n \geq 1\}$ such that $A \subseteq \bigcup_n A_n$ and $\Lambda^h(A_n) < \infty$ for each n. Since Λ^h is a σ-additive measure on the Borel σ-field of E, and since for any $A \subseteq E$, $\Lambda^h(A) = \inf\{\Lambda^h(B) : B \supseteq A$ and B is Borel$\}$, it is clear that any σ-finite set is thin.

The converse is true in some cases, for example if E is a compact subset of an Euclidean space. In the general case the problem is probably still open.

In any case, the class of σ-finite sets is a σ-ideal of thin sets.

<u>Example 1.4.</u> Suppose Ω is the family of right continuous maps from R_+ into E and $\{m_x : x \in E\}$ is the family of probability measures over Ω constructed from a Hunt Borel semigroup (for the notations, see end of Chapter 2; the reader is not obliged to know the definition of a Hunt Borel semigroup in order to understand something of this example). For each $x \in E$, m_x is what we know about the stochastic process $(X_t)_{t\in R_+}$ when $X_0(\omega) = x$ for each $\omega \in \Omega$. For $x \in E$, define a capacity I_x on E by

$$I_x(A) = m_x(\{\omega \in \Omega : X_t(\omega) \in A \text{ for some } t \in R_+\})$$

(The "going down on compacts" is a consequence of the fact that we are dealing with a Hunt semigroup.)

Since $x \to I_x(A)$ is an analytic function when A is analytic, we can define a capacity I_μ for each probability measure μ on E by $I_\mu(A) = \int I_x(A)d\mu(x)$. An analytic set A is called <u>polar</u> if $m_x(\{\omega : X_t(\omega) \in A$ for some $t > 0\}) = 0$, for every $x \in E$; in other words, it is a set which is (almost) never met by the process. To simplify matters, we suppose there is a probability measure λ on E such that A is polar iff A is

I_λ-null. This condition is in fact satisfied by a large class of Markov processes.

We call an analytic set semipolar[(1)] if $m_x(\{\omega : X_t(\omega) \in A$ for uncountably many $t\}) = 0$ for every $x \in E$; in other words, it is a set which is (almost) only countably met by the process[(2)]. Using Mokobodzki's theorem, it is possible to prove that an analytic set A is semipolar iff A is I_λ-thin and, for each $y \in A$, the set $\{y\}$ is semipolar.

§2. Thickness of a set. Let I be a capacity on E satisfying (a), (b), (c) at the beginning of the chapter. For an analytic subset A of E, the thickness $J(A)$ of A is defined as follows:

$$J(A) = \text{lub}\{t \geq 0 : \text{there exists an uncountable family } \{A_\gamma : \gamma \in \Gamma\}$$
$$\text{of disjoint analytic subsets of } A \text{ such that } I(A_\gamma) \geq t \text{ for all } \gamma \in \Gamma\}.$$

The thickness of an arbitrary $C \subseteq E$ is defined by $J(C) = \inf\{J(A) : A \text{ analytic}, A \supseteq C\}$.

Proposition 2.1.

(1) J is nondecreasing and going up.

(2) If A is analytic with $J(A) > t$, then there exist disjoint compact sets K_0 and K_1 such that $J(A \cap K_i) > t$, for $i = 0,1$.

Proof. The proof of (1) is similar to Proposition 1.1.

To prove (2), first note that by the capacitability theorem applied to I, there exists an uncountable family $\{K_\gamma : \gamma \in \Gamma\}$ of disjoint compact subsets of A, with $I(K_\gamma) > t$ for each $\gamma \in \Gamma$.

Now $\mathcal{K}(E)$ is a compact metrisable space and $\{K_\gamma : \gamma \in \Gamma\}$ is an uncountable subset of it. Hence this set has two distinct members K_α and K_β which are condensation points of it (i.e. any neighborhood of K_α or K_β in $\mathcal{K}(E)$ contains an uncountable number of K_γ's). To finish the proof, take K_0 and K_1 to be disjoint compact neighborhoods of K_α and K_β in E and note that for any neighborhood U of a compact set K in E, $\{L \in \mathcal{K}(E) : L \subseteq U\}$ is a neighborhood of K in $\mathcal{K}(E)$.

The main step in the proof that J is a caliber is a generalization of the classical Souslin theorem on uncountable analytic sets. We state it as a lemma.

---

(1) This is not the classical definition of a semipolar set in potential theory. However it is equivalent although this is difficult to prove.

(2) The class of semipolar sets is a σ-ideal of I_λ-thin sets.

Lemma 2.1. Let $\mathcal{C}$ be a map from $\mathcal{P}(E)$ into $\{0,1\}$ such that

(a) $\mathcal{C}$ is nondecreasing and going up.

(b) if A is analytic and $\mathcal{C}(A) = 1$, then there exist two disjoint compact sets K_0, K_1 such that $\mathcal{C}(A \cap K_i) = 1$ for $i = 0,1$.

Then if A is analytic and $\mathcal{C}(A) = 1$, there exist an upper semicontinuous map K from the Cantor space $\{0,1\}^N$ into $\mathcal{K}(E)$ such that

(1) the $K(\sigma)$ are disjoint.

(2) $\cup_\sigma K(\sigma)$ is compact and contained in A.

(3) for any σ and any open U such that $K(\sigma) \subseteq U$, $\mathcal{C}(U) = 1$.

Proof. Let A be an analytic set with $\mathcal{C}(A) = 1$. We can write $A = \theta(H_1,H_2,\ldots)$ where θ is an analytic operation and each H_n is a K_σ set. As we have seen in Chapter 2, we can suppose $H_n = N$ for all n. Let $\underline{m} = \{1,2,\ldots,m\}$.

We define, for each finite sequence s of 0's and 1's a $K_s \in \mathcal{K}(E)$ and for each natural number n, another natural number m_n such that the following holds.

(1) For all s, t, if t is an extension of s, then $K_t \subseteq K_s$.

(2) If s, t are incompatible, $K_t \cap K_s = \emptyset$.

(3) If $n = \ell(s)$, i.e. the length of s, and $A_n = \theta(\underline{m}_1,\ldots,\underline{m}_n,N,\ldots,N,\ldots)$ then $\mathcal{C}(A_n \cap K_s) = 1$ and $K_s \subseteq \overline{A}_n$.

Suppose we have constructed K_s and m_n for all s and n. Put $K(\sigma) = \bigcap_n K_{\sigma|n}$ where $\sigma|n$ denotes the finite sequence $\sigma(1)\sigma(2)\ \ldots\ \sigma(n)$. Then it results from (2) that the $K(\sigma)$ are disjoint and from (3) that for all σ, $K(\sigma) \subseteq \theta(\underline{m}_1,\underline{m}_2,\ldots,\underline{m}_n,\ldots)$, θ being right continuous on compacts, so that $K(\sigma) \subseteq A$. Now from (1) and (3) it follows that for any σ and any open $U \supseteq K(\sigma)$ there is some n such that $K_{\sigma|n} \subseteq U$. Thus $\mathcal{C}(U) = 1$ and K is upper semicontinuous, in particular $\cup_\sigma K_\sigma$ is compact. (Note that this can also be deduced from (2).)

We now proceed with the construction of the K_s and m_n by induction on $\ell(s)$.

First choose m_1 so that $\mathcal{C}(A_1) = 1$ where $A_1 = \theta(\underline{m}_1,N,N,\ldots)$. This is clearly possible. Since A_1 is analytic there exist, by hypothesis, two disjoint compact sets L_0 and L_1 such that $\mathcal{C}(A_1 \cap L_0) = \mathcal{C}(A_1 \cap L_1) = 1$.

Put $K_0 = \overline{A_1 \cap L_0}$, $K_1 = \overline{A_1 \cap L_1}$.

Now suppose K_s and m_n have been constructed for $\ell(s) \leq p$ and $n \leq p$. Choose m_{p+1} such that if $A_{p+1} = \theta(\underline{m}_1,\ldots,\underline{m}_p,\underline{m}_{p+1},N,\ldots)$, then $\mathcal{C}(A_{p+1} \cap K_s) = 1$ for all s with $\ell(s) \leq p$. This is possible since by hypothesis, $\mathcal{C}(A_p \cap K_s) = 1$ for each such s and there are only finitely many such s. Now A_{p+1} is analytic, so for a fixed s of length p, we can find disjoint compact sets L_{s0}, L_{s1} such that

$$C(A_{p+1} \cap K_s \cap L_{s0}) = C(A_{p+1} \cap K_s \cap L_{s1}) = 1 .$$

Put $K_{s0} = \overline{A_{p+1} \cap K_s \cap L_{s0}}$, $K_{s1} = \overline{A_{p+1} \cap K_s \cap L_{s1}}$.

This completes the construction.

Proposition 2.2. Let A be analytic with $J(A) > t$. Then there exist an upper semicontinuous map K from $\{0,1\}^N$ into $\mathcal{K}(E)$ such that each $K(\sigma) \subseteq A$ and $I(K(\sigma)) > t$ for all σ. In particular, $\bigcup_\sigma K(\sigma)$ is a compact subset of A and $J(\bigcup_\sigma K(\sigma)) > t$.

Proof. Choose $t' > t$ such that $J(A) > t'$ and put, for any $H \subseteq E$,

$$C(H) = 1 \quad \text{if} \quad J(H) > t'$$

$$= 0 \quad \text{otherwise} .$$

C satisfies the conditions of the lemma. Taking K as in the lemma, we have for any σ and any open $U \supseteq K(\sigma)$, $J(U) > t'$ so that $I(U) > t'$. Since I is a capacity, by right continuity, $I(K(\sigma)) \geq t'$.

Theorem 2.1. The thickness J is a caliber.

Proof. Let $\mathcal{D} \subseteq \mathfrak{M}^1(\mathcal{K}(E))$ consist of all probability measures λ on $\mathcal{K}(E)$ such that

(1) λ is diffuse

(2) there exists a compact subset Λ of $\mathcal{K}(E)$ made of disjoint compact subsets of E such that $\lambda(\Lambda) = 1$.

So we have $\lambda \in \mathcal{D}$ iff $(\forall K \in \mathcal{K}(E))(\lambda(\{K\}) = 0)$ and $(\exists \Lambda \in \mathcal{K}(\mathcal{K}(E)))((\lambda(\Lambda) = 1)$ and $(K_1 \in \Lambda$ and $K_2 \in \Lambda$ and $K_1 \neq K_2 \Rightarrow K_1 \cap K_2 = \emptyset))$.

It is easy to prove that $\mathcal{D}$ is a G_δ subset of the compact metrisable space $\mathfrak{M}^1(\mathcal{K}(E))$.

We claim that for an analytic subset A of E, $J(A) > t$ iff $(\exists \lambda \in \mathcal{D})\lambda(\{K \in \mathcal{K}(E) : I(K \cap A) > t\}) > 0$.

In Theorem 3.1 of Chapter 3, replacing F by $\mathcal{K}(E)$, J by I and A by $(A \times \mathcal{K}(E)) \cap \{(x,K) \in E \times \mathcal{K}(E) : x \in K\}$, we see that $\{K \in \mathcal{K}(E) : I(K \cap A) > t\}$ is an analytic subset of $\mathcal{K}(E)$.

Now to prove the "if" part, let $\lambda\{K \in \mathcal{K}(E) : I(K \cap A) > t\} > 0$ and Λ be the compact subset of $\mathcal{K}(E)$ corresponding to λ as in (2); then $\lambda(\{K \in \Lambda : I(K \cap A) > t\}) = \lambda(\{K \in \mathcal{K}(E) : I(K \cap A) > t\}) > 0$. Since λ is diffuse, the set $\{K \in \Lambda : I(K \cap A) > t\}$ is uncountable and so we have $J(A) > t$.

For the converse, let $K : \{0,1\}^N \to \mathcal{K}(E)$ be as in the preceding theorem and let μ be the image under K of the Lebesgue measure on $\{0,1\}^N$. Then μ is a diffuse probability measure on $\mathcal{K}(E)$ and $\mu(M) = 1$ where $M = \{K(\sigma) : \sigma \in \{0,1\}^N\}$. Now M is the image of $\{0,1\}^N$ under the Borel map K, hence M is analytic (in fact it is Borel, K being injective). Thus we can find a compact subset Λ of $\mathcal{K}(E)$ such that $\Lambda \subseteq M$ and $\mu(\Lambda) > 0$. Finally, let λ be the probability measure $\frac{1}{\mu(\Lambda)} \mu\big|_{\Lambda}$.

So we have proved that for analytic A

$$J(A) > t \leftrightarrow (\exists\lambda \in \mathcal{D})\lambda(\{K \in \mathcal{K}(E) : I(K \cap A) > t\}) > 0 .$$

It is easy to see that

$$(\exists\lambda \in \mathcal{D})\lambda(\{K \in \mathcal{K}(E) : I(K \cap A) > t\}) > 0 \leftrightarrow$$

$$(\exists\lambda \in \mathcal{D})\lambda(\{K \in \mathcal{K}(E) : I(K \cap A) > t\}) = 1 .$$

For any fixed $t \in R_+$, using the proof of Theorem 3.1 in Chapter 3 and the composition theorem, we have $A \to \{K : I(K \cap A) > t\}$ is an analytic operation.

Using the composition theorem again, we see that $A \to \sup_{\lambda\in\mathcal{D}} \lambda(\{K : I(K \cap A) > t\})$ is a caliber with value in $\{0,1\}$. Finally we get that, for fixed t, the function

$$J_t(A) = 1 \text{ if } J(A) > t$$

$$= 0 \text{ otherwise}$$

is a caliber. To finish the proof, note that for fixed A, $t \to J_t(A)$ is decreasing and that $J(A) = \int_0^\infty J_t(A)dt$ and approximate this integral by an increasing sequence of Riemann sums.

Corollary. If we set

$$\tilde{J}(A) = 1 \text{ if } A \text{ is thick}$$

$$= 0 \text{ otherwise ,}$$

then $\tilde{J}$ is a caliber.

Proof. Put $\tilde{J} = J_0$ where J_t is defined as in the theorem.

Structure of the σ-ideals of thin sets. Recall that a σ-ideal ζ of subsets of E is called a *σ-ideal of thin sets* if $\mathfrak{n} \subseteq \zeta \subseteq \mathfrak{m}$ where $\mathfrak{n}$ is the σ-ideal consisting of all null sets (i.e. sets of null capacity with respect to I) and $\mathfrak{M}$ is the σ-ideal consisting of all thin sets.

Theorem 2.2. Let ζ be a σ-ideal of thin sets. An analytic subset A of E belongs to ζ iff A is of the form $N \cup (\bigcup_n K_n)$ where K_n is a sequence of disjoint compact sets belonging to ζ and N is a null set.

Proof. The "if" part is trivial. For the "only if" let $A \in \zeta$. Consider a maximal family of disjoint compact subsets of A with a strictly positive capacity. Since A is thin, such a family is countable, say $\{K_n : n \geq 1\}$. Now $N = A - \bigcup_n K_n$ is analytic and does not contain a nonnull compact set. By the capacitability theorem, N is a null set.

Theorem 2.3. Let ζ be a σ-ideal of thin sets. An analytic subset A of E belongs to ζ iff every compact subset of A belongs to ζ.

Proof. The "only if" part is trivial. For the "if" part, suppose first A is thick. Then by Proposition 2.2, A contains a compact set which is thick and hence does not belong to ζ. Next suppose A is thin. Applying Theorem 2.2 to $\mathfrak{M}$, we find $A = N \cup (\bigcup_n K_n)$ where N is null and K_n is a thin compact set for $n \geq 1$. Now if every compact subset of A belongs to ζ, applying Theorem 2.2 to ζ, we get $A \in \zeta$.

References

N. Bourbaki [1975], Eléments de Mathématique, Topologie générale, 3ème édition (the printing is important!), Hermann, Paris 1975.

D. Busch [1973], Some problems connected with the Axiom of Determinacy, Ph.D. Thesis, Rockefeller University, 1973.

G. Choquet [1955], Theory of capacities, Ann. Inst. Fourier Grenoble, 5, 1955, p. 131-295.

G. Choquet [1959], Forme abstraite du théorème de capacitabilité, Ann. Inst. Fourier Grenoble, 9, 1959, p. 83-89.

R. O. Davies and C. A. Rogers [1969], The problem of subsets of finite positive measure, Bull. London Math. Soc., 1, 1969, p. 47-54.

C. Dellacherie [1972], Ensembles analytiques, capacités, mesures de Hausdorff, Lecture Notes in Mathematics, No. 295, Springer, Heidelberg 1972.

C. Dellacherie [1978], Un cours sur les ensembles analytiques, Proceedings of the Summer School of Analytic Sets, London 1978, to appear by Academic Press.

J. Jayne [1976], Structure of analytic Hausdorff spaces, Mathematika, **23**, 1976, p. 208-211.

C. Kuratowski [1958], Topologie, Volume I, 4ème édition, P.W.N., Warszawa 1958.

A. Louveau [1979], Familles séparantes pour les ensembles analytiques, C.R. Acad. Sc. Paris, **288**, 1979, p. 391-394.

G. Mokobodzki [1978], Ensembles à coupes dénombrables et capacités dominées par une mesure, Séminaire de Probabilités XII, Lecture Notes in Mathematics, No. 649, p. 492-511, Springer, Heidelberg 1978.

D. Preiss [1973], Metric spaces in which Prokhorov's theorem is not valid, Z. für Wahrscheinlichkeitstheorie, **27**, 1973, p. 109-116.

D. Shochat [1972], Capacitability of $\boldsymbol{\Sigma}^1_2$ sets, Ph.D. Dissertation, UCLA 1972.

M. Sion [1963], On capacitability and measurability, Ann. Inst. Fourier, Grenoble, **13**, 1963, p. 88-99.

HOMOGENEOUS TREES AND PROJECTIVE SCALES

Alexander S. Kechris*
Department of Mathematics
California Institute of Technology
Pasadena, California 91125

This exposition is a sequel to Kechris [1978]. Its main purpose is to show how set theoretical techniques, among them infinite exponent partition relations can be used to produce homogeneous trees for projective sets. The work here is again understood as being carried completely within $L[\mathbb{R}]$, with the hypothesis that AD + DC holds in this model. As applications, one has Kunen's important reduction of the problem of computing $\underset{\sim}{\delta}^1_5$ to the problem of computing certain ultrapowers of $\underset{\sim}{\delta}^1_3 = \omega_{\omega+1}$ and also a result of Martin on constructibility relative to subsets of $\omega_{\omega+1}$. An observation on the Victoria Delfino 3rd problem concludes the present paper.

Most of the results and constructions of trees presented in §§3,4,5 below are due to Kunen, and go back to his [1971a]. On the other hand, some of the effective calculations, in §§3,4, for the scales resulting from these trees need the recent results of Kechris and Martin [1978] and Harrington and Kechris [A].

§0. Introduction. In this introductory section we collect various notational conventions and some prerequisites needed to follow this paper.

0.1. Trees. If X is a set, X^ω is the set of all infinite sequences from X and $X^{<\omega}$ the set of all finite sequences from X. If $s,t \in X^{<\omega}$ and $f \in X^\omega$, then $s \subseteq t$ and $s \subseteq f$ denote the extension relation in each case. If $s = (x_0,\dots,x_{n-1})$, we let $s(i) \equiv s_i = x_i$ for $i < n$ and we put $\ell h s = n$. Sometimes we also write $(x_1,\dots,x_n)$ for $(y_0,\dots,y_{n-1})$, where $y_i = x_{i+1}$, $0 \leq i < n$. If $s = (x_0,\dots,x_{n-1})$ and $m \leq n$, then $s \upharpoonright m = (x_0,\dots,x_{m-1})$.

We reserve usually letters $\sigma,\tau,\dots$ for members of $\omega^{<\omega}$ and $u,v,w,\dots$ for members of $\mathrm{ORD}^{<\omega}$. As usual $\alpha,\beta,\gamma,\dots$ denote reals i.e. elements of

*The preparation of this paper was partially supported by NSF Grant MCS 76-17254 A01. The author is an A.P. Sloan Foundation Fellow. We would like to thank Y.N. Moschovakis for making a number of valuable suggestions for improving the presentation of this paper.

$\mathcal{R} = \omega^\omega$. We fix a recursive 1-1 correspondence $\tau_0, \tau_1, \tau_2, \ldots$ between ω and $\omega^{<\omega}$, such that $\tau_j \supsetneq \tau_i \Rightarrow j > i$ and if $\ell_i = \ell h \tau_i$ then $\ell_i \leq i$. Moreover we agree to take $\tau_0 = \emptyset$, $\tau_1 = (0)$.

By a tree on $\omega^k \times \lambda$, where $k \in \omega$ and $\lambda \in$ ORD, we mean a set T of $k+1$-tuples of the form $(\sigma_1, \ldots, \sigma_k, u) \in (\omega^{<\omega})^k \times \lambda^{<\omega}$, where each σ_i and u have all the same length and such that if $(\sigma_1, \ldots, \sigma_k, u) \in T$ and $n \leq \ell h \sigma_1$, then $(\sigma_1 \restriction n, \ldots, \sigma_k \restriction n, u \restriction n) \in T$. For such a tree T and for $\ell \leq k$ we let $T(\sigma_1, \ldots, \sigma_\ell) = \{(\sigma_{\ell+1}, \ldots, \sigma_k, u) : (\sigma_1, \ldots, \sigma_k, u) \in T\}$, $T(\alpha_1, \ldots, \alpha_\ell) = \bigcup_{n \in \omega} T(\alpha_1 \restriction n, \ldots, \alpha_\ell \restriction n)$ and finally $T^{\subseteq}(\sigma_1, \ldots, \sigma_\ell) = \bigcup_{n \leq \ell h \sigma_1} T(\sigma_1 \restriction n, \ldots, \sigma_\ell \restriction n)$. By $[T]$ we denote the set of all infinite branches through T i.e. $[T] = \{(\alpha_1, \ldots, \alpha_k, f) : \forall n(\alpha_1 \restriction n, \ldots, \alpha_k \restriction n, f \restriction n) \in T\}$. Let also $p[T] = \{(\alpha_1, \ldots, \alpha_k) : \exists f(\alpha_1, \ldots, \alpha_k, f) \in [T]\}$. For $X \subseteq \lambda$, $T \restriction X = \{(\sigma_1, \ldots, \sigma_k, u) \in T : u \in X^{<\omega}\}$.

A tree T on $\omega^k \times \lambda$ is <u>wellfounded</u> iff $[T] = \emptyset$. Equivalently T is wellfounded if the <u>Kleene-Brouwer ordering</u> $<_{KB}$ on $\bigcup_n (\omega^n)^k \times \lambda^n$ is a wellordering when restricted to T. Here $<_{KB}$ is defined as follows: Take the case $k = 1$ for notational simplicity:

$$((a_0, \ldots, a_{n-1}), (\xi_0, \ldots, \xi_{n-1})) <_{KB} ((b_0, \ldots, b_{m-1}), (\eta_0, \ldots, \eta_{m-1}))$$

$$\Leftrightarrow [(a_0, \ldots, a_{n-1}) \supsetneq (b_0, \ldots, b_{m-1}) \wedge$$

$$(\xi_0, \ldots, \xi_{n-1}) \supsetneq (\eta_0, \ldots, \eta_{m-1})] \vee$$

$$[\text{if } \ell \text{ is least such that } (a_\ell, \xi_\ell) \neq (b_\ell, \eta_\ell),$$

$$\text{then } \xi_\ell < \eta_\ell \text{ or } (\xi_\ell = \eta_\ell \text{ and } a_\ell < b_\ell)] .$$

(The use of the anti-lexicographical ordering of pairs $(a, \xi) \in \omega \times$ ORD will be convenient later on).

The usual rank function for a wellordering or wellfounded relation W will be denoted by rank_W and in case of a wellfounded tree T by rank_T.

0.2. <u>Scales.</u> If $P \subseteq \omega^k \times \mathcal{R}^m$ is a pointset, a <u>scale</u> on P is a sequence $\overline{\varphi} = \{\varphi_n\}$ of <u>norms</u> on P i.e. mappings into the ordinals, with the following property:

If $x_0, x_1, \ldots \in P$ and $x_i \to x$ and for all n, $\varphi_n(x_i)$ is eventually constant (as $i \to \infty$), say equal to λ_n, then $x \in P$ and $\varphi_n(x) \leq \lambda_n$. If each φ_n maps P into λ, we call $\overline{\varphi}$ a λ-scale. A scale $\overline{\varphi} = \{\varphi_n\}$ is <u>regular</u> if each norm φ_n maps P onto an ordinal.

Finally, if Γ is a pointclass and $\overline{\varphi}$ a scale on P, we call $\overline{\varphi}$ a Γ-<u>scale</u> if the following two relations are in Γ:

$$R(n,x,y) \Leftrightarrow x \in P \wedge [y \notin P \vee \varphi_n(x) \leq \varphi_n(y)] ,$$

$$Q(n,x,y) \Leftrightarrow x \in P \wedge [y \notin P \vee \varphi_n(x) < \varphi_n(y)] .$$

0.3. Indiscernibles. Assuming that $\forall \alpha (\alpha^{\#}$ exists), let for each real α $\mathcal{I}_\alpha$ be the class of Silver indiscernibles for $L[\alpha]$ and $\mathcal{U} = \bigcap_\alpha \mathcal{I}_\alpha = \{\underset{\sim}{u}_1, \underset{\sim}{u}_2, \ldots, \underset{\sim}{u}_\xi, \ldots\}_{\xi \in ORD}$ the class of uniform indiscernibles. Under AD, $\underset{\sim}{u}_n = \omega_n$ for $n < \omega$. By a result of Solovay each ordinal $< \underset{\sim}{u}_{n+1}$ can be written in the form $t^{L[\bar{\alpha}]}(\underset{\sim}{u}_1, \ldots, \underset{\sim}{u}_n)$, for some term t and some real α. For each $f : [\omega_1]^n \to \omega_1$ in $\tilde{L} = \bigcup_\alpha L[\alpha]$, where $[\omega_1]^n = \{(\xi_0, \ldots, \xi_{n-1}) \in \omega_1^n : \xi_0 < \xi_1 < \cdots < \xi_{n-1}\}$, define $\tilde{f}(\underset{\sim}{u}_1, \ldots, \underset{\sim}{u}_n) = t^{L[\alpha]}(\underset{\sim}{u}_1, \ldots, \underset{\sim}{u}_n)$, where $f(\xi_1, \ldots, \xi_n) = t^{L[\alpha]}(\xi_1, \ldots, \xi_n)$ for $\xi_1 < \cdots < \xi_n < \omega_1$. Thus every ordinal $< \underset{\sim}{u}_{n+1}$ has the form $\tilde{f}(\underset{\sim}{u}_1, \ldots, \underset{\sim}{u}_n)$ for some $f : [\omega_1]^n \to \omega_1$ in $\tilde{L}$. If now $H : \omega_1 \to \omega_1$ is in $\tilde{L}$ define $\tilde{H} : \underset{\sim}{u}_\omega \to \underset{\sim}{u}_\omega$ by $\tilde{H}(\tilde{f}(\underset{\sim}{u}_1, \ldots, \underset{\sim}{u}_n)) = \widetilde{H \circ f}(\underset{\sim}{u}_1, \ldots, \underset{\sim}{u}_n)$. For $X \subseteq \omega_1$ let also $\tilde{X} \subseteq \underset{\sim}{u}_\omega$ be defined by:

$\tilde{f}(\underset{\sim}{u}_1, \ldots, \underset{\sim}{u}_n) \in \tilde{X} \Leftrightarrow \exists C \subseteq \omega_1$ (C is closed unbounded(cub) in $\omega_1 \wedge$ for all $\xi_1 < \cdots < \xi_n$ in C, $f(\xi_1, \ldots, \xi_n) \in C$).

0.4. The work in this paper takes place in ZF + DC until otherwise specified.

§1. Π^1_1 sets; the tree S_1.

1.1. Definition of S_1. (a) Let $A \subseteq \mathbb{R}$ be a Π^1_1 set of reals. Then there is a recursive tree T on $\omega \times \omega$ such that

$$\alpha \in A \Leftrightarrow T(\alpha) \text{ is wellfounded} .$$

If $\sigma \in \omega^{<\omega}$ and $\ell h\sigma = n$ define the following ordering $<_\sigma$ on $\{0,1,\ldots,n-1\} = n$:

$$i <_\sigma j \overset{def}{\Leftrightarrow} \begin{array}{ll} & 1.\ \tau_i, \tau_j \notin T^{\subseteq}(\sigma) \wedge i < j \\ \vee & 2.\ \tau_i \notin T^{\subseteq}(\sigma) \wedge \tau_j \in T^{\subseteq}(\sigma) \\ \vee & 3.\ \tau_i, \tau_j \in T^{\subseteq}(\sigma) \wedge \tau_i <_{KB} \tau_j . \end{array}$$

Thus identifying τ_i with i here, we can visualize $<_\sigma$ as being the Kleene-Brouwer ordering of $T^{\subseteq}(\sigma) \cap n$, with the rest of n thrown in at the bottom with its natural ordering. Note now the following:

(i) If $\sigma \neq \emptyset$, 0 is the top element of $<_\sigma$.

(ii) $\sigma \subseteq \sigma' \Rightarrow <_\sigma \subseteq <_{\sigma'}$ (i.e. $<_\sigma$ is a subordering of $<_{\sigma'}$).

That (i) holds comes from the fact that $\emptyset = \tau_0$ is the top element of $<_{KB}$. That (ii) holds is an immediate consequence of the remark that for $i < \ell h\sigma$, $\sigma \subseteq \sigma'$

$$\tau_i \in T^{\subseteq}(\sigma) \Leftrightarrow \tau_i \in T^{\subseteq}(\sigma') .$$

This is of course because always $\ell_i = \ell h\tau_i \leq i$, thus $\tau_i \in T^{\subseteq}(\sigma) \Leftrightarrow (\sigma \restriction \ell_i, \tau_i) \in T \Leftrightarrow (\sigma' \restriction \ell_i, \tau_i) \in T \Leftrightarrow \tau_i \in T^{\subseteq}(\sigma')$.

Put now for each $\alpha \in \mathcal{R}$:

$$<_\alpha =^{def} \bigcup_n <_{\alpha \restriction n} ,$$

so that $<_\alpha$ is a linear ordering of (all of) ω, with top element 0 again. Note that $<_\alpha$ is just the Kleene-Brouwer ordering on $T(\alpha)$ with the rest of ω thrown in at the bottom with its natural ordering. Thus

$$\begin{aligned} \alpha \in A &\Leftrightarrow T(\alpha) \text{ is wellfounded} \\ &\Leftrightarrow \langle T(\alpha), <_{KB} \rangle \text{ is wellordered} \\ &\Leftrightarrow <_\alpha \text{ is a wellordering} . \end{aligned}$$

(b) Define now:

$$\begin{aligned} S_1(A;T) = S_1 =^{def} \{(\sigma,u) : \sigma \in \omega^{<\omega},\ u \in \omega_1^{<\omega} \wedge \\ \ell h\sigma = \ell hu (= n, \text{ say}) \wedge \\ u : n \to \omega_1 \text{ is order preserving} \\ \text{relative to } <_\sigma \text{ i.e. for } 0 \leq i, \\ j < n : i <_\sigma j \Leftrightarrow u_i < u_j\} . \end{aligned}$$

Then we obviously have that:

$$\begin{aligned} \alpha \in A &\Leftrightarrow \exists f (\alpha, f) \in [S_1] \\ &\Leftrightarrow S_1(\alpha) \text{ is not wellfounded} . \end{aligned}$$

1.2. Scales for Π_1^1 sets. (a) If J is a tree on λ we shall say that J has an honest leftmost branch if there is a branch $f \in [J]$ such that for all branches $g \in [J]$:

$$\forall i, \quad f(i) \leq g(i) .$$

Every non-wellfounded tree J has a leftmost branch $h \in [J]$, which is by definition characterized by the property that for $g \in [J]$:

$$h \leq_{\ell ex} g \quad \text{i.e.} \quad h = g \vee \exists i[h(i) < g(i) \wedge \forall j < i(h(j) = g(j))] .$$

But only special trees J have honest leftmost branches. We shall show that those of the form $S_1(\alpha)$ are among them.

Indeed, let for $\alpha \in A$:

$$f_\alpha(i) = \text{rank}_{<_\alpha}(i) .$$

Clearly $f_\alpha \in [S_1(\alpha)]$. On the other hand if $g \in [S_1(\alpha)]$, then $g : \omega \to \omega_1$ is such that

$$i <_\alpha j \Leftrightarrow g(i) < g(j) ,$$

thus

$$f_\alpha(i) = \text{rank}_{<_\alpha}(i) \leq g(i) ,$$

and we are done.

(b) Define now for $\alpha \in A$:

$$\varphi_i(\alpha) = f_\alpha(i) .$$

Then $\{\varphi_i\}$ is a scale on A (and thus an ω_1-scale). Indeed, letting $\overline{\varphi}(\alpha) = (\varphi_0(\alpha), \varphi_1(\alpha), \ldots)$ and assuming that for all n, $\alpha_n \in A$ while $\alpha_n \to \alpha$ and $\overline{\varphi}(\alpha_n) \to g$ (i.e. $\varphi_i(\alpha_n) = g(i)$, for all large enough n) we conclude that $(\alpha, g) = \lim(\alpha_n, \overline{\varphi}(\alpha_n)) \in [S_1]$ and moreover $f_\alpha(i) = \varphi_i(\alpha) \leq g(i)$.

Now by the usual argument (see for example Kechris and Moschovakis [1978]) one can show that if for $\alpha \in A$ we put

$$\psi_i(\alpha) = \langle \varphi_0(\alpha), \varphi_i(\alpha) \rangle ,$$

where $\langle \xi, \eta \rangle \stackrel{\text{def}}{=}$ the ordinal attached to the pair (ξ, η) in the lexicographical ordering of $\omega_1 \times \omega_1$, then $\{\psi_i\}$ is a Π^1_1-scale on A. Thus we have shown that Π^1_1 has the scale property.

1.3. <u>Homogeneity properties of</u> S_1. (a) Fix now $\sigma \in \omega^{<\omega}$, $\sigma \neq \emptyset$. Let $\ell h \sigma = n$ and define

$$\pi_\sigma : n \to n$$

to be the unique permutation of n defined by:

$$i <_\sigma j \Leftrightarrow \pi_\sigma(i) < \pi_\sigma(j) .$$

In particular,

$$\pi_\sigma(0) = n - 1 .$$

Then note that if for any $A \subseteq \mathrm{ORD}$ and any $\eta \in \mathrm{ORD}$ we let

$$[A]^\eta \stackrel{\mathrm{def}}{=} \text{the set of all increasing maps from } \eta \text{ into } A ,$$

we have

$$(\sigma, v) \in S_1 \Leftrightarrow \ell h v = n \wedge$$
$$\exists u \in [\omega_1]^n (v = u \circ \pi_\sigma, \text{ i.e.}$$
$$v = (u_{\pi_\sigma(0)}, u_{\pi_\sigma(1)}, \ldots, u_{\pi_\sigma(n-1)})) .$$

Thus

$$S_1(\sigma) = ([\omega_1]^{\ell h \sigma})\pi_\sigma \stackrel{\mathrm{def}}{=} \{(\eta_{\pi_\sigma(0)}, \ldots, \eta_{\pi_\sigma(n-1)}) : \eta_0 < \eta_1 < \cdots < \eta_{n-1} < \omega_1\} ,$$

so that $S_1(\sigma)$ is just a "permutation" of $[\omega_1]^{\ell h \sigma}$.

Since $\pi_\sigma(0) = n - 1$ we have the following important "boundedness" property:

$$(\xi_0, \ldots, \xi_{n-1}) \in S_1(\sigma) \Leftrightarrow \xi_0 > \xi_1, \xi_2, \ldots, \xi_{n-1} .$$

This will be quite useful in §2.

(b) Assume now $H : \omega_1 \to \omega_1$ is an increasing map i.e. $H \in [\omega_1]^{\omega_1}$. For any $u = (\xi_0, \ldots, \xi_{n-1}) \in \omega_1^n$ we put also $H(u) = (H(\xi_0), \ldots, H(\xi_{n-1}))$ and for any $(\sigma, u) \in \omega^n \times \omega_1^n$ we let again

$$H(\sigma, u) = (\sigma, H(u)) .$$

The claim now is that

$$H : S_1 \to S_1 ,$$

i.e. S_1 is invariant under H, so that in particular for any α:

$$H : S_1(\alpha) \to S_1(\alpha) .$$

This is of course because if $(\sigma,v) \in S_1$ then for some $u \in [\omega_1]^{\ell h\sigma}$, $v = u \circ \pi_\sigma$ so that $H(v) = H(u) \circ \pi_\sigma$, where, since H is increasing, $H(u) \in [\omega_1]^{\ell h\sigma}$. Thus $(\sigma, H(v)) \in S_1$. Since $(\sigma,u) \mapsto H(\sigma,u)$ clearly preserves the relation of proper extension between sequences we have, letting $D =$ range(H):

$S_1(\alpha)$ is not wellfounded $\Rightarrow S_1 \upharpoonright D(\alpha)$ is not wellfounded.

So we have shown that for each $D \subseteq \omega_1$, D uncountable, we have for all α:

$$\alpha \in A \Leftrightarrow S_1(\alpha) \text{ is not wellfounded}$$

$$\Leftrightarrow S_1 \upharpoonright D(\alpha) \text{ is not wellfounded},$$

so that the non-wellfoundedness of $S_1(\alpha)$ depends only on its restriction to any uncountable subset of ω_1. This too will be useful in §2.

Note. The construction of S_1 is due to Shoenfield [1961]. The homogeneity properties of S_1 have been studied and used by Solovay, Mansfield [1971] and Martin [A].

§2. Π^1_2 sets; the tree S_2. Assume from now on and for the rest of this paper that $\forall\alpha(\alpha^\#$ exists).

2.1. Definition of S_2. (a) Let $A \subseteq \mathcal{R}$ be Π^1_2. Then for some Π^1_1 set $B \subseteq \mathcal{R} \times \mathcal{R}$ we have

$$\alpha \in A \Leftrightarrow \neg \exists\beta\, B(\alpha,\beta)$$

$$\Leftrightarrow \neg \exists\beta\, \exists f(\alpha,\beta,f) \in [S_1]$$

$$\Leftrightarrow S_1(\alpha) \text{ is wellfounded},$$

where S_1 is the tree associated with B as in §1. Strictly speaking we have talked in §1 only about subsets of $\mathcal{R}$ but it is obvious how to modify this discussion so that it applies to Π^1_1 subsets of any $\mathcal{R} \times \mathcal{R} \times \cdots \times \mathcal{R}$. Thus the tree S_1 associated with B will be a tree on $\omega \times \omega \times \omega_1$, so that $S_1(\alpha)$ is a tree on $\omega \times \omega_1$. A typical element of S_1 will be a triple (σ,τ,u), where $\ell h\sigma = \ell h\tau = \ell hu = n$ and $u : n \to \omega_1$ is order preserving relative to $<_{\sigma,\tau}$, which is defined as in 1.1 by replacing σ by σ, τ everywhere.

Let now for $\sigma \neq \emptyset$, $\ell h\sigma = n$:

$$S_1^*(\sigma) =^{\text{def}} \{(\tau_i,u) : (\sigma \upharpoonright \ell_i, \tau_i, u) \in S_1 \wedge 1 \leq i \leq n\}.$$

Then $\langle S_1^*(\sigma), <_{KB}\rangle$ is clearly a wellordering as $S_1^*(\sigma)$ contains only sequences of bounded length. We shall denote it by W_σ and its order type by ρ_σ. Note that

$$\emptyset \neq \sigma \subseteq \sigma' \Rightarrow W_\sigma \subseteq W_{\sigma'} .$$

If $S_1(\alpha)$ is wellfounded, let again W_α denote the wellordering $\langle S_1(\alpha), <_{KB}\rangle$ and ρ_α its order type. As $S_1(\alpha) = \bigcup_{n\geq 1} S_1^*(\alpha \restriction n)$ we have $W_\alpha = \bigcup_{n\geq 1} W_{\alpha\restriction n}$.

One should note now that for each $\sigma \neq \emptyset$

$$\rho_\sigma = \omega_1 .$$

Clearly $\rho_\sigma \geq \omega_1$ since $S_1^*(\sigma)$ is uncountable, as $S_1^*(\sigma) \supseteq \{((0),u) : (\sigma \restriction 1,(0),u) \in S_1\} = \{((0),(\xi)) : \xi < \omega_1\}$, since $\tau_1 = (0)$, $\ell_1 = 1$. On the other hand if $(\tau_i,u),(\tau_j,v) \in S_1^*(\sigma)$, say with $v = (\xi_0,\ldots,\xi_{\ell_i-1})$, $u = (\eta_0,\ldots,\eta_{\ell_j-1})$ then since $\ell_j \geq 1$ (as we took $j \geq 1$) we must have $\eta_0 > \eta_1,\ldots,\eta_{\ell_j-1}$ so that if $(\tau_j,u) <_{KB} (\tau_i,v)$ then $\eta_0 \leq \xi_0$ (recall the use of antilexicographical ordering of pairs in $<_{KB}$ here) so that $\eta_0,\eta_1,\ldots,\eta_{\ell_j-1} \leq \xi_0$ i.e. there are only countably many predecessors of (τ_i,v). Thus $\rho_\sigma \leq \omega_1$. (The reason we have excluded $(\emptyset,\emptyset)$ from $S_1^*(\sigma)$ was precisely so that $\rho_\sigma = \omega_1$; otherwise it would have been $\omega_1 + 1$, which would be technically awkward). Similarly if $S_1(\alpha)$ is wellfounded, $\rho_\alpha = \omega_1$.

Let as usual

$$\widetilde{L} = \bigcup_{\alpha\in\mathcal{R}} L[\alpha] .$$

Let also for each wellordering W and set $A \subseteq$ ORD, $[A]^W$ be the set of all increasing mappings from (the domain of) W into A. As in this notation, we sometimes do not distinguish between W and its domain, when convenient. To each $h \in [\omega_1]^{W_\sigma} \cap \widetilde{L}$, where $\sigma \neq \emptyset$, we will assign a tuple of ordinals $p_\sigma(h) = (\xi_1,\ldots,\xi_{\ell h\sigma})$ as follows:

Since h maps $S_1^*(\sigma)$ into ω_1, it splits into $\ell h\sigma = n$ many maps $h_1,\ldots,h_n$, where $\text{domain}(h_i) = S_1(\sigma \restriction \ell_i,\tau_i)$ and

$$h_i(u) = h(\tau_i,u) .$$

Now h_i can be identified with the map $h_i' : [\omega_1]^{\ell_i} \to \omega_1$ given by

$$h_i'(v) = h_i(v \circ \pi_{\sigma\restriction \ell_i,\tau_i}) ,$$

since $S_1(\sigma \upharpoonright \ell_i, \tau_i) = [\omega_1]^{\ell_i} \pi_{\sigma \upharpoonright \ell_i, \tau_i}$, as we saw in 1.3.(a). Put now

$$p_\sigma(h) = (\tilde{h}'_i(\underset{\sim}{u}_1, \ldots, \underset{\sim}{u}_{\ell_i}))_{1 \leq i \leq n} .$$

We are now ready to define:

$$(\sigma, u) \in S_2 \overset{\text{def}}{\Leftrightarrow} \exists h \in [\omega_1]^{W_\sigma} \cap \tilde{L}(p_\sigma(h) = u) \vee (\sigma = u = \emptyset) .$$

(Again S_2 depends on A, B, S_1 but we won't indicate this explicitly).

We first verify that this in indeed a tree: Let $(\sigma, u) \in S_2$ and $1 \leq n \leq \ell h\sigma = \ell hu$. Put $\sigma \upharpoonright n = \sigma'$. We can find $h \in [\omega_1]^{W_\sigma} \cap \tilde{L}$ such that $p_\sigma(h) = u$. Let $h' \in [\omega_1]^{W_{\sigma'}} \cap \tilde{L}$ be defined by $h' = h \upharpoonright W_{\sigma'}$. Note here that S_1, W_σ, $W_{\sigma'}$ are in L so that $h' \in \tilde{L}$. Then clearly $p_{\sigma'}(h') = u \upharpoonright n$, therefore $(\sigma \upharpoonright n, u \upharpoonright n) \in S_2$.

(b) We shall verify now that

$$\alpha \in A \Leftrightarrow S_2(\alpha) \text{ is not wellfounded} .$$

Indeed, if $\alpha \in A$, $S_1(\alpha)$ is wellfounded so let $h : W_\alpha \to \omega_1$ be the rank function of W_α, i.e. the unique isomorphism between W_α and ω_1. Clearly $h \in L[\alpha]$. Then if for each $n \geq 1$, $h^n = h \upharpoonright W_{\alpha \upharpoonright n}$ we have that $p_{\alpha \upharpoonright n}(h^n) \in S_2(\alpha \upharpoonright n)$ and also $p_{\alpha \upharpoonright 1}(h^1) \subseteq p_{\alpha \upharpoonright 2}(h^2) \subseteq \cdots$, so $S_2(\alpha)$ is not wellfounded. Conversely, assume $(\eta_1, \eta_2, \ldots)$ is an infinite branch through $S_2(\alpha)$. Then for each $n \geq 1$, let ${}^n h \in [\omega_1]^{W_{\alpha \upharpoonright n}} \cap \tilde{L}$ be such that $p_{\alpha \upharpoonright n}({}^n h) = (\eta_1, \ldots, \eta_n)$. Then clearly for all $i \geq 1$ and all $n, m \geq i$, $\widetilde{{}^n h'_i}(\underset{\sim}{u}_1, \ldots, \underset{\sim}{u}_{\ell_i}) = \widetilde{{}^m h'_i}(\underset{\sim}{u}_1, \ldots, \underset{\sim}{u}_{\ell_i})$. So find $C^i_{n,m}$ a cub subset of ω_1 such that for all $v \in [C^i_{n,m}]^{\ell_i}$:

$${}^n h'_i(v) = {}^m h'_i(v) .$$

This allows us to define the following map q from $S_1 \upharpoonright C(\alpha)$ into the ordinals, where

$$C = \bigcap_{1 \leq i \leq n, m} C^i_{n,m} :$$

$$q(\tau_i, u) = {}^n h'_i(v) = {}^n h(\tau_i, u) ,$$

where $n \geq i$ and $u = v \circ \pi_{\alpha \upharpoonright \ell_i, \tau_i}$. We now claim that this is an order preserving map from $\langle S_1 \upharpoonright C(\alpha), <_{KB} \rangle$ into the ordinals. Indeed, if

$(\tau_i,u),(\tau_j,w) \in S_1 \restriction C(\alpha)$ and

$$(\tau_i,u) <_{KB} (\tau_j,w)$$

then $(\tau_i,u),(\tau_j,w) \in S_1^*(\alpha \restriction n)$, for $n \geq i,j$, thus

$$^n h(\tau_i,u) < {}^n h(\tau_j,w)$$

$\therefore$

$$q(\tau_i,u) < q(\tau_j,w) .$$

So $S_1 \restriction C(\alpha)$ is wellfounded and therefore $S_1(\alpha)$ is wellfounded by 1.3.(b) i.e. $\alpha \in A$.

2.2 Scales for Π^1_2 sets. We claim now that for each $\alpha \in A$, $S_2(\alpha)$ has an honest leftmost branch, Indeed in the notation of 2.1.(b) if $h : W_\alpha \to \omega_1$ is as defined there and we let for each $i > 0$, $h_i(u) = h(\tau_i,u)$, for $u \in S_1(\alpha \restriction \ell_i,\tau_i)$ and $\xi_i = \widetilde{h_i'}(\underset{\sim}{u}_1,\dots,\underset{\sim}{u}_{\ell_i})$, then clearly $(\xi_1,\xi_2,\dots) \in [S_2(\alpha)]$. Moreover if $(\eta_1,\eta_2,\dots)$ is any branch of $[S_2(\alpha)]$, then again in the notation of 2.1.(b) q is an order preserving map from $S_1 \restriction C(\alpha)$ into ω_1. Let $H : \omega_1 \to C$ be the normal function enumerating C. Then by 1.3 $H : S_1(\alpha) \to S_1 \restriction C(\alpha)$ and of course H preserves $<_{KB}$ so $h(\tau_i,u) \leq q(\tau_i,H(u))$. Now let $D \subseteq C$ be cub such that $H(\xi) = \xi$ for $\xi \in D$. Then $h(\tau_i,u) \leq q(\tau_i,u)$, for $u \in D^{\ell_i}$, thus $h_i'(v) \leq {}^n h_i'(v)$, for $v \in [D]^{\ell_i}$ $\therefore$ $\xi_i = \widetilde{h_i'}(\underset{\sim}{u}_1,\dots,\underset{\sim}{u}_{\ell_i}) \leq \widetilde{{}^n h_i'}(\underset{\sim}{u}_1,\dots,\underset{\sim}{u}_{\ell_i}) = \eta_i$ and we are done.

We can now define for each $\alpha \in A$ and each $i \geq 1$

$$\varphi_i(\alpha) = f_\alpha(i) ,$$

where f_α is the leftmost branch of $S_2(\alpha)$ as above. Clearly $\overline{\varphi}(\alpha) = (\varphi_1(\alpha),\varphi_2(\alpha),\dots)$ is a scale on A. We want to show actually that it is a Δ^1_3-scale. For this note that

$$\varphi_i(\alpha) = \widetilde{h_i'}(\underset{\sim}{u}_1,\dots,\underset{\sim}{u}_{\ell_i}) ,$$

where $h_i'(v) = \mathrm{rank}_{W_\alpha}(v \circ \pi_{\alpha \restriction \ell_i,\tau_i})$, for $v \in [\omega_1]^{\ell_i}$. Thus for $\alpha,\beta \in A$

$$\varphi_i(\alpha) \leq \varphi_i(\beta) \Leftrightarrow \mathrm{rank}_{W_\alpha}(v \circ \pi_\alpha \restriction \ell_i, \tau_i)$$

$$\leq \mathrm{rank}_{W_\beta}(v \circ \pi_\beta \restriction \ell_i, \tau_i),$$

for all $v \in [C]^{\ell_i}$, where C is some cub subset of ω_1

$\Leftrightarrow L[\alpha,\beta] \models \theta_i(\alpha,\beta,v)$, for all $v \in [C]^{\ell_i}$, where $C \subseteq \omega_1$ is some cub set ,

(here θ_i is some formula recursively determined from i),

$$\Leftrightarrow L[\alpha,\beta] \models \theta_i(\alpha,\beta,\underset{\sim}{u}_1,\ldots,\underset{\sim}{u}_{\ell_i})$$

$$\Leftrightarrow \langle\alpha,\beta\rangle^{\#}(\ulcorner\theta_i\urcorner) = 0 ,$$

which is obviously Δ^1_3.

<u>Remark</u>. Note that we can also describe $\varphi_i(\alpha)$ as follows:

Let $\hat{S}_1 = \{(\sigma,\tau,u) : \ell h\sigma = \ell h\tau = \ell h u$ (= say, n)

$\wedge\ \sigma,\tau \in \omega^{<\omega} \wedge u \in \mathrm{ORD}^{<\omega} \wedge$

$u : n \to \mathrm{ORD}$ is order preserving

relative to $<_{\sigma,\tau}\}$.

Thus $\hat{S}_1$ is the "liftup" of S_1 to all ordinals. Clearly $\hat{S}_1 \restriction \omega_1 = S_1$ and $\hat{S}_1$ is a definable class in L. Now by an easy indiscernibility argument we have for all $i \geq 1$:

$$\varphi_i(\alpha) = f_\alpha(i) = \mathrm{rank}_{\langle\hat{S}_1(\alpha),<_{KB}\rangle}(\tau_i,(\underset{\sim}{u}_1,\ldots,\underset{\sim}{u}_{\ell_i}) \circ \pi_\alpha \restriction \ell_i,\tau_i) .$$

2.3. <u>Homogeneity properties of</u> S_2. <u>The tree</u> S_2^-. <u>From now on and for the rest of this paper (except for portion of 2.5) assume full</u> AD. (What we have in mind of course is that we work completely inside $L[\mathbb{R}]$ granting that $L[\mathbb{R}] \models$ AD, so that every one of our results below which is sufficiently absolute holds also in the real world.)

(a) For each $A \subseteq \mathrm{ORD}$ and W a wellordering, let $A \not\models^W$ denote the subset of $[A]^W$ defined as follows (where $W = \langle S,<\rangle$):

$$h \in A \not\models^W \Leftrightarrow h \in [A]^W \text{ and}$$

(i) $\forall x \in S(h(x) > \sup\{h(y) : y < x\})$

(ii) There is $\{\xi_n^x\}_{x\in S}$ with $\xi_0^x < \xi_1^x < \cdots \to h(x)$, for all $x \in S$.

According to a result of Martin (see Kechris [1978] for example) we have for any W of order type ω_1 and any $X \subseteq [\omega_1]^W$, that there is $C \subseteq \omega_1$ cub such that

$$C \models^W \subseteq X \quad \text{or} \quad C \models^W \subseteq {\sim}X .$$

This clearly defines a measure (i.e. countably additive ultrafilter) on $\omega_1 \models^W$.

Since W_σ has order type ω_1 and since $p_\sigma : [\omega_1]^{W_\sigma} \xrightarrow{\text{onto}} S_2(\sigma)$, this induces a measure on $S_2(\sigma)$, when $\sigma \neq \emptyset$. For the purpose of making this measure more explicit we shall actually consider the subtree S_2^- of S_2 defined by

$$(\sigma,u) \in S_2^- \Leftrightarrow \exists h \in \omega_1 \models^{W_\sigma}(p_\sigma(h) = u) \vee (\sigma = u = \emptyset)$$

It is trivial of course to check that also,

$$\alpha \in A \Leftrightarrow S_2^-(\alpha) \text{ is not wellfounded} .$$

Let $\bar{u}_\sigma$ be the above mentioned measure on $\omega_1 \models^{W_\sigma}$ i.e.

$$X \in \bar{u}_\sigma \Leftrightarrow \exists C \subseteq \omega_1 \ (C \text{ cub} \wedge C \models^{W_\sigma} \subseteq X)$$

and let u_σ be the measure on $S_2^-(\sigma)$ induced by p_σ i.e. $u_\sigma = (p_\sigma)_* \bar{u}_\sigma$ or explicitly, for $A \subseteq S_2^-(\sigma)$:

$$A \in u_\sigma \Leftrightarrow \exists C \subseteq \omega_1[C \text{ cub} \wedge p_\sigma[C \models^{W_\sigma}] \subseteq A] .$$

We shall actually show that u_σ is generated by the sets of the form

$$S_2^-(\sigma) \cap \tilde{C}^{\ell h\sigma} ,$$

where $C \subseteq \omega_1$ is cub. In other words, we claim that for $A \subseteq S_2^-(\sigma)$:

$$A \in u_\sigma \Leftrightarrow \exists C \subseteq \omega_1 \ [C \text{ cub} \wedge S_2^-(\sigma) \cap \tilde{C}^{\ell h\sigma} \subseteq A] .$$

For that is clearly enough to prove the following:

<u>Lemma</u>. For any uncountable $C \subseteq \omega_1$,

$$p_\sigma[C \models^{W_\sigma}] = S_2^-(\sigma) \cap \tilde{C}^{\ell h\sigma} .$$

Proof. ($\subseteq$): If $h : W_\sigma \to C$ is order preserving, then

$$p_\sigma(h) = (\tilde{h}_i'(\underset{\sim}{u}_1,\ldots,\underset{\sim}{u}_{\ell_i}))_{1\leq i\leq \ell h\sigma} ,$$

where

$$h_i'(v) = h(\tau_i, v \circ \pi_{\sigma\restriction \ell_i,\tau_i}) \in C \therefore \tilde{h}_i'(\underset{\sim}{u}_1,\ldots,\underset{\sim}{u}_{\ell_i}) \in \tilde{C} \therefore p_\sigma(h) \in S_2^-(\sigma) \cap \tilde{C}^{\ell h\sigma}.$$

($\supseteq$) Let $(\xi_1,\ldots,\xi_n) \in S_2^-(\sigma) \cap \tilde{C}^{\ell h\sigma}$. Then for some $h \in \omega_1 \models^{W}\sigma$, $p_\sigma(h) = (\xi_1,\ldots,\xi_n)$, i.e. $\xi_i = \tilde{h}_i'(\underset{\sim}{u}_1,\ldots,\underset{\sim}{u}_{\ell_i})$, where $h_i'(v) = h(\tau_i, v \circ \pi_{\sigma\restriction \ell_i,\tau_i})$, for $v \in [\omega_1]^{\ell_i}$. Moreover, $\tilde{h}_i'(\underset{\sim}{u}_1,\ldots,\underset{\sim}{u}_{\ell_i}) \in \tilde{C}$ $\therefore$ there is $D \subseteq C$, D cub such that

$$h_i'(v) = h(\tau_i, v \circ \pi_{\sigma\restriction \ell_i,\tau_i}) \in C$$

for all $v \in [D]^{\ell_i}$. Let $H : \omega_1 \to D$ be the normal enumeration of D. Put

$$g(\tau_i,u) = h(\tau_i,H(u)) .$$

Then $g \in [C]^{W_\sigma}$. Also if $g_i(u) = g(\tau_i,u)$ and $g_i'(v) = g_i(v \circ \pi_{\sigma\restriction \ell_i,\tau_i})$, for $v \in [\omega_1]^{\ell_i}$, then $g_i'(v) = h_i'(v)$ for $v \in [E]^{\ell_i}$, where E is cub and $\xi \in E \Rightarrow H(\xi) = \xi$. Thus $\xi_i = \tilde{h}_i'(\underset{\sim}{u}_1,\ldots,\underset{\sim}{u}_{\ell_i}) = \tilde{g}_i'(\underset{\sim}{u}_1,\ldots,\underset{\sim}{u}_{\ell_i}) \in C$.

So it only remains to show that actually $g \in C \models^{W_\sigma}$. Recalling the definition of $C \models^{W_\sigma}$ before, it is clear that condition (ii) holds for g. So it is enough to verify (i), i.e. that for $x \in S_1^*(\sigma)$ we have

$$g(x) > \sup\{g(y) : y \in S_1^*(\sigma) \wedge y <_{KB} x\} .$$

If $x = (\tau_i,u)$, then

$$g(\tau_i,u) = h(\tau_i,H(u)) > \sup\{h(\tau_j,w) : (\tau_j,w) \in S_1^*(\sigma) \wedge (\tau_j,w) <_{KB} (\tau_i,H(u))\}$$

$$\geq \sup\{h(\tau_j,H(v)) : (\tau_j,v) \in S_1^*(\sigma) \wedge (\tau_j,v) <_{KB} (\tau_i,u)\}$$

$$= \sup\{g(\tau_j,v) : (\tau_j,v) \in S_1^*(\sigma) \wedge (\tau_j,v) <_{KB} (\tau_i,u)\}$$

and we are done.

(b) We examine now a further homogeneity property of S_2^-. Let $H : \omega_1 \to \omega_1$ be normal. Then if $h \in \omega_1 \models^{W_\sigma}$ clearly $H \circ h \in \omega_1 \models^{W_\sigma}$ too (this does not happen in general if H is just increasing). Now it is easy to check that

$$p_\sigma(H \circ h) = \widetilde{H}(p_\sigma(h)) .$$

Indeed, if $p_\sigma(h) = (\xi_1,\ldots,\xi_n)$, where $\xi_i = \widetilde{h}'_i(\underset{\sim}{u}_1,\ldots,\underset{\sim}{u}_{\ell_i})$, then

$$\widetilde{H}(\xi_i) = \widetilde{H \circ h'_i}(\underset{\sim}{u}_1,\ldots,\underset{\sim}{u}_{\ell_i}) = \widetilde{(H \circ h)'_i}(\underset{\sim}{u}_1,\ldots,\underset{\sim}{u}_{\ell_i}) .$$

Thus $\widetilde{H}$ maps S_2^- into S_2^- (in the usual sense that if $(\tau,u) \in S_2^-$, then $(\tau,\widetilde{H}(u)) \in S_2^-$), where

$$\widetilde{H}(\xi_1,\ldots,\xi_n) = (\widetilde{H}(\xi_1),\ldots,\widetilde{H}(\xi_n)) .$$

As $\widetilde{H}$ obviously preserves the proper extension relation among sequences we have for any cub $C \subseteq \omega_1$:

$$\alpha \in A \Leftrightarrow S_2^-(\alpha) \text{ is not wellfounded}$$

$$\Leftrightarrow S_2^- \restriction \widetilde{C}(\alpha) \text{ is not wellfounded} .$$

<u>Remark</u>. It is easy also to check that S_2 is preserved under any $\widetilde{H}$, where $H : \omega_1 \to \omega_1$ is just order preserving, and so for any unbounded $C \subseteq \omega_1$: $\alpha \in A \Leftrightarrow S_2(\alpha)$ is not wellfounded $\Leftrightarrow S_2 \restriction \widetilde{C}(\alpha)$ is not wellfounded.

2.4. <u>Some definability estimates for S_2^-</u>. Consider the structure

$$\mathfrak{D}_3 = \langle \underset{\sim}{u}_\omega, \leq, \{\underset{\sim}{u}_n\}_{n<\omega} \rangle ,$$

as in Kechris-Martin [1978]. (Recall that since we are working with AD, $\underset{\sim}{u}_n = \omega_n$, $\forall n \leq \omega$.) We want to show that:

(i) S_2^- is Δ_1^1 on $\mathfrak{D}_3$.

(ii) The second order relation

$$A \subseteq S_2^-(\sigma) \wedge A \in \mathcal{U}_\sigma$$

is also Δ_1^1 on $\mathfrak{D}_3$.

Now (ii) follows immediately from (i) as for $A \subseteq S_2^-(\sigma)$:

$$A \in \mathcal{U}_\sigma \Leftrightarrow \exists C(C \subseteq \omega_1 \text{ is cub} \wedge \tilde{C}^{\ell h\sigma} \cap S_2^-(\sigma) \subseteq A)$$
$$\Leftrightarrow \forall C(C \subseteq \omega_1 \text{ is cub} \Rightarrow \tilde{C}^{\ell h\sigma} \cap A \neq \emptyset) .$$

(It is needed of course to verify here that the map

$$X \mapsto \tilde{X}$$

for $X \subseteq \omega_1$, is also Δ_1^1, or equivalently that the second order relation

$$X \subseteq \omega_1 \wedge \xi \in \tilde{X}$$

is Δ_1^1 on $\mathfrak{D}_3$. This follows from the fact that for $X \subseteq \omega_1$:

$$\xi \in \tilde{X} \Leftrightarrow \exists\alpha\, \exists \text{term } t \text{ such that}$$
$$[t^{L[\alpha]}(\underset{\sim}{u}_1,\ldots,\underset{\sim}{u}_{r(t)}) = \xi$$
$$\wedge \exists C \subseteq \omega_1 (C \text{ cub}$$
$$\wedge \forall v \in [C]^{r(t)}(t^{L[\alpha]}(v) \in X))]$$
$$\Leftrightarrow \forall\alpha\forall \text{term } t[t^{L[\alpha]}(\underset{\sim}{u}_1,\ldots,\underset{\sim}{u}_{r(t)}) = \xi$$
$$\Rightarrow \forall C \subseteq \omega_1 (C \text{ cub} \Rightarrow \exists v \in [C]^{r(t)}(t^{L[\alpha]}(v) \in X))] .$$

So it is enough to prove (i). For that let us say that a sequence $(\xi_1,\ldots,\xi_n)$, where $\xi_i < \underset{\sim}{u}_{\ell_i+1}$, has the <u>gap property</u> if there are f_i with $\tilde{f}_i(\underset{\sim}{u}_1,\ldots,\underset{\sim}{u}_{\ell_i}) = \xi_i$ and there is cub $D \subseteq \omega_1$ such that for all $1 \leq i \leq n$ and all $v \in [D]^{\ell_i}$:

$$f_i(v) > \sup\{f_j(w) : f_j(w) < f_i(v) \wedge w \in [D]^{\ell_j}\} .$$

Note that this is independent of the particular choice of f_i's which represent the ξ_i's. Now we have the equivalence (for $\sigma \neq \emptyset$)

$$(\sigma,(\xi_1,\ldots,\xi_n)) \in S_2^- \Leftrightarrow \ell h\sigma = n$$
$$\&\ (a)\quad \xi_i < \underset{\sim}{u}_{\ell_i+1} \wedge \text{cofinality}(\xi_i) = \omega$$
$$\&\ (b)\quad (\xi_1,\ldots,\xi_n) \text{ has the gap property}$$
$$\&\ (c)\quad \text{for any (all) } f_1,\ldots,f_n \text{ such that}$$

$\tilde{f}_i(\underset{\sim}{u}_1,\dots,\underset{\sim}{u}_{\ell_i}) = \xi_i$, if we let

$h(\tau_i, v \circ \pi_{\sigma\restriction \ell_i,\tau_i}) = f_i(v)$, then for

some cub set $C \subseteq \omega_1$, $h \restriction (S_1^* \restriction C(\sigma))$

is order preserving relative to $<_{KB}$.

The proof of this equivalence is similar to the proof of the Lemma in 2.3.

Note now that (a) is easily Δ_1^1 on $\mathfrak{D}_3$, while in (c) C can be taken to be equal to I_α (= set of Silver indescernibles of $L[\alpha]$ below ω_1), for any α such that $f_1,\dots,f_n \in L[\alpha]$, thus (c) is also Δ_1^1 in $\mathfrak{D}_3$. It only remains to check that (b) is Δ_1^1 in $\mathfrak{D}_3$. For that is clearly enough to show that in the definition of gap property one can take D to be I'_α = all limit points of I_α, for any real α such that $f_1,\dots,f_n \in L[\alpha]$. This is exactly what is proved by an indiscernibility argument in Part A8, Lemma 1 of Solovay [1978] and we won't repeat the argument.

2.5. <u>An alternative tree</u> S_2^+. We shall now describe an alternative version of a tree for a given Π_2^1 set A. This version is relevant to Martin's recent proof of Determinacy($\underset{\sim}{\Pi}_2^1$) from very large cardinals. We assume only $\forall\alpha\,(\alpha^\#$ exists) until further notice.

In the notation of 2.1 define (for $\sigma \neq \emptyset$)

$(\sigma,(\xi_1,\dots,\xi_n)) \in S_2^+ \Leftrightarrow \ell h\sigma = n$

& (i) $\xi_i < \underset{\sim}{u}_{\ell_i+1}$

& (ii) there are $f_i : [\omega_1]^{\ell_i} \to \omega_1$ in $\tilde{L}$ such that

$\tilde{f}_i(\underset{\sim}{u}_1,\dots,\underset{\sim}{u}_{\ell_i}) = \xi_i$ and if we let

$h(\tau_i, v \circ \pi_{\sigma\restriction \ell_i,\tau_i}) = f_i(v)$,

then for some cub $C \subseteq \omega_1$ and all $u \in C^{\ell_j}$,

$w \in C^{\ell_i}$, if $(\tau_j,u),(\tau_i,w) \in S_1^*(\sigma)$ then:

$(\tau_j,u) <^1 (\tau_i,w) \Rightarrow h(\tau_j,u) < h(\tau_i,w)$,

where

$(\tau,u) <^1 (\tau',u') \Leftrightarrow \tau$ is a 1-point extension of τ' and u is

a 1-point extension of u' .

Clearly, S_2 is a subtree of S_2^+.

Now it is not hard to check as in 2.1 that

$$\alpha \in A \Leftrightarrow S_2^+(\alpha) \text{ is not wellfounded },$$

and that for each $\alpha \in A$, $S_2^+(\alpha)$ has an honest leftmost branch, say f_α, and that if $\overline{\varphi}(\alpha) = f_\alpha$ then $\overline{\varphi}$ is a $\underset{\sim}{\Delta}_3^1$-scale on A.

Our next goal will be to get an explicit description of the condition $(\xi_1,\dots,\xi_n) \in S_2^+(\sigma)$ in terms of the ordinals themselves instead of their representing functions.

For that recall that for $(\sigma,(\xi_1,\dots,\xi_n)) \in S_2^+$ there must be functions $f_i \in \widetilde{L}$, $f_i : [\omega_1]^{\ell_i} \to \omega_1$ such that $\tilde{f}_i(\underset{\sim}{u}_1,\dots,\underset{\sim}{u}_{\ell_i}) = \xi_i$ and for some cub $C \subseteq \omega_1$ and all $v \in [C]^{\ell_j}$, $v' \in [C]^{\ell_i}$ we have

$$(*) \qquad (\tau_j, v \circ \pi_{\sigma\restriction \ell_j,\tau_j}) <^1 (\tau_i, v' \circ \pi_{\sigma\restriction \ell_i,\tau_i}) \Rightarrow f_j(v) < f_i(v') .$$

Now the hypothesis of $(*)$ implies that τ_j is a 1-point extension of τ_i, so that in particular $\ell_j = \ell_i + 1$ and moreover if we put for simplicity $\pi_j = \pi_{\sigma\restriction \ell_j,\tau_j}$ and similarly for π_i, we must also have that

$$v_{\pi_j(k)} = v'_{\pi_i(k)}, \qquad \forall k < \ell_i$$

or equivalently

$$v_{\pi_j \circ \pi_i^{-1}(k)} = v'_k, \qquad \forall k < \ell_i .$$

Note now that from the following commutative diagram of order preserving maps:

$$\begin{array}{ccc} \langle \ell_j, <_{\sigma\restriction \ell_j,\tau_j}\rangle & \xrightarrow{\ \pi_j\ } & \langle \ell_j, <\rangle \\ \uparrow \text{inclusion} & & \uparrow \ \pi_j \circ \pi_i^{-1} \\ \langle \ell_i, <_{\sigma\restriction \ell_i,\tau_i}\rangle & \xrightarrow[\ \pi_i\ ]{} & \langle \ell_i, <\rangle \end{array}$$

we must have that

$$\pi_j \circ \pi_i^{-1} : \ell_i \to \ell_j$$

is order preserving i.e. for some

$$m = m(\sigma,i,j) \leq \ell_i$$

$$\pi_j \circ \pi_i^{-1}(k) = \begin{cases} k, & \text{if } 0 \leq k < m \\ k+1, & \text{if } m \leq k < \ell_i - 1 . \end{cases}$$

Thus $(v'_0,\ldots,v'_{\ell_i-1}) = (v_0,\ldots,\hat{v}_m,\ldots,v_{\ell_j-1})$, where $\hat{v}_m$ signifies the fact that v_m is omitted.

Recall now that for each $m \geq 1$ we have the following embedding

$$j_m : \underset{\sim}{u}_\omega \to \underset{\sim}{u}_\omega ,$$

where

$$j_m(\underset{\sim}{u}_n) = \begin{cases} \underset{\sim}{u}_n, & \text{if } n < m , \\ \underset{\sim}{u}_{n+1}, & \text{if } n \geq m , \end{cases}$$

and $j_m(\tilde{f}(\underset{\sim}{u}_1,\ldots,\underset{\sim}{u}_t)) = \tilde{f}(j_m(\underset{\sim}{u}_1),\ldots,j_m(\underset{\sim}{u}_t))$. Then an easy indiscernibility argument plus the above analysis easily yields that (for $\sigma \neq \emptyset$)

$$(**) \qquad (\sigma,(\xi_1,\ldots,\xi_n)) \in S_2^+ \Leftrightarrow \ell h\sigma = n$$

$$\& \ (\text{i}) \ \ \xi_i < \underset{\sim}{u}_{\ell_i+1}$$

$$\& \ (\text{ii}) \ \ \text{For all } 1 \leq i, j \leq n:$$

$$\tau_j <^1 \tau_i \Rightarrow \xi_j < j_{m(\sigma,i,j)+1}(\xi_i) .$$

This is the particular form that is relevant in Martin's proof.

From this explicit form of S_2^+ and using again full AD one can easily check that each $S_2^+(\sigma)$ is a finite union of Kunen sets $A^t_{m,n}$, where $m = \max_{1\leq i\leq n} \ell_i$ and $n = \ell h\sigma$. The notion and the notation involved here is as in Solovay [1978]. Now each of these $A^t_{m,n}$ carries a canonical measure generated by the sets of the form $\tilde{C}^n \cap A^t_{m,n}$, with $C \subseteq \omega_1$ cub; see again Solovay [1978]. This establishes a homogeneity property of S_2^+. It is relevant to notice here that $S_2^-(\sigma)$ is exactly one of the sets $A^t_{m,n}$ that get into $S_2^+(\sigma)$, so that the passage from S_2^+ to S_2^- has the effect of canonically choosing from the finitely many Kunen sets $A^t_{m,n}$ involved in each $S_2^+(\sigma)$, exactly one which is then equal to $S_2^-(\sigma)$. Although one could write a description of each $S_2^-(\sigma)$ using the embeddings j_m as in (**) it would be a bit messy and not as elegant or useful as (**) itself.

As a final comment we mention that it would be easy to show again that S_2^+ has also the following homogeneity property: For all unbounded $C \subseteq \omega_1$:

$$\alpha \in A \Leftrightarrow S_2^+(\alpha) \text{ is not wellfounded}$$

$$\Leftrightarrow S_2^+ \upharpoonright \tilde{C}(\alpha) \text{ is not wellfounded} .$$

Note. The construction of S_2 is due to Mansfield [1971], Martin [A] following work of Martin-Solovay [1969]. The homogeneity properties of these trees have been studied and used by Kunen [1971] and Martin.

§3. Π_3^1 sets; the tree S_3.

3.1. Definition of S_3. (a) Let now $A \subseteq \mathbb{R}$ be Π_3^1. Then for some $B \in \Pi_2^1$

$$\alpha \in A \Leftrightarrow \neg \exists \beta\, B(\alpha,\beta)$$

$$\Leftrightarrow \neg \exists \beta\, \exists f(\alpha,\beta,f) \in [S_2^-]$$

$$\Leftrightarrow S_2^-(\alpha) \text{ is wellfounded} .$$

Let again for $\sigma \neq \emptyset$, $lh\sigma = n$:

$$S_2^*(\sigma) = \{(\tau_i,u) : (\sigma \upharpoonright \ell_i,\tau_i,u) \in S_2^- \wedge i < n\} .$$

Note that we allow here $(\emptyset,\emptyset) \in S_2^*(\sigma)$, while we have excluded it from $S_1^*(\sigma)$: Again for each such σ let W_σ denote the wellordering $\langle S_2^*(\sigma),<_{KB}\rangle$ and ρ_σ its order type. Then ρ_σ is a successor ordinal (as $(\emptyset,\emptyset)$ is the top element of it) and $\rho_\sigma < \underset{\sim}{u}_{k+1}$, where $k = \max\{\ell_i + 1 : i < n\}$. If $\alpha \in A$ let W_α be the wellordering $\langle S_2^-(\alpha),<_{KB}\rangle$ and ρ_α its order type. Again $W_\alpha = \bigcup_{n\geq 1} W_{\alpha\upharpoonright n}$. Note also that $\rho_\alpha < (\underset{\sim}{u}_\omega)^+ = \omega_{\omega+1}$.

Remark. We have been using the same notation W_σ, ρ_σ in both §2 and in the present §3, although it would be more accurate to distinguish them by superscripts: W_σ^1, ρ_σ^1, W_σ^2, ρ_σ^2. However, in §3 we will be only using the present $W_\sigma = W_\sigma^2$, $\rho_\sigma = \rho_\sigma^2$, so there will be no danger of confusion. Similar remarks will apply also to W_α, ρ_α, p_σ (to be defined below) and in the subsequent sections.

Given now $h \in [\omega_{\omega+1}]^{W_\sigma}$, let h_i, for $i < n = lh\sigma$, be defined as follows: The domain of h_i is $S_2^-(\sigma \upharpoonright \ell_i,\tau_i)$ and for $v \in S_2^-(\sigma \upharpoonright \ell_i,\tau_i)$:

$$h_i(v) = h(\tau_i,v) .$$

Thus for $i = 0$, h_0 is the single ordinal $h(\emptyset,\emptyset)$. Recall now from §2 that $S_2^-(\sigma \upharpoonright \ell_i, \tau_i)$ carries the measure

$$\mathcal{U}_{\sigma\upharpoonright \ell_i,\tau_i} \equiv \mathcal{U}_{\sigma,i} \quad \text{(for simplicity)}$$

and let

$$\xi_0 = h_0$$

$$\xi_i = [h_i]_{\mathcal{U}_{\sigma,i}} .$$

Finally put

$$p_\sigma(h) = (\xi_0,\xi_1,\ldots,\xi_{n-1}) .$$

Then define the tree S_3 by:

$$(\sigma,u) \in S_3 \Leftrightarrow \exists h \in [\omega_{\omega+1}]^{W_\sigma}(p_\sigma(h) = u) \vee (\sigma = u = \emptyset) .$$

(b) We show now that

$$\alpha \in A \Leftrightarrow S_3(\alpha) \text{ is not wellfounded} .$$

First, if $\alpha \in A$ then $S_2^-(\alpha)$ is wellfounded so let $h : W_\alpha \to \omega_{\omega+1}$ be the unique isomorphism between W_α and an initial segment of $\omega_{\omega+1}$, which is of course equal to ρ_α. Let $h_i(v) = h(\tau_i,v)$, for $i > 0$ & $v \in S_2(\alpha \upharpoonright \ell_i,\tau_i)$ and let $\xi_i = [h_i]_{\mathcal{U}_{\alpha\upharpoonright n,i}}$ for any $n > i$. Let also $\xi_0 = h(\emptyset,\emptyset)$. Then $(\xi_0,\ldots,\xi_{n-1}) = p_{\alpha\upharpoonright n}(h^n)$, where $h^n = h \upharpoonright W_{\alpha\upharpoonright n} \in [\omega_{\omega+1}]^{W_{\alpha\upharpoonright n}}$ so that $(\xi_0,\ldots,\xi_{n-1}) \in S_3(\alpha \upharpoonright n)$ i.e. $(\xi_0,\xi_1,\ldots) \in S_3(\alpha)$. Conversely, assume $(\eta_0,\eta_1,\ldots) \in [S_3(\alpha)]$. For each $n > 0$, let ${}^n h \in [\omega_{\omega+1}]^{W_{\alpha\upharpoonright n}}$ be such that $p_{\alpha\upharpoonright n}({}^n h) = (\eta_0,\eta_1,\ldots,\eta_{n-1})$. Then for all $n > 0$, ${}^n h(\emptyset,\emptyset) = \eta_0$ and for all $i > 0$ and all $m,n > i$

$$[{}^m h_i]_{\mathcal{U}_{\alpha\upharpoonright m,i}} = [{}^n h_i]_{\mathcal{U}_{\alpha\upharpoonright n,i}} ;$$

where $\mathcal{U}_{\alpha\upharpoonright m,i} = \mathcal{U}_{\alpha\upharpoonright n,i}$ of course. By the results in 2.3, we can now find $C^i_{n,m} \subseteq \omega_1$ cub such that

$$v \in \widetilde{C^i_{n,m}}^{\ell_i} \cap S_2^-(\alpha \upharpoonright \ell_i,\tau_i) \Rightarrow {}^m h_i(v) = {}^n h_i(v) .$$

Let $C = \bigcap_{0<i<n,m} C^i_{n,m}$. Then $C \subseteq \omega_1$ is cub and the conclusion of (*) holds for all $0 < i < n,m$ and all $v \in \tilde{C}^{\ell_i} \cap S_2^-(\alpha \restriction \ell_i, \tau_i)$. This allows us to define the following map q from $S_2^- \restriction \tilde{C}(\alpha)$ into the ordinals:

$$q(\tau_i, v) = {}^n h_i(v) ,$$

where $n > i$. We now claim that this is an order preserving map from $\langle S_2^- \restriction \tilde{C}(\alpha), <_{KB}\rangle$ into the ordinals. Indeed, if $(\tau_i, u), (\tau_j, w) \in S_2^- \restriction \tilde{C}(\alpha)$ and

$$(\tau_i, u) <_{KB} (\tau_j, w) ,$$

then $(\tau_i, u), (\tau_j, w) \in S_2^*(\alpha \restriction n)$ for $n > i,j$, thus

$${}^n h(\tau_i, u) < {}^n h(\tau_j, w)$$

i.e.

$${}^n h_i(u) < {}^n h_j(w)$$

$\therefore$

$$q(\tau_i, u) < q(\tau_j, w) .$$

So $S_2^- \restriction \tilde{C}(\alpha)$ is wellfounded, thus $\alpha \in A$ by 2.3(b).

(c) We note also the following two basic properties of S_3:

(i) S_3 is a tree on $\omega \times \omega_{\omega+1}$ i.e. all the ordinals occurring in it are $< \omega_{\omega+1}$. This is because by a result of Kunen (see [1971a]) is $\mathcal{U}$ is a measure on any set I of cardinality $< \omega_{\omega+1}$ then for any $f : I \to \omega_{\omega+1}$, $[f]_{\mathcal{U}} < \omega_{\omega+1}$.

(ii) Let for any measure $\mathcal{U}$ on a set I, $i^{\mathcal{U}} : \mathrm{ORD} \to \mathrm{ORD}$ be the embedding it generates i.e.

$$i^{\mathcal{U}}(\xi) = \sup\{[f]_{\mathcal{U}} \mid f : I \to \xi\} .$$

$$= [C_\xi]_{\mathcal{U}}, \text{ for } C_\xi \text{ the constant } \xi \text{ function} .$$

Then we claim that if $\ell h\sigma = n \neq 0$:

$$(\sigma, (\xi_0, \ldots, \xi_{n-1}) \in S_3 \Rightarrow \xi_i < i^{\mathcal{U}_{\sigma,i}}(\xi_0), \text{ for } i > 0 .$$

This is because for $i > 0$, $\xi_i = [h_i]_{\mathcal{U}_{\sigma,i}}$, where $h : W_\sigma \to \omega_{\omega+1}$ is order preserving and $h_i(v) = h(\tau_i, v)$ so that

$$\xi_0 = h(\emptyset, \emptyset) > h_i(v), \text{ for all } i > 0 ,$$

as $(\emptyset,\emptyset) >_{KB} (\tau_i, v)$. Thus

$$ {}_i u_{\sigma,i}(\xi_0) > [h_i]_{u_{\sigma,i}} \, . $$

This is analogous to a property of S_1 which we established in §1, and it will be useful in §4.

3.2. <u>Scales for Π^1_3 sets.</u> As usual we verify now that if $\alpha \in A$ then $S_3(\alpha)$ has an honest leftmost branch. For that in the notation of 3.1.(b) if $h : W_\alpha \to \omega_{\omega+1}$ is as defined there and we let h_i and ξ_i be again as defined there, we have $(\xi_0, \xi_1, \ldots) \in [S^-_3(\alpha)]$. Moreover if $(\eta_0, \eta_1, \ldots)$ is any branch of $S^-_3(\alpha)$ then, again in the notation of 3.1.(b), q is an order preserving map from $S^-_2 \restriction \tilde{C}(\alpha)$ into $\omega_{\omega+1}$. Let $H : \omega_1 \to C$ be the normal function enumerating C. Then by 2.3, $\tilde{H} : S^-_2(\alpha) \to S^-_2 \restriction \tilde{C}(\alpha)$, and of course $\tilde{H}$ preserves $<_{KB} \therefore h(\tau_i, u) \leq q(\tau_i, \tilde{H}(u))$. Now let $D \subseteq C$ be cub sich that $H(\xi) = \xi$ for $\xi \in D$. Then $\tilde{H}(u) = u$ for $u \in \tilde{D}^{<\omega} \therefore h(\tau_i, u) \leq q(\tau_i, u)$ for $u \in \tilde{D}^{\ell_i} \therefore h_i(u) \leq {}^n h_i(u)$ for $u \in \tilde{D}^{\ell_i} \cap S^-_2(\alpha \restriction n, \tau_i)$, if $n > i \therefore [h_i]_{u_{\alpha \restriction n,i}} = \xi_i \leq [{}^n h_i]_{u_{\alpha\restriction n,i}} = \eta_i$ for $i > 0$. Also $\xi_0 = h(\emptyset,\emptyset) \leq q(\emptyset,\emptyset) = {}^n h(\emptyset,\emptyset) = \eta_0$ i.e. $\xi_i \leq \eta_i$, $\forall i \geq 0$ and we are done.

This implies now that if for each $\alpha \in A$ we put

$$ \varphi_i(\alpha) = f_\alpha(i) \, , $$

where f_α is the leftmost branch of $S_3(\alpha)$, then $\overline{\varphi} = \{\varphi_i\}$ is an $\omega_{\omega+1}$-scale on A. By modifying this slightly (for reasons that will become apparent in a moment) we will obtain a Π^1_3-scale on A. Indeed put for $\alpha \in A$:

$$ \psi_i(\alpha) = \langle \varphi_0(\alpha), \overline{\alpha}(i), \varphi_i(\alpha) \rangle \, , $$

where $\langle \xi, \eta, \theta \rangle$ refers to the ordinal associated to the triple (ξ, η, θ) in the lexicographical ordering of $(\omega_{\omega+1})^3$. Now we claim that $\overline{\psi} = \{\psi_i\}$ is a Π^1_3-scale on A. For that just note that for $\alpha, \beta \in A$:

$$ \begin{aligned} \psi_i(\alpha) \leq \psi_i(\beta) \Leftrightarrow \varphi_0(\alpha) < \varphi_0(\beta) &\vee [\varphi_0(\alpha) = \varphi_0(\beta) \wedge \overline{\alpha}(i) < \overline{\beta}(i)] \\ &\vee [\varphi_0(\alpha) = \varphi_0(\beta) \wedge \overline{\alpha}(i) = \overline{\beta}(i) \\ &\wedge \varphi_i(\alpha) \leq \varphi_i(\beta)] \, . \end{aligned} $$

So if $\beta \in A$:

$$\alpha \in A \wedge \psi_i(\alpha) \leq \psi_i(\beta) \Leftrightarrow [\alpha \in A \wedge \varphi_0(\alpha) < \varphi_0(\beta)] \vee$$
$$[(\alpha \in A \wedge \varphi_0(\alpha) = \varphi_0(\beta)) \wedge \bar{\alpha}(i) < \bar{\beta}(i)] \vee$$
$$[(\alpha \in A \wedge \varphi_0(\alpha) = \varphi_0(\beta)) \wedge \bar{\alpha}(i) = \bar{\beta}(i)$$
$$\wedge \varphi_i(\alpha) \leq \varphi_i(\beta)] ,$$

for which we have to calculate that it is Δ^1_3 uniformly in β. But by the results of Kechris-Martin [1978] it is enough to show that it is Δ^1_1 over $\mathfrak{Q}_3$, uniformly in β. To check this notice that

$$\alpha \in A \wedge \varphi_0(\alpha) \leq \varphi_0(\beta) \Leftrightarrow \text{There is an embedding from } \langle S^-_2(\alpha), <_{KB}\rangle \text{ into } \langle S^-_2(\beta), <_{KB}\rangle$$
$$\Leftrightarrow \langle S^-_2(\alpha), <_{KB}\rangle \text{ is a wellordering and there is no embedding of } \langle S^-_2(\beta), <_{KB}\rangle \text{ into a proper initial segment of } \langle S^-_2(\alpha), <_{KB}\rangle .$$

Since in 2.4 we have shown that S^-_2 is Δ^1_1 in $\mathfrak{Q}_3$ the above equivalences show that "$\alpha \in A \wedge \varphi_0(\alpha) \leq \varphi_0(\beta)$" is Δ^1_1 in $\mathfrak{Q}_3$, uniformly in β. A similar calculation applies to the predicate "$\alpha \in A \wedge \varphi_0(\alpha) = \varphi_0(\beta)$." So to complete this proof it is sufficient to check that in case $\alpha \in A$, $\beta \in A$ and $\bar{\alpha}(i) = \bar{\beta}(i)$ the predicate

$$\varphi_i(\alpha) \leq \varphi_i(\beta), \quad \text{for } i > 0$$

is (uniformly in i, α, β) Δ^1_1 in $\mathfrak{Q}_3$. Recall that for $i > 0$

$$\varphi_i(\alpha) = f_\alpha(i) = [\lambda u.\ \mathrm{rank}_{W_\alpha}(\tau_i, u)]_{\mathcal{U}_{\alpha \restriction \ell_i, i}} ;$$

here the u varies over $S^-_2(\alpha \restriction \ell_i, \tau_i)$. Since $\alpha \restriction i = \beta \restriction i$ and $i \geq \ell_i$ we clearly have that $S^-_2(\alpha \restriction \ell_i, \tau_i) = S^-_2(\beta \restriction \ell_i, \tau_i)$ and $\mathcal{U}_{\alpha \restriction \ell_i, i} = \mathcal{U}_{\beta \restriction \ell_i, i}$. Then

$$\varphi_i(\alpha) \leq \varphi_i(\beta) \Leftrightarrow \{u \in S^-_2(\alpha \restriction \ell_i, \tau_i) : \mathrm{rank}_{W_\alpha}(\tau_i, u) \leq \mathrm{rank}_{W_\beta}(\tau_i, u)\} \in \mathcal{U}_{\alpha \restriction \ell_i, i} .$$

Since as above the relation

$$\mathrm{rank}_{W_\alpha}(\tau_i, u) \leq \mathrm{rank}_{W_\beta}(\tau_i, u)$$

is Δ^1_1 in $\mathfrak{D}_3$, uniformly in all the parameters involved, the results in 2.4 imply that "$\varphi_i(\alpha) \leq \varphi_i(\beta)$" is Δ^1_1 in $\mathfrak{D}_3$, uniformly in α, β as above.

3.3. <u>Homogeneity properties of</u> S_3. <u>The tree</u> S_3^-. (a) By analogy with the work in 2.3 we define a subtree S_3^- of S_3 as follows:

$$(\sigma, u) \in S_3^- \Leftrightarrow \exists h \in \omega_{\omega+1} \models^{W_\sigma} (p_\sigma(h) = u) \vee \sigma = u = \emptyset .$$

Again we have: $\alpha \in A \Leftrightarrow S_3^-(\alpha)$ is not wellfounded.

By a result of Kunen [1971a] we have

$$\omega_{\omega+1} \to (\omega_{\omega+1})^\rho_\nu, \qquad \forall \rho, \nu < \omega_{\omega+1} .$$

This implies that we have the following $\omega_{\omega+1}$-additive measure $\bar{\mathcal{V}}_\sigma$ on $\omega_{\omega+1} \models^{W_\sigma}$:

$$X \in \bar{\mathcal{V}}_\sigma \Leftrightarrow \exists C \subseteq \omega_{\omega+1} (C \text{ cub} \wedge C \models^{W_\sigma} \subseteq X) .$$

Since

$$p_\sigma : \omega_{\omega+1} \models^{W_\sigma} \xrightarrow{\text{onto}} S_3^-(\sigma) ,$$

this induces the $\omega_{\omega+1}$-additive measure

$$(p_\sigma)_* \bar{\mathcal{V}}_\sigma = \mathcal{V}_\sigma$$

on $S_3^-(\sigma)$, for $\sigma \neq \emptyset$. Thus for $A \subseteq S_3^-(\sigma)$:

$$A \in \mathcal{V}_\sigma \Leftrightarrow \exists C \subseteq \omega_{\omega+1} (C \text{ cub} \wedge p_\sigma[C \models^{W_\sigma}] \subseteq A) .$$

We shall now try to get a more explicit form of this measure $\mathcal{V}_\sigma$. This will be based on the analog of the Lemma in 2.3.(a).

Let $\mathcal{U}$ be a measure on a set I of cardinality $< \omega_{\omega+1}$. Then as we mentioned before $i^{\mathcal{U}}(\omega_{\omega+1}) = \omega_{\omega+1}$, where $i^{\mathcal{U}}$ is the associated embedding generated by $\mathcal{U}$. For each $X \subseteq \omega_{\omega+1}$ let $i^{\mathcal{U}}(X)$ be the image of X under this embedding i.e.

$$i^{\mathcal{U}}(X) = \{[f]_{\mathcal{U}} : \{t : f(t) \in X\} \in \mathcal{U}\}.$$

Then since $i^{\mathcal{U}}(\omega_{\omega+1}) = \omega_{\omega+1}$, clearly $i^{\mathcal{U}}(X) \subseteq \omega_{\omega+1}$. (<u>Caution</u>. In general $i^{\mathcal{U}}[X] = \{i^{\mathcal{U}}(\xi) : \xi \in X\} \subsetneqq i^{\mathcal{U}}(X)$). Now we have

Lemma. For any unbounded $C \subseteq \omega_{\omega+1}$:

$$p_\sigma[C \models^{W_\sigma}] = S_3^-(\sigma) \cap (C \times i^{u_{\sigma,1}}(C) \times i^{u_{\sigma,2}}(C) \times \cdots \times i^{u_{\sigma,n-1}}(C)) ,$$

where $\ell h\sigma = n > 0$.

Proof. First let $h \in C \models^{W_\sigma}$. Then, if $p_\sigma(h) = (\xi_0,\ldots,\xi_{n-1})$, $\xi_0 = h(\emptyset,\emptyset) \in C$, and if $h_i(u) = h(\tau_i,u)$ for $i > 0$, then $\xi_i = [h_i]_{u_{\sigma,i}}$ so that (since $h_i(u) \in C$) $\xi_i \in i^{u_{\sigma,i}}(C)$. Thus $(\xi_0,\ldots,\xi_{n-1}) \in S_3^-(\sigma) \cap (C \times i^{u_{\sigma,1}}(C) \times \cdots \times i^{u_{\sigma,n-1}}(C))$. The proof of the converse is very similar to the proof of the Lemma in 2.3.(a) and we omit the details. ⊣

Thus the measure ν_σ on each $S_3^-(\sigma)$ (for $\sigma \neq \emptyset$) is generated by the sets of the form

$$S_3^-(\sigma) \cap \left(C \times \prod_{0<i<\ell h\sigma} i^{u_{\sigma,i}}(C)\right)$$

for C cub, $C \subseteq \omega_{\omega+1}$, i.e.

$$A \in \nu_\sigma \Leftrightarrow \exists C \subseteq \omega_{\omega+1}\left[C \text{ cub} \wedge A \supseteq S_3^-(\sigma) \cap \left(C \times \prod_{0<i<\ell h\sigma} i^{u_{\sigma,i}}(C)\right)\right] .$$

This bears some resemblance to the corresponding result about the generation of the measures u_σ on $S_2^-(\sigma)$.

(b) Finally we establish the usual further homogeneity of S_3^-. Let $H : \omega_{\omega+1} \to \omega_{\omega+1}$ be a function. If u is a measure on I, where I has cardinality $< \omega_{\omega+1}$, we let $i^u(H)$ be the image of H under i^u i.e. $i^u(H) : \omega_{\omega+1} \to \omega_{\omega+1}$ and

$$i^u(H)([f]_u) = [H \circ f]_u .$$

Then if $\text{range}(H) = C$, $i^u(H) : \omega_{\omega+1} \xrightarrow{\text{onto}} i^u(C)$.

Assume now $H : \omega_{\omega+1} \to \omega_{\omega+1}$ is order preserving and for any $\sigma \neq \emptyset$, if $\ell h\sigma = n$, define for $u = (\xi_0,\ldots,\xi_{n-1})$:

$$H^\sigma(u) = (H(\xi_0), i^{u_{\sigma,1}}(H)(\xi_1),\ldots,i^{u_{\sigma,n-1}}(H)(\xi_{n-1})) .$$

Then it easy to check that

$$(\sigma,u) \in S_3 \Rightarrow (\sigma,H^\sigma(u)) \in S_3 ,$$

while if H is also normal,

$$(\sigma,u) \in S_3^- \Rightarrow (\sigma,H^\sigma(u)) \in S_3^- .$$

In particular, if $\alpha \in \mathbb{R}$ and we let $H^\alpha = \bigcup_{n\geq 1} H^{\alpha\restriction n}$ so that $H^\alpha(u) = H^{\alpha\restriction n}(u)$, where $n > \ell h u$, then H^α maps $S_3^-(\alpha)$ into $S_3^-(\alpha)$. Thus if for each $X \subseteq \omega_{\omega+1}$ we let

$$S_3^- \restriction X^\alpha = \{(\sigma,(\xi_0,\ldots,\xi_{n-1})) \in S_3^- : \xi_0 \in X \wedge \text{for } 0 < i < n,$$
$$\xi_i \in i^{u_{\alpha\restriction n,i}}(X)\} ,$$

then for any cub $C \subseteq \omega_{\omega+1}$ we have

$$\alpha \in A \Leftrightarrow S_3^-(\alpha) \text{ is not wellfounded}$$
$$\Leftrightarrow S_3^- \restriction C^\alpha(\alpha) \text{ is not wellfounded} .$$

(For S_3 we have this for any unbounded C).

Note. The construction of S_3 (and $-$) is due to Kunen [1971a]. The calculation of a Π_3^1 scale from this tree is originally due to Martin by a different argument than the one we gave in 3.2.

§4. Π_4^1 sets; the tree S_4.

4.1. Definition of S_4. (a) Let $A \subseteq \mathbb{R}$ be a Π_4^1 set of reals. Then for some $B \in \Pi_3^1$,

$$\alpha \in A \Leftrightarrow \neg\exists\beta\, B(\alpha,\beta)$$
$$\Leftrightarrow \neg\exists\beta\, \exists f(\alpha,\beta,f) \in [S_3^-]$$
$$\Leftrightarrow S_3^-(\alpha) \text{ is wellfounded} .$$

Again for each $\sigma \neq \emptyset$, $\ell h\sigma = n$ we let

$$S_3^*(\sigma) = \{(\tau_i,u) : (\sigma \restriction \ell_i,\tau_i,u) \in S_3^- \wedge 1 \leq i \leq n\} ,$$

and we define W_σ to be $\langle S_3^*(\sigma),<_{KB}\rangle$ and ρ_σ, W_α (for $\alpha \in A$), as in 2.1. Note that $\rho_\sigma = \omega_{\omega+1}$ and for $\alpha \in A$, $\rho_\alpha = \omega_{\omega+1}$. This is because of 3.1.(c), (ii).

Now for each $h \in [\omega_{\omega+1}]^{W_\sigma}$, let h_i, for $1 \leq i \leq n = \ell h \sigma$, be defined as follows:

$$\text{domain}(h_i) = S_3^-(\sigma \restriction \ell_i, \tau_i)$$

and for $v \in S_3^-(\sigma \restriction \ell_i, \tau_i)$

$$h_i(v) = h(\tau_i, v) .$$

Recall that $S_3^-(\sigma \restriction \ell_i, \tau_i)$ carries the measure

$$\nu_{\sigma \restriction \ell_i, \tau_i} \equiv \nu_{\sigma, i} \quad \text{(for simplicity)}$$

and put

$$\xi_i = [h_i]_{\nu_{\sigma, i}} .$$

Finally, let

$$p_\sigma(h) = (\xi_1, \ldots, \xi_n) = ([h_i]_{\nu_{\sigma, i}})_{1 \leq i \leq n}$$

and define

$$(\sigma, u) \in S_4 \Leftrightarrow \exists h \in [\omega_{\omega+1}]^{W_\sigma}(p_\sigma(h) = u) \vee (\sigma = u = \emptyset) .$$

It is easy now to verify as in 2.1 that

$$\alpha \in A \Leftrightarrow S_4(\alpha) \text{ is not wellfounded} .$$

Put

$$\lambda_5 = \sup\{i^{\nu_{\sigma, i}}(\omega_{\omega+1}) : \sigma \neq \emptyset,\ 1 \leq i \leq \ell h \sigma\} .$$

Then clearly S_4 is a tree on $\omega \times \lambda_5$.

4.2. Scales for Π^1_4 sets. Again as in 2.2 we shall verify that for each $\alpha \in A$, $S_4(\alpha)$ has an honest leftmost branch. For that let $h : W_\alpha \to \omega_{\omega+1}$ be the rank function of W_α. For each $i \geq 1$ let h_i be the function on $S_3^-(\alpha \restriction \ell_i, \tau_i)$ given by $h_i(v) = h(\tau_i, v)$ and let $\xi_i = [h_i]_{\nu_{\alpha \restriction \ell_i, i}}$. Clearly $(\xi_1, \xi_2, \ldots)$ is a branch through $S_4(\alpha)$. Now let $(\eta_1, \eta_2, \ldots)$ be a branch through $S_4(\alpha)$. We want to show that $\xi_i \leq \eta_i$, $\forall i \geq 1$. Since $(\eta_1, \eta_2, \ldots, \eta_n) \in S_4(\alpha \restriction n)$ let ${}^n h \in [\omega_{\omega+1}]^{W_{\alpha \restriction n}}$ be such that $p_{\alpha \restriction n}[{}^n h] = (\eta_1, \ldots, \eta_n)$. Let ${}^n h_i(v) = {}^n h(\tau_i, v)$. Then for $n, m \geq i$,

$[{}^n h_i]_{U_{\alpha\restriction \ell_i, i}} = [{}^m h_i]_{U_{\alpha\restriction \ell_i, i}}$ $\therefore$ there is a cub $C^i_{n,m} \subseteq \omega_{\omega+1}$ such that ${}^n h_i(v) = {}^m h_i(v)$ for $v \in (C^i_{n,m} \times \prod_{0<j<\ell_i} i^{U_{\alpha\restriction \ell_i, \tau_i, j}}(C^i_{n,m})) \cap S_3^-(\alpha \restriction \ell_i, \tau_i)$. Let $C = \bigcap_{1<i<n,m} C^i_{n,m}$. Then $C \subseteq \omega_{\omega+1}$ is cub and ${}^n h_i(v) = {}^m h_i(v)$ for all $v \in (C \times \prod_{0<j<\ell_i} i^{U_{\alpha\restriction \ell_i, \tau_i, j}}(C)) \cap S_3^-(\alpha \restriction \ell_i, \tau_i)$ and all $n,m \geq i \geq 1$. Then we can define the following function q from

$$S_3^- \restriction C^\alpha(\alpha) = \{(\tau_i, (\xi_0, \ldots, \xi_{\ell_i - 1})) : (\alpha \restriction \ell_i, \tau_i, (\xi_0, \ldots, \xi_{\ell_i - 1})) \in S_3^- \wedge \xi_0 \in C \wedge \xi_j \in i^{U_{\alpha\restriction \ell_i, \tau_i, j}}(C) \text{ for } 0 < j < \ell_i\},$$

into the ordinals:

$$q(\tau_i, v) = {}^n h_i(v), \quad \text{for any } n \geq i .$$

As usual q is order preserving from $\langle S_3^- \restriction C^\alpha(\alpha), <_{KB}\rangle$ into the ordinals. Let now H be the normal function enumerating C. Then if $i \geq 1$ and we let

$$H^{\alpha\restriction \ell_i, \tau_i}(\xi_0, \ldots, \xi_{\ell_i - 1}) = (H(\xi_0), i^{U_{\alpha\restriction \ell_i, \tau_i, 1}}(H)(\xi_1), \ldots, i^{U_{\alpha\restriction \ell_i, \tau_i, \ell_i - 1}}(H)(\xi_{\ell_i - 1})),$$

the map

$$(\tau_i, v) \mapsto (\tau_i, H^{\alpha\restriction \ell_i, \tau_i}(v))$$

maps $\langle S_3^-(\alpha), <_{KB}\rangle$ in an order preserving way into $\langle S_3^- \restriction C^\alpha(\alpha), <_{KB}\rangle$ so that if $h : \rho_\alpha \to \omega_{\omega+1}$ is as before then $h(\tau_i, v) \leq q(\tau_i, H^{\alpha\restriction \ell_i, \tau_i}(v)) = {}^n h_i(H^{\alpha\restriction \ell_i, \tau_i}(v))$, for $n \geq i$. So $h_i(v) \leq {}^n h_i(H^{\alpha\restriction \ell_i, \tau_i}(v))$, for $n \geq i$. Now we can find a cub $D \subseteq \omega_{\omega+1}$ such that $H \restriction D = \mathrm{id} \restriction D$. Then for $v = (\xi_0, \ldots, \xi_{\ell_i - 1}) \in D \times \prod_{0<j<\ell_i} i^{U_{\alpha\restriction \ell_i, \tau_i, j}}(D)$, $h_i(v) \leq {}^n h_i(v)$. Indeed if $j > 0$ and $\xi_j \in i^{U_{\alpha\restriction \ell_i, \tau_i, j}}(D)$ we have $\xi_j = [f]_{U_{\alpha\restriction \ell_i, \tau_i, j}}$, where $f(x) \in D$ a.e. (mod $U_{\alpha\restriction \ell_i, \tau_i, j}$). But then $H(f(x)) = f(x)$ a.e.

$(\mathrm{mod}\ \mathcal{U}_{\alpha\restriction \ell_i,\tau_i,j})$ $\therefore$ $\xi_j = [f]_{\mathcal{U}_{\alpha\restriction \ell_i,\tau_i,j}} = [H\circ f]_{\mathcal{U}_{\alpha\restriction \ell_i,\tau_i,j}} =$

$i^{\mathcal{U}_{\alpha\restriction \ell_i,\tau_i,j}}(H)([f]_{\mathcal{U}_{\alpha\restriction \ell_i,\tau_i,j}}) = i^{\mathcal{U}_{\alpha\restriction \ell_i,\tau_i,j}}(H)(\xi_j)$. Similarly for $j = 0$.

Thus $\xi_i = [h_i]_{\mathcal{V}_{\alpha\restriction \ell_i,i}} \leq [^n h_i]_{\mathcal{V}_{\alpha\restriction \ell_i,i}} = \eta_i$.

If now for each $\alpha \in A$ we let f_α be the leftmost branch of $S_4(\alpha)$ and

$$\varphi_i(\alpha) = f_\alpha(i) = \xi_i = [h_i]_{\mathcal{V}_{\alpha\restriction \ell_i,i}}, \quad \text{(in the preceding notation)},$$

then we can verify again that

$$\psi_i(\alpha) = \langle \bar{\alpha}(i), \varphi_i(\alpha)\rangle$$

is a Δ^1_5-scale on A. Indeed, for $\alpha,\beta \in A$:

$$\psi_i(\alpha) \leq \psi_i(\beta) \Leftrightarrow \bar{\alpha}(i) < \bar{\beta}(i) \vee \{\bar{\alpha}(i) = \bar{\beta}(i) \wedge [\lambda v.\mathrm{rank}_{W_\alpha}(\tau_i,v)]_{\mathcal{V}_{\alpha\restriction \ell_i,i}}$$

$$\leq [\lambda v.\mathrm{rank}_{W_\beta}(\tau_i,v)]_{\mathcal{V}_{\alpha\restriction \ell_i,i}}\}$$

$$\Leftrightarrow \bar{\alpha}(i) < \bar{\beta}(i) \vee \{\bar{\alpha}(i) = \bar{\beta}(i) \wedge \exists C \subseteq \omega_{\omega+1}[C \text{ cub } \wedge$$

$$\forall(\xi_0,\ldots,\xi_{\ell_i-1})[[(\alpha\restriction \ell_i,(\xi_0,\ldots,\xi_{\ell_i-1})) \in S_3^- \wedge$$

$$\xi_0 \in C \wedge \forall j[0 < j < \ell_i \Rightarrow \xi_i \in i^{\mathcal{U}_{\alpha\restriction \ell_i,\tau_i,j}}(C)]]$$

$$\Rightarrow \mathrm{rank}_{W_\alpha}(\tau_i,(\xi_0,\ldots,\xi_{\ell_i-1})) \leq \mathrm{rank}_{W_\beta}(\tau_i,(\xi_0,\ldots,\xi_{\ell_i-1}))]]\}.$$

This relation can be verified to be Σ^1_1 over the structure $\langle \omega_{\omega+1},<,S_2^-,S_3^-\rangle$, by using the results of Kechris-Martin [1978]. By the Moschovakis Coding Lemma (see Moschovakis [1970] or Kechris [1978]) and the techniques of Harrington-Kechris [A] one can verify then that every relation on reals which is Σ^1_1 on $\langle \omega_{\omega+1},<,S_2^-,S_3^-\rangle$ is Σ^1_5 (and conversely), so "$\psi_i(\alpha) \leq \psi_i(\beta)$" is also Σ^1_5. Similarly "$\psi_i(\alpha) < \psi_i(\beta)$" is Σ^1_5 and we are done.

4.3. <u>Homogeneity properties of</u> S_4. Unfortunately not much can be said at this time about the homogeneity properties of S_4 as the combinatorial property

$$\omega_{\omega+1} \to (\omega_{\omega+1})^{\omega_{\omega+1}}$$

is still an open question (recall that the fact that $\omega_1 \to (\omega_1)^{\omega_1}$ is the key to establishing the homogeneity properties of S_2).

Note. The construction of S_4 is due to Kunen [1971a].

§5. On $\underset{\sim}{\delta}^1_5$. Recall that we have defined in §4:

$$\lambda_5 = \sup\{i^{\nu_{\sigma,i}}(\omega_{\omega+1}) : \sigma \neq \emptyset,\ 1 \leq i \leq \ell h\sigma\} .$$

The following result reduces the problem of computing $\underset{\sim}{\delta}^1_5$ to the problem of computing these $i^{\nu_{\sigma,i}}(\omega_{\omega+1})$.

Theorem (Kunen [1971a]). $\underset{\sim}{\delta}^1_5 = (\lambda_5)^+ =$ smallest cardinal $> \lambda_5$.

Proof. By the results in §4, every Π^1_4, and thus every Σ^1_5 set, is λ_5-Souslin i.e. it can be written in the form

$$\exists f(\alpha,f) \in S ,$$

where S is a tree on $\omega \times \lambda_5$. Thus be the Kunen-Martin Theorem (see Martin [A]) $\underset{\sim}{\delta}^1_5 \leq (\lambda_5)^+$. So it is enough to prove that $\lambda_5 < \underset{\sim}{\delta}^1_5$ i.e. that for each σ, i as above $i^{\nu_{\sigma,i}}(\omega_{\omega+1}) < \underset{\sim}{\delta}^1_5$. Put $\nu_{\sigma,i} \equiv \nu$. Then ν is a measure on $S_3^-(\sigma \restriction \ell_i, \tau_i) \equiv I$. Now the relation

$$f < g \Leftrightarrow f,g : I \to \omega_{\omega+1} \wedge [f]_\nu < [g]_\nu$$

can be easily seen to be Σ^1_1 on the structure $\langle \omega_{\omega+1}, <, S_3^- \rangle$. One can now use the Moschovakis Coding Lemma to code functions $f : I \to \omega_{\omega+1}$ by reals. Say ε codes f_ε. Then as in 4.2 one can verify that $\prec$ is Σ^1_5 in the codes i.e. the following relation is Σ^1_5:

$$\varepsilon \prec^* \delta \Leftrightarrow \varepsilon,\ \delta \text{ code functions } f_\varepsilon,\ f_\delta \text{ (resp.) from}$$

$$I \text{ into } \omega_{\omega+1} \wedge f_\varepsilon \prec f_\delta .$$

As $\prec^*$ is a wellfounded relation of rank $i^{\nu_{\sigma,i}}(\omega_{\omega+1})$ (since $\text{rank}_{\prec^*}(\varepsilon) = \text{rank}_{\prec}(f_\varepsilon) = [f_\varepsilon]_\nu$) we have that $i^{\nu_{\sigma,i}}(\omega_{\omega+1}) < \underset{\sim}{\delta}^1_5$. ⊣

§6. Homogeneous trees in general. We shall discuss now a general notion of homogeneity shared by the trees constructed before. We shall also formulate the type of tree construction utilized in §2-4 as a general transfer theorem for homogeneous trees.

A tree T on $\omega \times \lambda$ is <u>homogeneous</u> if for each $\sigma \neq \emptyset$ in $\omega^{<\omega}$ there is a measure μ_σ on $T(\sigma)$ with the following two properties:

(i) Let for $\sigma' \supseteq \sigma$, $\pi_{\sigma'\sigma} : T(\sigma') \to T(\sigma)$ be the restriction map:

$$\pi_{\sigma'\sigma}(u) = u \upharpoonright \ell h\sigma .$$

Then $(\pi_{\sigma'\sigma})_* \mu_{\sigma'} = \mu_\sigma$ (i.e. for $X \subseteq T(\sigma)$, $X \in \mu_\sigma \Leftrightarrow \pi^{-1}_{\sigma'\sigma}[X] \in \mu_{\sigma'}$).

(ii) If $T(\alpha)$ is not wellfounded and for each $n \geq 1$, $X_n \subseteq T(\alpha \upharpoonright n)$ and $X_n \in \mu_{\alpha \upharpoonright n}$, then there is $f \in \lambda^\omega$ with $f \upharpoonright n \in X_n$, for all n.

The basic way in which homogeneous trees have been obtained in this paper is as follows:

Suppose T is a tree on $\omega \times \lambda$. Suppose also that there is an ordinal κ and for each $\sigma \neq \emptyset$ a wellordering W_σ of order type $\rho_\sigma \leq \kappa$ and a map

$$p_\sigma : [\kappa]^{W_\sigma} \xrightarrow{\text{onto}} T(\sigma) ,$$

such that

$$\sigma \subseteq \sigma' \Rightarrow W_\sigma \subseteq W_{\sigma'}$$

and moreover the following two conditions hold:

(i)' If $\sigma \subseteq \sigma'$, then for $h \in [\kappa]^{W_{\sigma'}}$, $p_{\sigma'}(h) \upharpoonright \ell h\sigma = p_\sigma(h \upharpoonright W_\sigma)$.

(ii)' If $T(\alpha)$ is not wellfounded, then the union

$$W_\alpha = \bigcup W_{\alpha \upharpoonright n}$$

is a wellordering of order type $\rho_\alpha \leq \kappa$.

Then granting that for each σ

$$\kappa \to (\kappa)^{\omega \cdot \rho_\sigma} ,$$

we have that T is homogeneous.

Indeed, let ν_σ be the following measure on $[\kappa]^{W_\sigma}$:

$$\nu_\sigma(X) = 1 \Leftrightarrow \exists C \subseteq \kappa [C \text{ cub} \wedge C^{W_\sigma} \subseteq X] .$$

Then let

$$\mu_\sigma = (p_\sigma)_* \nu_\sigma .$$

To check property (i) of homogeneity we use (i)': Indeed let $X \subseteq T(\sigma)$. If $\mu_\sigma(X) = 1$, then there is $C \subseteq \kappa$, C cub with $C \models^{W_\sigma} \subseteq p_\sigma^{-1}[X]$. If now $h \in C \models^{W_{\sigma'}}$, $h \restriction W_\sigma \in C \models^{W_\sigma} \therefore p_\sigma(h \restriction W_\sigma) = p_{\sigma'}(h) \restriction \ell h\sigma \in X \therefore$ $p_{\sigma'}(h) \in \pi_{\sigma'\sigma}^{-1}[X] \therefore \mu_{\sigma'}(\pi_{\sigma'\sigma}^{-1}[X]) = 1$, so that $(\pi_{\sigma'\sigma})_* \mu_{\sigma'}(X) = 1$.

For (ii) we use of course (ii)': Let $T(\alpha)$ be not wellfounded and let X_n be such that $\mu_{\alpha \restriction n}(X_n) = 1$. Pick C_n cub in κ with $C_n \models^{W_{\alpha \restriction n}} \subseteq p_{\alpha \restriction n}^{-1}[X_n]$. Let $C = \bigcap_n C_n$, so that C is cub in κ. Let, since $\rho_\alpha \leq \kappa$, $h \in C \models^{W_\alpha}$. If $h_n = h \restriction W_{\alpha \restriction n}$, then $h_n \in C \models^{W_{\alpha \restriction n}}$, so that $p_{\alpha \restriction n}(h_n) \in X_n$. Moreover if $n < m$ then $p_{\alpha \restriction n}(h_n)$ is an initial segment of $p_{\alpha \restriction m}(h_m)$, thus there is $f \in \lambda^\omega$ with $f \restriction n = p_{\alpha \restriction n}(h_n)$ so that $f \restriction n \in X_n$ and we are done.

It is now easy to see that S_1 is an example of such a tree with $\kappa = \omega_1$, $W_\sigma = \langle \ell h\sigma, <_\sigma \rangle$, $\rho_\sigma = \ell h\sigma$ and $p_\sigma(u) = u \circ \pi_\sigma$. Moreover, by their construction, S_2, S_3, S_4 are all of that form with W_σ, ρ_σ, π_σ as given in §§2-4.

Note. Similar definitions of homogeneity apply to trees on $\omega^k \times \lambda$.

We shall now state and prove a general transfer theorem for homogeneous trees. We say below that $A \subseteq \mathcal{R}$ admits the tree T if $A = p[T]$. Similarly for $A \subseteq \mathcal{R} \times \mathcal{R}$ etc.

Transfer Theorem for Homogeneous Trees (Kunen, Martin). Assume $B \subseteq \mathcal{R}^2$ admits the homogeneous tree T (on some $\omega^2 \times \lambda$). Then put

$$\alpha \in A \Leftrightarrow \neg \exists \beta \, B(\alpha, \beta),$$

and let ρ_σ be the order type of the wellordering $W_\sigma = \langle T^*(\sigma), <_{KB} \rangle$, where $T^*(\sigma) = \{(\tau_i, u) : (\sigma \restriction \ell_i, \tau_i, u) \in T \wedge 1 \leq i \leq \ell h\sigma\}$ and let ρ_α be the order type of the wellordering $W_\alpha = \bigcup_n W_{\alpha \restriction n}$, where $\alpha \in A$. Assume that there is

$$\kappa \geq \max\{\sup \rho_\sigma : \sigma \neq \emptyset \wedge \sigma \in \omega^{<\omega}\}, \sup\{\rho_\alpha : \alpha \in A\}\}$$

with

$$\kappa \to (\kappa)^{\omega \cdot \rho_\sigma}$$

for all $\sigma \in \omega^{<\omega}$, $\sigma \neq \emptyset$. Then A admits a homogeneous tree $\hat{T}$ (on some $\omega \times \hat{\lambda}$).

Proof. Let $\mu_{\sigma,\tau}$ be the measure on $T(\sigma,\tau)$, where $\ell h\sigma = \ell h\tau \neq 0$. Let $\hat{\lambda} = \sup\{i^{\mu_{\sigma,\tau}}(\kappa) : \ell h\sigma = \ell h\tau \neq 0\}$. Then define a tree $\hat{T}$ on $\omega \times \hat{\lambda}$ and maps

$$p_\sigma : [\kappa]^{W_\sigma} \xrightarrow{\text{onto}} \hat{T}(\sigma)$$

as follows:

Given $h \in [\kappa]^{W_\sigma}$, let for $1 \leq i \leq \ell h\sigma$, $h_i(v) = h(\tau_i, v)$ so that $h_i : T(\sigma \restriction \ell_i, \tau_i) \to \kappa$. Abbreviate

$$\mu_{\sigma,i} \equiv \mu_{\sigma \restriction \ell_i, \tau_i}$$

and put

$$p_\sigma(h) = ([h_i]_{\mu_{\sigma,i}})_{1 \leq i \leq \ell h\sigma}$$

and

$$(\sigma, u) \in \hat{T} \Leftrightarrow \exists h \in [\kappa]^{W_\sigma}(p_\sigma(h) = u) \vee (\sigma = u = \emptyset) .$$

First note that if $\sigma \subseteq \sigma'$ and $h \in [\kappa]^{W_{\sigma'}}$, then $p_{\sigma'}(h) \restriction \ell h\sigma = ([h_i]_{\mu_{\sigma',i}})_{1 \leq i \leq \ell h\sigma} = p_\sigma(h \restriction W_\sigma)$, so that $\hat{T}$ is indeed a tree and condition (i)' before is satisfied.

If now $\hat{T}(\alpha)$ is not wellfounded, then as we will see in a moment $\alpha \in A$ $\therefore T(\alpha)$ is wellfounded, therefore $W_\alpha = \bigcup_n W_{\alpha \restriction n}$ is a wellordering and $\rho_\alpha \leq \kappa$, thus (ii)' is also satisfied. Since we have assumed that $\kappa \to (\kappa)^{\omega \cdot \rho_\sigma}$ we have by our preceding discussion that $\hat{T}$ is homogeneous.

So it only remains to show that

$$\alpha \in A \Leftrightarrow \hat{T}(\alpha) \text{ is not wellfounded} .$$

The direction $\Rightarrow$ is clear. To prove the other direction, assume $(\eta_1, \eta_2, \dots) \in [\hat{T}(\alpha)]$. Then for each $n \geq 1$, we can find ${}^n h \in [\kappa]^{W_{\alpha \restriction n}}$ such that if we let ${}^n h_i(v) = {}^n h(\tau_i, v)$ for $1 \leq i \leq n$, then $\eta_i = [{}^n h_i]_{\mu_{\alpha \restriction n, i}}$. Since $[{}^n h_i]_{\mu_{\alpha \restriction n,i}} = [{}^{n'} h_i]_{\mu_{\alpha \restriction n',i}}$ for $n, n' \geq i$ let $Z_i \subseteq T(\alpha \restriction \ell_i, \tau_i)$ have $\mu_{\alpha \restriction \ell_i, \tau_i}$-measure 1 and be such that ${}^n h_i(v) = {}^{n'} h_i(v)$ for all $n, n' \geq i$ and all $v \in Z_i$.

If now $\alpha \notin A$, towards a contradiction, find β with $T(\alpha, \beta)$ not wellfounded. Let for $k \geq 1$, $\tau_{i_k} = \beta \restriction k$ so that $i_1 < i_2 < \cdots$. Let $X_k = Z_{i_k} \subseteq T(\alpha \restriction k, \beta \restriction k)$. By the homogeneity of T there is $f \in [T(\alpha, \beta)]$ with $f \restriction k \in X_k$ for all $k \geq 1$. Then obviously $(\tau_{i_1}, f \restriction 1) >_{KB} (\tau_{i_2}, f \restriction 2) >_{KB} \cdots$. But for each $k \geq 1$, if n is large enough, then ${}^n h_{i_k}(f \restriction k) = \theta_k$

is independent of n and also $\theta_k > \theta_{k+1}$, so that $\theta_1 > \theta_2 > \cdots$, a contradiction. ⊣

Let now $\kappa^{\mathbb{R}}$ be the least non-hyperprojective ordinal or equivalently the ordinal of the smallest admissible set containing the reals. Then by Kechris-Kleinberg-Moschovakis-Woodin [A], there are arbitrarily large $\lambda < \kappa^{\mathbb{R}}$ with $\lambda \to (\lambda)^{\lambda}$ (and also this holds for $\lambda = \kappa^{\mathbb{R}}$). So it follows that every projective set admits a homogeneous tree or $\omega \times \lambda$ for some $\lambda < \kappa^{\mathbb{R}}$.

Moschovakis has pointed out that in the preceding theorem it is enough to assume that B admits only a weakly homogeneous tree (to conclude again under the appropriate assumptions that A carries a homogeneous tree). Here a tree T on $\omega \times \lambda$ is called weakly homogeneous if for each $\sigma \neq \emptyset$ there is a partition

$$T(\sigma) = \cup_{i \in I_\sigma} K_{\sigma,i}$$

where each I_σ is a countable set such that the following hold:

(a) If $\sigma \subseteq \sigma'$ and $T(\sigma) = \cup_i K_{\sigma,i}$, $T(\sigma') = \cup_j K_{\sigma',j}$ then for every j there is an i such that $K_{\sigma',j} \upharpoonright \ell h\sigma \equiv \{v \upharpoonright \ell h\sigma : v \in K_{\sigma',j}\} \subseteq K_{\sigma,i}$.

(b) Each $K_{\sigma,i}$ carries a measure $\mu_{\sigma,i}$ with the following property: If $T(\alpha)$ is not wellfounded and for each $n > 0$, $i \in I_{\alpha\upharpoonright n}$, $X_{\alpha\upharpoonright n,i} \subseteq K_{\alpha\upharpoonright n,i}$ and $X_{\alpha\upharpoonright n,i}$ has $\mu_{\alpha\upharpoonright n,i}$-measure 1, then there is $f \in \lambda^{\omega}$ such that for each $n > 0$,

$$f \upharpoonright n \in \bigcup_i X_{\alpha\upharpoonright n,i} \, .$$

The proof is similar to the one given before and we leave it to the reader. Notice also the simple fact that if $B \subseteq \mathcal{R}^2$ admits a weakly homogeneous tree then so does

$$C = \{\alpha : \exists\beta\, B(\alpha,\beta)\} \, .$$

Note. The concepts and results in this section originate in Kunen [1971a] and Martin [1977].

§7. A result of Martin on subsets of $\underset{\sim}{\delta}^1_3$. Let $\mathcal{P} \subseteq \mathcal{R}$ be a universal Π^1_3 set of reals i.e. assume that $\mathcal{P} \in \Pi^1_3$ and for each $A \subseteq \mathcal{R}$ in Π^1_3 there is $n \in \omega$ such that $\alpha \in A \Leftrightarrow n^\frown\alpha = (n,\alpha(0),\alpha(1),\dots) \in \mathcal{P}$. Let $\overline{\varphi} = \{\varphi_n\}$ be a regular Π^1_3-scale on $\mathcal{P}$ - i.e. each φ_n maps $\mathcal{P}$ onto an initial segment of ordinals (therefore, as is well known, $\mathrm{range}(\varphi_n) = \underset{\sim}{\delta}^1_3 = \omega_{\omega+1}$). The tree associated with this scale is defined by

$$T_3(\overline{\varphi}) = \{(\alpha \restriction n, (\varphi_0(\alpha), \varphi_1(\alpha), \ldots, \varphi_{n-1}(\alpha))) : \alpha \in \mathcal{P},\ n \in \omega\} .$$

Thus $T_3(\overline{\varphi})$ is a tree on $\omega \times \omega_{\omega+1}$. Also $\mathcal{P} = p[T_3(\overline{\varphi})]$ and for every $\alpha \in \mathcal{P}$ $T_3(\overline{\varphi})(\alpha)$ has an honest leftmost branch, namely $\overline{\varphi}(\alpha)$. Note also that there is a function $K : \omega_{\omega+1} \to \omega_{\omega+1}$ such that

$$(\sigma, (\xi_0, \ldots, \xi_{n-1})) \in T_3(\overline{\varphi}) \wedge \xi_0 \leq \xi \Rightarrow \xi_0, \xi_1, \ldots, \xi_{n-1} \leq K(\xi) .$$

Indeed, if $(\sigma, (\xi_0, \ldots, \xi_{n-1})) \in T_3(\overline{\varphi})$ then for some $\alpha \supseteq \sigma$, $\varphi_0(\alpha) = \xi_0, \ldots, \varphi_{n-1}(\alpha) = \xi_{n-1}$. Thus $\max\{\xi_0, \xi_1, \ldots, \xi_{n-1}\} \leq \sup\{\varphi_n(\alpha) : n \in \omega \wedge \alpha \in \mathcal{P} \wedge \varphi_0(\alpha) \leq \xi\} =^{\text{def}} K(\xi)$. That $K(\xi) < \omega_{\omega+1}$ follows from the fact that for each $\xi < \omega_{\omega+1} = \underset{\sim}{\delta}^1_3$, $\{\alpha : \alpha \in \mathcal{P} \wedge \varphi_0(\alpha) \leq \xi\}$ is $\underset{\sim}{\Delta}^1_3$, so by boundedness $\{\varphi_n(\alpha) : n \in \omega \wedge \alpha \in \mathcal{P} \wedge \varphi_0(\alpha) \leq \xi\}$ is bounded below $\underset{\sim}{\delta}^1_3$.

If $X \subseteq \omega_{\omega+1}$, then we say that X is Σ^1_n <u>in the codes</u> or just Σ^1_n if

$$X^* =^{\text{def}} \{\alpha \in \mathcal{P} : \varphi_0(\alpha) \in X\}$$

is Σ^1_n. One can use the results of Harrington-Kechris [A] to show that this is independent of the choice of $\mathcal{P}$, φ_0, where φ_0 is any regular Π^1_3-norm on a universal Π^1_3 set $\mathcal{P}$, provided that $n \geq 4$.

Let also S_3 be the tree associated with $\mathcal{P}$ as in §3.1. Thus again $\mathcal{P} = p[S_3]$. The result below provides an analog of Theorem 1 in Kechris-Moschovakis [1972].

<u>Theorem</u>. (Martin [1977]). If $X \subseteq \omega_{\omega+1}$ is Σ^1_4, then $X \in L[S_3, T_3(\overline{\varphi})]$.

<u>Proof</u>. Say, putting $T_3(\overline{\varphi}) \equiv T_3$,

$$\beta \in X^* \Leftrightarrow \exists\gamma(n_0{}^\frown\langle\beta,\gamma\rangle \in \mathcal{P})$$

$$\Leftrightarrow \exists\gamma\, \exists p(n_0{}^\frown\langle\beta,\gamma\rangle, p) \in [T_3] ,$$

and

$$\alpha \in \mathcal{P} \wedge \beta \in \mathcal{P} \wedge \varphi_0(\alpha) \leq \varphi_0(\beta) \Leftrightarrow n_1{}^\frown\langle\alpha,\beta\rangle \in \mathcal{P}$$

$$\Leftrightarrow \exists f(n_1{}^\frown\langle\alpha,\beta\rangle, f) \in [S_3] ,$$

where as usual $\langle\alpha,\beta\rangle = (\alpha(0), \beta(0), \alpha(1), \beta(1), \ldots)$. Consider then the following game G_ξ, for $\xi < \omega_{\omega+1}$:

I		II				
$\alpha(0)$	$h(0)$					
		$\beta(0)$	$g(0)$	$\gamma(0)$	$p(0)$	$f(0)$
$\alpha(1)$	$h(1)$					
		$\beta(1)$	$g(1)$	$\gamma(1)$	$p(1)$	$f(1)$
$\vdots$				$\vdots$		
α	h	β	g	γ	p	f ,

(where $h,g,p,f \in \omega^{\omega}_{\omega+1}$), whose payoff set is defined as follows:

We say that I has played correctly up to his m^{th} move, for $m \geq 1$, if $(\alpha \restriction m, h \restriction m) \in T_3 \wedge h(0) \leq \xi$. Then note that also $\forall i < m(h(i) \leq K(\xi))$. We say that II has played correctly up to his m^{th} move, for $m \geq 1$ again, if $(\beta \restriction m, g \restriction m) \in T_3 \wedge g(0) \leq \xi \wedge (n_0 ^\frown \langle\beta,\gamma\rangle \restriction m, p \restriction m) \in T_3 \wedge (n_1 ^\frown \langle\alpha,\beta\rangle \restriction m, f \restriction m) \in S_3$.

Now II wins iff for all $m \geq 1$: I has played correctly up to his m^{th} move $\Rightarrow$ II has played correctly up to his m^{th} move.

Clearly this is a closed game for player II and it is in $L[S_3, T_3]$, uniformly on ξ. So it is enough (by the absoluteness of closed games) to show that

$$\xi \in X \Leftrightarrow \text{II has a winning strategy in } G_\xi .$$

($\Leftarrow$). Say s is a winning strategy for II in G_ξ. Let I play (α, f) where $\alpha \in P$, $\varphi_0(\alpha) = \xi$ ahd $h = \overline{\varphi}(\alpha)$. Then I plays always correctly, so if II following his winning strategy s produces (β,g,γ,p,f) he must have played also always correctly $\therefore (\beta,g) \in [T_3] \wedge g(0) \leq \xi \wedge (n_0 ^\frown \langle\beta,\gamma\rangle, p) \in [T_3] \wedge (n_1 ^\frown \langle\alpha,\beta\rangle, f) \in [S_3] \therefore \beta \in P$, $\varphi_0(\beta) \leq g(0) \leq \xi$ (as $\overline{\varphi}(\beta)$ is the honest leftmost branch of $T_3(\beta)$), $\beta \in X^*$ and $\varphi_0(\alpha) \leq \varphi_0(\beta) \therefore \varphi_0(\beta) = \varphi_0(\alpha) = \xi$ and $\xi \in X$.

($\Rightarrow$). Assume now I has a winning strategy t in G_ξ but, towards a contradiction, that $\xi \in X$. Then fix $\beta \in P$ with $\varphi_0(\beta) = \xi$, $g = \overline{\varphi}(\beta)$ and γ, p so that $(n_0 ^\frown \langle\beta,\gamma\rangle, p) \in [T_3]$. Let for each $\sigma \in \omega^{<\omega}$, $\sigma \neq \phi$ ν_σ be the measure on $S_3(\sigma)$ as in 3.3. Note that since $\omega_{\omega+1} \to (\omega_{\omega+1})^\rho_\nu$, $\forall \rho < \omega_{\omega+1}$, $\forall \nu < \omega_{\omega+1}$, we actually have that ν_σ is $\omega_{\omega+1}$-additive. Of course these measures satisfy the homogeneity conditions (i), (ii) of §6.

Define now inductively values $\alpha(0), h(0), \alpha(1), h(1), \ldots$ and sets $X_1, X_2, \ldots$ as follows (recall below that $n_1 ^\frown \langle\alpha,\beta\rangle = (n_1, \alpha(0), \beta(0), \ldots)$:

First $\alpha(0)$, $h(0)$ are the values called by t in I's initial move, which is clearly correct. Now for each $(f(0)) \in S_3((n_1))$, if II plays $\beta(0)$, $g(0)$, $\gamma(0)$, $p(0)$, $f(0)$ in his first move, I answers by t to play correctly $\alpha(1)$, $h(1)$. In particular $h(1) < K(\xi)$ so by the $\omega_{\omega+1}$-additivity of $\mathcal{U}_{(n_1)}$ let X_1 be in $\mathcal{U}_{(n_1)}$ and such that $\alpha(1)$, $h(1)$ are always the same for $(f(0)) \in X_1$. This is our $\alpha(1)$, $h(1)$. Then for each $(f(0),f(1)) \in S_2((n_1,\alpha(0))$ if II next plays $\beta(1)$, $g(1)$, $\gamma(1)$, $p(1)$, $f(1)$ I answers following t to play $\alpha(2)$, $h(2)$ which, by an argument exactly as before, is the same for all $(f(0),f(1)) \in X_2$ for some $X_2 \in \mathcal{U}_{(n_1,\alpha(0))}$ etc. Now as $(\alpha,h) \in [T_3]$ and $h(0) \leq \xi$ clearly $\varphi_0(\alpha) \leq \xi$ so $\varphi_0(\alpha) \leq \varphi_0(\beta)$, thus $S_3(n_1 ^\frown \langle\alpha,\beta\rangle)$ is not wellfounded. Since $X_k \in \mathcal{U}_{n_1 ^\frown \langle\alpha,\beta\rangle \restriction k}$ for all $k \geq 1$ we have by condition (ii) of homogeneity that there is f such that $f \restriction k \in X_k$ for each $k \geq 1$. If II plays now β, g, γ, p, f, he plays always correctly and if I follows t he plays α, h so that he also plays correctly. But then II won, a contradiction. ⊣

<u>Corollary</u>. If $X \subseteq \omega_{\omega+1}$ then there is $\alpha \in \mathbb{R}$ such that $X \in L[X_3, T_3(\overline{\varphi}), \alpha]$.

<u>Proof</u>. By the Moschovakis' Coding Lemma every $X \subseteq \omega_{\omega+1}$ is $\Sigma^1_4(\alpha)$ for some $\alpha \in \mathbb{R}$. ⊣

§8. On the Victoria Delfino Third Problem. Let $\mathcal{P}$ be a universal Π^1_3 set of reals and $\overline{\varphi}$ a regular Π^1_3-scale on it. Let $T_3(\overline{\varphi})$ be its associated tree as in §7. The Victoria Delfino Third Problem (see Kechris-Moschovakis (eds) [1978]) is the question:

Is $L[T_3(\overline{\varphi})]$ independent of $\mathcal{P}$, $\overline{\varphi}$?

Let also $\tilde{L}[T_3(\overline{\varphi})] = \bigcup_{\alpha \in \mathbb{R}} L[T_3(\overline{\varphi}),\alpha]$. Surely the independence of $\tilde{L}[T_3(\overline{\varphi})]$ from $\mathcal{P}$, $\overline{\varphi}$ would be very strong evidence for an affirmative answer to the above problem. So the following result, despite its dependence on an unproven yet hypothesis is of interest here. Its proof uses methods of Kunen; see Kunen [1971b] and Kechris [1978].

<u>Theorem</u>. Assume $\omega_{\omega+1} \to (\omega_{\omega+1})^{\omega_{\omega+1}}$. Then $\text{power}(\omega_{\omega+1}) \subseteq \tilde{L}[T_3(\overline{\varphi})]$, so in particular $\tilde{L}[T_3(\overline{\varphi})]$ is independent of $\mathcal{P}$, $\overline{\varphi}$.

<u>Proof</u>. The heart of the proof is the following:

<u>Lemma A</u>. If $f : \omega_{\omega+1} \to \omega_{\omega+1}$, then there is $g \in \tilde{L}[T_3(\overline{\varphi})]$ such that $f(\xi) \leq g(\xi)$, $\forall \xi < \omega_{\omega+1}$.

From that it follows that if $C \subseteq \omega_{\omega+1}$ is cub, then there is $\overline{C} \subseteq C$, $\overline{C}$ cub such that $\overline{C} \in \tilde{L}[T_3(\overline{\varphi})]$. Indeed, let $f : \omega_{\omega+1} \to C$ be the increasing enumeration of C and let g be as in Lemma A. Then if $\overline{C} = \{\xi < \omega_{\omega+1} : \xi$ is limit $\wedge \forall \eta < \xi (g(\eta) < \xi)\}$, $\overline{C}$ is cub and if $\xi \in \overline{C}$ then $\forall \eta < \xi (f(\eta) \leq g(\eta) < \xi)$ $\therefore$ $f(\xi) = \xi \in C$ i.e. $\overline{C} \subseteq C$. Now we have

Lemma B. If $\omega_{\omega+1} \to (\omega_{\omega+1})^{\omega_{\omega+1}}$, and for every cub $C \subseteq \omega_{\omega+1}$ there is $\bar{C} \subseteq C$, $\bar{C}$ cub such that $\bar{C} \in \tilde{L}[T_3(\bar{\varphi})]$, then $\mathrm{power}(\omega_{\omega+1}) \subseteq \tilde{L}[T_3(\bar{\varphi})]$.

Proof of Lemma B. Consider the following partition of $[\omega_{\omega+1}]^{\omega_{\omega+1}}$:

$$f \in X \Leftrightarrow f \in \tilde{L}[T_3(\bar{\varphi})] .$$

Then let C be cub such that $C \models^{\omega_{\omega+1}} \subseteq X$ or $C \models^{\omega_{\omega+1}} \subseteq \sim X$. By our hypothesis C can be assumed to be in $\tilde{L}[T_3(\bar{\varphi})]$ so that we must have $C \models^{\omega_{\omega+1}} \subseteq \tilde{L}[T_3(\bar{\varphi})]$. Let $\{\eta_\theta : \theta < \omega_{\omega+1}\}$ be the increasing enumeration of C and put $H = \{\eta_{\theta+\omega} : \theta < \omega_{\omega+1}\}$. Clearly $[H]^{\omega_{\omega+1}} \subseteq \tilde{L}[T_3(\bar{\varphi})]$. Let now A be an arbitrary unbounded subset of $\omega_{\omega+1}$. Let also f_H be the increasing enumeration of H. Then $f_H[A] \subseteq H$, so if g is the increasing enumeration of $f_H[A]$, $g \in \tilde{L}[T_3(\bar{\varphi})]$. But $\xi \in A \Leftrightarrow f_H(\xi) \in f_H(A) \Leftrightarrow f_H(\xi) \in \mathrm{range}(g)$ $\therefore$ $A \in L[f_H, g] \subseteq \tilde{L}[T_3(\bar{\varphi})]$ and we are done.

So we only have to prove Lemma A above. For that we will play a Solovay-type game which is a variant of a game of Kunen, see Kunen [1971b].

Let $\mathcal{S}'$ be a Π^1_2 subset of $\mathcal{R} \times \mathcal{R}$, such that if $\mathcal{S}(\alpha) \Leftrightarrow \exists\beta\, \mathcal{S}'(\alpha,\beta)$, then $\mathcal{S} \in \Sigma^1_3 - \underset{\sim}{\Pi}^1_3$. Let

$$\mathcal{S}'(\alpha,\beta) \Leftrightarrow n_0{}^\frown\langle\alpha,\beta\rangle \in \mathcal{P} ,$$

so that

$$\begin{aligned} \exists\beta\, \mathcal{S}'(\alpha,\beta) &\Leftrightarrow \exists\beta(n_0{}^\frown\langle\alpha,\beta\rangle \in \mathcal{P}) \\ &\Leftrightarrow \exists f\, \exists\beta\, \forall k(n_0{}^\frown\langle\alpha,\beta\rangle \restriction k, f \restriction k) \in T_3(\bar{\varphi}) \\ &\Leftrightarrow \exists g\, \forall m(\alpha \restriction m, f \restriction m) \in U , \end{aligned}$$

where

$$\begin{aligned} ((a_0,\ldots,a_{m-1}),(\xi_0,\ldots,\xi_{m-1})) \in U &\Leftrightarrow (\xi_0)_0,\ldots,(\xi_{m-1})_0 \in \omega \\ &\wedge (n_0{}^\frown(a_0,(\xi_0)_0,\ldots,a_{m-1},(\xi_{m-1})_0) \restriction m, ((\xi_0)_1,\ldots,(\xi_{m-1})_1)) \in T_3(\bar{\varphi}) \end{aligned}$$

where $\xi \mapsto ((\xi)_0,(\xi_1))$ is some simple 1-1 correspondence of $\omega_{\omega+1}$ with $(\omega_{\omega+1})^2$. Clearly $U \in L[T_3(\bar{\varphi})]$ and if

$$\mathcal{S}(\alpha) \Leftrightarrow \exists\beta\, \mathcal{S}'(\alpha,\beta) ,$$

then $\alpha \in \mathcal{S} \Leftrightarrow U(\alpha)$ is not wellfounded. Now we claim that U is a tree on $\omega \times \lambda$, for some $\lambda < \omega_{\omega+1}$. Indeed in the notation above if $(\bar{a},\bar{\xi}) \in U$, then there is a γ such that $n_0{}^\frown\gamma \in \mathcal{P}$, $\gamma = \langle\alpha,\beta\rangle$, $\bar{a} \subseteq \alpha$ and $\varphi_i(n_0{}^\frown\gamma) = (\xi_i)_1$ for $i < m = \ell h\bar{a}$. Thus $(\alpha,\beta) \in \mathcal{S}'$. But $\mathcal{S}'' = \{n_0{}^\frown\langle\alpha,\beta\rangle : (\alpha,\beta) \in \mathcal{S}'\}$ is a Π^1_2 subset of $\mathcal{P}$, so (by boundedness) there is $\mu < \omega_{\omega+1}$ such that $n_0{}^\frown\gamma \in \mathcal{S}'' \Rightarrow \varphi_i(n_0{}^\frown\gamma) < \mu$ for all i $\therefore$ $(\xi_i)_1 < \mu$, $\forall i < m$ $\therefore$ there is $\lambda < \omega_{\omega+1}$ such that $\xi_i < \lambda$ and we are done.

Thus if $U(\alpha)$ is wellfounded, $\text{rank}(U(\alpha)) < \omega_{\omega+1}$. Moreover since $\mathcal{S} \notin \underset{\sim}{\Delta}^1_3$,

$$\sup\{\text{rank}(U(\alpha)) : U(\alpha) \text{ is wellfounded}\} = \omega_{\omega+1} .$$

Otherwise for some $\rho < \omega_{\omega+1}$

$$\alpha \in \mathcal{S} \Leftrightarrow \neg\,(\text{rank}(U(\alpha)) < \rho) ,$$

therefore by Martin's result (see Martin [A]) that $\underset{\sim}{\Delta}^1_3$ is closed under $< \omega_{\omega+1}$ intersections and unions, $\mathcal{S} \in \underset{\sim}{\Delta}^1_3$, a contradiction.

After these preliminaries consider the following game associated with each $f : \omega_{\omega+1} \to \omega_{\omega+1}$:

I	II
w	α

II wins iff $[w \in \mathcal{P} \Rightarrow U(\alpha)$ is wellfounded $\wedge$ $\text{rank}(U(\alpha)) > f(\varphi_0(w))]$.

By a simple boundedness argument and the above remarks I cannot have a winning strategy in this game, so II has a winning strategy s. Define then the following tree $\mathcal{J}$ on $\omega \times \omega_{\omega+1} \times \omega \times \lambda$:

$$(\tau,u,a,v) \in \mathcal{J} \Leftrightarrow (\tau,u) \in T_3(\overline{\varphi}) \wedge (a,v) \in U$$
$$\wedge\ a \text{ is the result of II playing according to s when I plays } \tau .$$

Clearly $\mathcal{J} \in L[T_3(\overline{\varphi}),s]$. Moreover $\mathcal{J}$ is wellfounded, since if $(w,f,\alpha,g) \in [\mathcal{J}]$ then $(w,f) \in [T_3(\overline{\varphi})]$ $\therefore w \in \mathcal{P}$, and $(\alpha,g) \in U$ $\therefore U(\alpha)$ is not wellfounded and also α is the result of a run in which II follows s against I playing w, a contradiction.

Fix now $\xi < \omega_{\omega+1}$. Let $w \in \mathcal{P}$ be such that $\varphi_0(w) = \xi$. Let α be the result of II playing according to s while I plays this w. Finally let $f = \overline{\varphi}(w)$. Then the map

$$v \mapsto (w \restriction \ell h v, f \restriction \ell h v, \alpha \restriction \ell h v, v)$$

is an embedding of $U(\alpha)$ into $\mathcal{J}$. But notice that actually this maps $U(\alpha)$ into

$$\mathcal{J}_{(\xi)} =^{\text{def}} \{(\tau,u,a,v) \in \mathcal{J} : u(0) \leq \xi\} .$$

Thus $\text{rank}(U(\alpha)) \leq \text{rank}(\mathcal{J}_{(\xi)})$. But also $f(\xi) = f(\varphi_0(w)) < \text{rank}(U(\alpha)) \therefore$

$$f(\xi) < \text{rank}(\mathcal{J}_{(\xi)}) =^{\text{def}} g(\xi) .$$

As $g \in L[T_3(\overline{\varphi}),s]$ it will be enough to show that $\text{rank}(\mathcal{J}_{(\xi)}) < \omega_{\omega+1}$. But recall from §7 that some $K(\xi) < \omega_{\omega+1}$:

$$w \in \mathcal{P} \wedge \varphi_0(w) \leq \xi \Rightarrow \forall i[\varphi_i(w) \leq K(\xi)] .$$

Thus if $(\tau,u,a,v) \in \mathcal{J}$, then $u(i) \leq K(u(0))$, since $(\tau,u) \in T_3(\overline{\varphi})$, so there is $w \in \mathcal{P}$ with $\tau \subseteq w$ and $\overline{\varphi}(w) \upharpoonright \ell h u = u \therefore \varphi_0(w) = u(0)$ and $u(i) = \varphi_i(w) \leq K(u(0))$. So

$$\mathcal{J}_{(\xi)} \subseteq \mathcal{J} \upharpoonright K(\xi) ,$$

thus

$$f(\xi) \leq g(\xi) = \text{rank}(\mathcal{J}_{(\xi)}) \leq \text{rank}(\mathcal{J} \upharpoonright K(\xi)) < \omega_{\omega+1} ,$$

where $\mathcal{J} \upharpoonright \theta = \{(\tau,u,a,v) \in \mathcal{J} : u \in \theta^{<\omega}\}$. This completes the proof.

References

L. A. Harrington and A. S. Kechris [A], On the determinacy of games on ordinals, to appear.

A. S. Kechris [1978], AD and projective ordinals, Cabal Seminar 76-77, Lecture Notes in Mathematics, Vol. 689, Springer-Verlag 1978.

A. S. Kechris, E. M. Kleinberg, Y. N. Moschovakis, H. Woodin [A], The axiom of determinacy, strong partition properties and nonsingular measures, this volume.

A. S. Kechris and D. A. Martin [1978], On the theory of Π^1_3 sets of reals, Bull. Amer. Math. Soc. 84 (1978), 149-151.

A. S. Kechris and Y. N. Moschovakis (eds.) [1978], Cabal Seminar 76-77, Lecture Notes in Mathematics, Vol. 689, Springer-Verlag 1978.

A. S. Kechris and Y. N. Moschovakis [1978], Notes on the theory of scales, Cabal Seminar 76-77, Lecture Notes in Mathematics, Vol. 689, Springer-Verlag 1978.

A. S. Kechris and Y. N. Moschovakis [1972], Two theorems about projective sets, Israel J. of Math., 12 (1972), 391-399.

K. Kunen [1971a], On $\underset{\sim}{\delta}^1_5$, circulated note, August 1971.

K. Kunen [1971b], Some singular cardinals, circulated note, September 1971.

R. Mansfield [1971], A Souslin operation on Π^1_2, Israel J. of Math., $\underset{\sim}{9}$ (1971), 367-379.

D. A. Martin [1977], On subsets of $\underset{\sim}{\delta}^1_3$, circulated note, January 1977.

D. A. Martin [A], Projective sets and cardinal numbers: some questions related to the continuum problem, J. of Symbolic Logic, to appear.

D. A. Martin and R. M. Solovay [1969], A basis theorem for Σ^1_3 sets of reals, Ann. of Math, $\underset{\sim}{89}$ (1969), 138-160.

Y. N. Moschovakis [1970], Determinacy and prewellorderings of the continuum, Math. Logic and Foundations of Set Theory, Ed. by Y. Bar Hillel, North Holland, 1970, 24-62.

J. R. Shoenfield [1961], The problem of predicativity, Essays on the Foundations of Mathematics, Magnes Press, Hebrew Univ. Jerusalem, 1961, 132-139.

R. M. Solovay [1978], A Δ^1_3 coding of the subsets of ω_ω, Cabal Seminar 76-77, Lecture Notes in Mathematics, Vol. 689, Springer-Verlag 1978.

THE AXIOM OF DETERMINACY, STRONG PARTITION PROPERTIES AND NONSINGULAR MEASURES

Alexander S. Kechris[(1)]
Department of Mathematics
California Institute of Technology
Pasadena, California 91125

Eugene M. Kleinberg[(2)]
Department of Mathematics
State University of New York
Buffalo, New York

Yiannis N. Moschovakis[(3)]
Department of Mathematics
University of California
Los Angeles, California 90024

W. Hugh Woodin
Department of Mathematics
University of California
Berkeley, California 94720

In this paper we study the relationship between AD and strong partition properties of cardinals as well as some consequences of these properties themselves.

Let us say that an uncountable cardinal κ has <u>the strong partition property</u> if

$$(\forall\mu < \kappa)[\kappa \rightarrow (\kappa)^{\kappa}_{\mu}] ,$$

i.e. if for every partition of all increaseng κ-term sequences into fewer than κ parts there is a set $C \subseteq \kappa$ of cardinality κ such that all the increasing sequences into C lie in the same part of the partition. We will show in §1 that the axiom of determinacy (AD) implies the existence of unboundedly many cardinals with the strong partition property below Θ, where

[(1)]Research partially supported by NSF Grant MCS79-20465. The author is an A. P. Sloan Foundation Fellow.

[(2)]Research partially supported by NSF Grant MCS78-03744.

[(3)]Research partially supported by NSF Grant MCS78-02989.

Θ = supremum of the ranks of the prewellorderings of the continuum .

In §2, we will show that conversely, if there is a cardinal κ with the strong partition property above an ordinal λ, then every λ-Suslin set is determined. Combining these two results we obtain an elegant purely set-theoretic characterization of AD within $\mathbb{R}^+$, the smallest admissible set containing the continuum; namely, in $\mathbb{R}^+$, AD holds if and only if the power set of every ordinal exists, and there are arbitrarily large cardinals with the strong partition property.

In §3 we will strengthen the main result of §1 to obtain from AD unboundedly many cardinals κ below Θ, such that not only κ has the strong partition property but also $\{\lambda < \kappa : \lambda$ has the strong partition property$\}$ is stationary in κ (thus κ is also Mahlo). Finally, in §4 we will show that every Mahlo cardinal with the strong partition property carries a normal measure concentrating on regular cardinals. Up until now all the normal measures on cardinals κ, produced by AD, were of the "singular" type i.e. for some regular $\lambda < \kappa$ they concentrated on the ordinals of cofinality λ.

In matters of descriptive set theory, we will follow in general the terminology and notation of Moschovakis [1980], which we will cite as DST. We refer to Kleinberg [1977] for information concerning partition properties.

Our work in this paper takes place in ZF + DC with all other hypotheses (mainly AD) stated explicitly.

§1. A partition theorem. In the main result of this section, we will establish that AD implies the existence of many cardinals with the strong partition property.

As usual, a space will be any product

$$\mathfrak{X} = X_1 \times \cdots \times X_n$$

of copies of ω and $\mathfrak{N} = {}^{\omega}\omega$, a pointset is any subset of a space and a pointclass is any collection Γ of pointsets. A Γ-norm on a pointset $P \subseteq \mathfrak{X}$ is any function

$$\varphi : P \to \text{Ordinals}$$

such that both $\leq^*_\varphi$ and $<^*_\varphi$ are in Γ, where

$$x \leq^*_\varphi y \Leftrightarrow x \in P \,\&\, [y \notin P \vee \varphi(x) \leq \varphi(y)] ,$$

$$x <^*_\varphi y \Leftrightarrow x \in P \,\&\, [y \notin P \vee \varphi(x) < \varphi(y)] .$$

We say that Γ has the prewellordering property if every pointset in Γ admits a Γ-norm.

Recall from DST that a Spector pointclass closed under $\forall^{\mathcal{N}}$ is a pointclass Γ which is closed under recursive substitutions, &, $\vee$, $\exists^{\omega}$, $\forall^{\omega}$ and $\forall^{\mathcal{N}}$, which is ω-parametrized and which has the prewellordering property. We will need here a slightly stronger notion.

A partial function

$$f : \mathcal{X} \to \omega$$

is Γ-recursive (or in Γ), if

$$\text{Graph}(f) = \{(x,i) : f(x) = i\}$$

is in Γ. We say that Γ is closed under Kleene's ^3E (deterministic quantification on $\mathcal{N}$), if whenever $f : \mathcal{X} \times \mathcal{N} \to \omega$ is a Γ-recursive partial function, then the realtion

$$P(x) \Leftrightarrow (\forall\alpha)[f(x,\alpha)\downarrow] \ \& \ (\exists\alpha)[f(x,\alpha) = 0]$$

is in Γ.

If Γ is closed under Kleene's ^3E, then (trivially) Γ is closed under $\forall^{\mathcal{N}}$ and Δ, $\underset{\sim}{\Delta}$ are both closed under $\forall^{\mathcal{N}}$ and $\exists^{\mathcal{N}}$; and if Γ is closed under both $\forall^{\mathcal{N}}$ and $\exists^{\mathcal{N}}$, then (again trivially) Γ is closed under Kleene's ^3E. By Moschovakis [1967], if A is any pointset then the envelope

$$\text{Env}(A,{}^3E) = \text{all pointsets which are Kleene-semirecursive in } {}^3E \text{ and (the characteristic function of) } A$$

is a Spector pointclass closed under ^3E, which furthermore contains both A and its complement and is not closed under $\exists^{\mathcal{N}}$. These Kleene envelopes are very important for the applications of our present results, although one need not know any recursion in type-3 in order to follow the proofs in §1.

We can now state our main result in this section.

1.1. Theorem. Assume AD, let Γ be a Spector pointclass closed under ^3E and let

$$\kappa = o(\underset{\sim}{\Delta})$$
$$= \text{supremum of the ranks of prewellorderings of } \mathcal{N} \text{ in } \underset{\sim}{\Delta}$$

be the ordinal associated with Γ; then κ has the strong partition property.

This result implies that (granting AD),

$$\kappa({}^3E) = \text{the Kleene ordinal of the continuum}$$
$$= \text{the ordinal associated with } \mathrm{Env}({}^3E)$$

has the strong partition property, as does

$$\kappa^{\mathbb{R}} = \text{the closure ordinal of the continuum}$$
$$= \sup\{\xi : \xi \text{ in the rank of a hyperprojective prewellordering of } \mathcal{N}\}.$$

It also implies that (under AD again), there are arbitrarily large cardinals with the strong partition property below Θ.

One very important problem is whether the projective ordinals $\underset{\sim}{\delta}^1_{2n+1}$, $n \geq 1$ (in particular $\underset{\sim}{\delta}^1_3$), have the strong partition property, granting AD. Clearly our Theorem 1.1 does not tell us anything in this case, since the Spector pointclasses $\Gamma = \Pi^1_{2n+1}$ are not closed under ${}^3E$.

Using a method of Martin (see Kechris [1978], Lemma 11.1) the proof of 1.1 is reduced to the problem of finding an appropriate coding (by elements of $\mathcal{N}$) of functions $f : \kappa \to \kappa$. Let us first reformulate Martin's Lemma in a form convenient for our purposes here.

Suppose Γ is a Spector pointclass closed under $\forall^{\mathcal{N}}$ and with associated ordinal κ, let

$$\varphi : S \twoheadrightarrow \kappa$$

be a Γ-norm which maps some set $S \subseteq \mathcal{N}$ in Γ onto κ and suppose that for each $\varepsilon \in \mathcal{N}$ we have a partial function

$$f_\varepsilon : \kappa \to \kappa ;$$

we will say that φ and $\{f_\varepsilon : \varepsilon \in \mathcal{N}\}$ define a <u>good</u> <u>coding</u> <u>in</u> Γ of the functions on κ to κ if conditions (i)-(iii) below hold, where for ordinals $\xi, \theta < \kappa$ we let

$$C_\xi(\varepsilon) \Leftrightarrow f_\varepsilon(\xi)\downarrow ,$$
$$C_{\xi,\theta}(\varepsilon) \Leftrightarrow (\forall \xi' \leq \xi)[f_\varepsilon(\xi')\downarrow \;\&\; f_\varepsilon(\xi') \leq \theta] ,$$

(i) For each fixed $\xi, \theta < \kappa$, the relation $C_{\xi,\theta}$ is in $\underset{\sim}{\Delta}$.

(ii) There is some relation $V(\varepsilon,\alpha,\beta)$ in $\neg\underset{\sim}{\Gamma}$ which computes the values of each f_ε relative to the norm φ in the following sense:

$$\alpha \in S \;\&\; f_\varepsilon(\varphi(\alpha))\downarrow \Rightarrow (\exists\beta)V(\varepsilon,\alpha,\beta)$$
$$\& \; (\forall\beta)\{V(\varepsilon,\alpha,\beta) \to [\beta \in S \;\&\; f_\varepsilon(\varphi(\alpha)) = \varphi(\beta)]\} .$$

(iii) For every total function $f : \kappa \to \kappa$, there is some $\varepsilon \in \mathcal{N}$ such that

$f = f_\varepsilon$.

1.2. Lemma. Assume AD and let Γ be a Spector pointclass closed under $\forall^{\mathcal{N}}$ with ordinal κ, which admits a good coding of the functions on κ to κ; then κ has the strong partition property.

Proof. Notice first that a good coding in Γ of the functions on κ to κ has the following additional boundedness property.

(ii') If $A \in \neg\underset{\sim}{\Gamma}$ and for some $\xi < \kappa$, $A \subseteq C_\xi$, then

$$\sup\{f_\varepsilon(\xi) : \varepsilon \in A\} < \kappa .$$

This is because by the definitions, if $\alpha \in S$ and $\varphi(\alpha) = \xi$, then

$$\sup\{f_\varepsilon(\xi) : \varepsilon \in A\} \leq \sup\{\varphi(\beta) : (\exists\varepsilon)[\varepsilon \in A \;\&\; V(\varepsilon,\alpha,\beta)]\}$$

and the set $\{\beta : (\exists\varepsilon)[\varepsilon \in A \;\&\; V(\varepsilon,\alpha,\beta)]\}$ is a subset of S which lies in $\neg\underset{\sim}{\Gamma}$, so that the standard boundedness argument for Spector pointclasses closed under $\forall^{\mathcal{N}}$ applies (see 4C.11 of DST).

Using this observation, we can prove the lemma by a small modification of the argument given in 11.1 of Kechris [1978], whose notation we will use.

Let $\{A_\eta\}_{\eta<\mu}$, where $\mu < \kappa$, be a partition of $[\kappa]^\kappa$ into μ many pieces. For each $\eta < \mu$ consider the game presented there with A replaced by $\sim A_\eta$. Call it G_η. If for some $\eta < \mu$ I has a winning strategy in G_η, then clearly there is a homogeneous set landing in A_η, so we are done. So assume II has a winning strategy in each G_η, $\eta < \mu$, towards a contradiction. By the Coding Lemma 7D.5 of DST let $\{S_\eta\}_{\eta<\mu}$ be a sequence of sets such that

(a) $S_\eta \neq 0$

(b) $\sigma \in S_\eta \Rightarrow \sigma$ is a winning strategy in G_η,

(c) $\bigcup_{\eta<\mu} S_\eta \in \neg\underset{\sim}{\Gamma}$,

and put

$$G(\xi,\theta) = \sup\{f_{\sigma(\varepsilon)}(\xi) + 1 : \sigma \in \textstyle\bigcup_{\eta<\mu} S_\eta \wedge \varepsilon \in C_{\xi,\theta}\} .$$

By boundedness again $G(\xi,\theta) < \kappa$, and from this it is easy to find a closed unbounded set $D \subseteq \kappa$ such that

$$D^{\underline{\kappa}} = D^{\underline{\ }} \subseteq \sim A_\eta, \qquad \forall\eta < \mu ;$$

thus $D^{\underline{\ }} \subseteq \bigcap_{\eta<\mu} \sim A_\eta = \emptyset$, a contradiction. ⊣

Thus to prove Theorem 1.1 it will be sufficient to show that if Γ is a Spector pointclass closed under Kleene's ${}^3E$, then Γ admits a good coding of the functions on its ordinal κ.

The key to this proof is a strong (uniform) version of the Coding Lemma (I), 7D.5 of DST, which is implicit in the proof of that result, particularly as that was described in the original paper Moschovakis [1970].

Consider then the customary <u>language of analysis</u> (or second-order number theory), where we have variables $n, k, i, \ldots$ over ω and $\alpha, \beta, \gamma, \ldots$ over $\mathcal{N}$, symbols for $0, 1, =, +$ and $\cdot$ on ω and application "$\alpha(t)$" of variables over $\mathcal{N}$ on terms. We obtain an extension $L(\chi)$ of this language by adding new prime formulas of the form

$$\chi(\alpha, \beta) \simeq m \;;$$

in the intended interpretation, χ will denote a partial function from $\mathcal{N} \times \mathcal{N}$ into ω in the obvious way. We will often denote a formula of $L(\chi)$ by a symbol such as

$$\varphi(\chi, x_1, \ldots, x_n) \;,$$

which shows explicitly the occurrence of χ; we will then use (ambiguously) the same symbol $\varphi(\chi, x_1, \ldots, x_n)$ to denote the relation on $\chi, x_1, \ldots, x_n$ defined by the formula.

Put

$\Sigma^1_1(\chi)$ = the smallest collection of formulas of $L(\chi)$ which contains all ordinary Σ^1_1 formulas (with no χ) and the prime formula "$\chi(\alpha, \beta) \simeq m$" and which is closed under the positive operations &, $\vee$, $\exists^\omega$, $\forall^\omega$ and $\exists^{\mathcal{N}}$.

It is important that we do not allow the negative formula $\neg\,(\chi(\alpha, \beta) \simeq m)$ in $\Sigma^1_1(\chi)$.

1.3. <u>Lemma.</u> For each $\Sigma^1_1(\chi)$ formula $\varphi(\chi, x_1, \ldots, x_n)$, there are (ordinary) Σ^1_1 formulas $\rho(x_1, \ldots, x_n)$ and $\sigma(x_1, \ldots, x_n, \alpha, \beta, \gamma)$ such that for all $\chi, x_1, \ldots, x_n$,

$$\varphi(\chi, x_1, \ldots, x_n) \Leftrightarrow \rho(x_1, \ldots, x_n)$$
$$\vee\, (\exists\alpha)(\exists\beta)(\exists\gamma)\{(\forall t)[\chi((\alpha)_t, (\beta)_t) \simeq \gamma(t)] \;\&\; \sigma(x_1, \ldots, x_n, \alpha, \beta, \gamma)\} \;.$$

<u>Proof</u> of this is very easy, as in 7D.7 of DST or Lemma 1 of Moschovakis [1970] - where the argument is not completely correct, as it does not work for the empty partial function χ. ⊣

From this lemma and the known parametrization theorems for Σ^1_1, we can obtain easily a uniform (in χ) parametrization theorem for $\Sigma^1_1(\chi)$.

1.4. Lemma. For each product

$$\mathfrak{X} = X_1 \times \cdots \times X_n$$

of copies of ω and $\mathcal{N}$, there is a fixed $\Sigma^1_1(X)$ formula

$$G^{\mathfrak{X}}(X,\varepsilon,x) \Leftrightarrow G(X,\varepsilon,x) ,$$

where ε varies over $\mathcal{N}$, x varies over $\mathfrak{X}$ and the following are true.

(i) If $\varphi(X,x)$ is any $\Sigma^1_1(X)$ formula with x varying over $\mathfrak{X}$, then for some recursive ε and all X and x,

$$\varphi(X,x) \Leftrightarrow G(X,\varepsilon,x) .$$

(ii) For each pair of spaces $\mathcal{Y}$, $\mathfrak{X}$, there is a recursive function

$$S^{\mathfrak{X},\mathcal{Y}} = S : \mathcal{N} \times \mathcal{Y} \to \mathcal{N} ,$$

so that for all X, x, y (omitting superscripts)

$$G(X,\varepsilon,y,x) \Leftrightarrow G(X,S(\varepsilon,y),x) . \qquad \dashv$$

After these preliminary lemmas, fix a prewellordering $\leq$ on a subset S of $\mathcal{N}$ with rank function

$$\rho : S \twoheadrightarrow \lambda ,$$

and let

$$f : \lambda^n \to \mathrm{Power}(\mathcal{Y})$$

be any function which assigns to n-tuples from λ subsets (possibly empty) of a space $\mathcal{Y}$. A choice set for f is any subset

$$C \subseteq \mathcal{N}^n \times \mathcal{Y} ,$$

such that

(i) $(\alpha_1,\ldots,\alpha_n,y) \in C \Rightarrow \alpha_1,\ldots,\alpha_n \in S$ & $y \in f(\rho(\alpha_1),\ldots,\rho(\alpha_n))$,

(ii) $f(\xi_1,\ldots,\xi_n) \neq \emptyset \Rightarrow$ for each $\alpha_1,\ldots,\alpha_n \in S$ with $\rho(\alpha_1) = \xi_1,\ldots,\rho(\alpha_n) = \xi_n$, there is some y such that $(\alpha_1,\ldots,\alpha_n,y) \in C$.

In effect, C assigns to each $\xi_1,\ldots,\xi_n < \lambda$ such that $f(\xi_1,\ldots,\xi_n) \neq \emptyset$ a non-empty subset of $f(\xi_1,\ldots,\xi_n)$.

If $\alpha \notin S$ put by convention

$$\rho(\alpha) = \infty > \lambda$$

and consider the partial function $\chi = \chi(\leq)$ encoding $\leq$:

$$\chi(\alpha,\beta) \simeq \begin{cases} 1, & \text{if } \alpha \in S \ \& \ \rho(\alpha) \leq \rho(\beta) \\ 0, & \text{if } \beta \in S \ \& \ \rho(\beta) < \rho(\alpha) . \end{cases}$$

For each $\mu < \lambda$ consider also the approximations

$$\chi_\mu(\alpha,\beta) \simeq \begin{cases} 1, & \text{if } \alpha \in S \ \& \ \rho(\alpha) \leq \mu \ \& \ \rho(\alpha) \leq \rho(\beta) \\ 0, & \text{if } \beta \in S \ \& \ \rho(\beta) \leq \mu \ \& \ \rho(\beta) < \rho(\alpha) . \end{cases}$$

Thus $\chi_\mu \subseteq \chi_{\mu'}$ if $\mu \leq \mu' < \lambda$ and

$$\bigcup_{\mu<\lambda} \chi_\mu = \chi .$$

If $f : \lambda^n \to \text{Power}(\mathcal{Y})$ as above and $\mu < \lambda$ we let f_μ be the restriction of f to μ,

$$f_\mu(\xi_1,\ldots,\xi_n) = \begin{cases} f(\xi_1,\ldots,\xi_n) & \text{if } \xi_1,\ldots,\xi_n \leq \mu , \\ \emptyset & \text{otherwise} . \end{cases}$$

1.5. <u>The Uniform Coding Lemma.</u> Assume AD, let $\leq$ be a prewellordering on a subset S of $\mathcal{N}$ with rank function

$$\rho : S \twoheadrightarrow \lambda$$

and associated partial functions χ_μ $(\mu < \lambda)$, let

$$f : \lambda^n \to \text{Power}(\mathcal{Y})$$

be a function and let $G(\chi,\varepsilon,\alpha_1,\ldots,\alpha_n,y)$ be the universal $\Sigma^1_1(\chi)$ formula of Lemma 1.4. There is a fixed $\varepsilon^* \in \mathcal{N}$ such that for all $\mu < \lambda$ the relation

$$C_\mu(\alpha_1,\ldots,\alpha_n,y) \Leftrightarrow G(\chi_\mu,\varepsilon^*,\alpha_1,\ldots,\alpha_n,y)$$

is a choice set for f_μ.

<u>Proof</u> of this is a modification of the proof of 7D.5 of DST.

Take $n = 1$ for convenience and in the notation of 1.5, call ε^* a <u>uniform</u> code of a choice set for f (relative to $\leq$). Assume towards a contradiction that this lemma fails and pick the least λ such that for some $\leq$ and some f as above there is no uniform code ε^* for f. It is easy to check that λ is limit. Fix such a counterexample $\leq$ and f for λ now. Consider then the game where I plays α and II plays β and

II wins iff [α is not a uniform code for any f_ξ, $\xi < \lambda$ (relative to $\leq \restriction \xi = \{(\alpha,\beta) : \alpha \leq \beta \wedge \rho(\beta) \leq \xi\}$)] $\vee$ [α is a uniform code for some f_ξ and there is $\eta > \xi$ such that β is a uniform code for $f_\eta]$.

The proof can be now completed as in 7D.5 of DST. $\dashv$

We can finally produce the desired coding of functions $f : \kappa \to \kappa$ by elements of $\mathcal{N}$.

1.6. Lemma. Assume AD, let Γ be a Spector pointclass closed under ^3E and let κ be the ordinal of Γ. Then Γ admits a good coding of the functions on κ to κ.

Proof. Choose a set $S \subseteq \mathcal{N}$ in Γ and a Γ-norm

$$\varphi : S \twoheadrightarrow \kappa$$

and let $\leq \, = \, \leq_\varphi$ be the associated prewellordering, so that in the notation we have established, for $\mu < \kappa$,

$$\chi_\mu(\alpha,\beta) \simeq \begin{cases} 1, & \text{if } \alpha \in S \ \& \ \varphi(\alpha) \leq \mu \ \& \ \varphi(\alpha) < \varphi(\beta), \\ 0, & \text{if } \beta \in S \ \& \ \varphi(\beta) \leq \mu \ \& \ \varphi(\beta) < \varphi(\alpha). \end{cases}$$

(Again $\varphi(\alpha) = \infty$, if $\alpha \notin S$.) Let $G(\chi,\varepsilon,\alpha,\beta)$ be the universal $\Sigma^1_1(\chi)$ formula of Lemma 1.4, and put for $\xi < \kappa$:

$$(*) \qquad \begin{aligned} f_\varepsilon(\xi)\downarrow \Leftrightarrow {} & \forall\alpha\{[\alpha \in S \ \& \ \varphi(\alpha) = \xi] \Rightarrow \\ & \exists\beta G(\chi_\xi,\varepsilon,\alpha,\beta)\} \ \& \\ & \forall\alpha\forall\alpha'\forall\beta\forall\beta'\{[\alpha,\alpha' \in S \wedge \varphi(\alpha) = \varphi(\alpha') = \xi \\ & \& \ G(\chi_\xi,\varepsilon,\alpha,\beta) \ \& \ G(\chi_\xi,\varepsilon,\alpha',\beta')] \Rightarrow \\ & [\beta \in S \ \& \ \beta' \in S \ \& \ \varphi(\beta) = \varphi(\beta')]\}\ ; \end{aligned}$$

if $f_\varepsilon(\xi)\downarrow$ put

$f_\varepsilon(\xi)$ = the unique ζ such that for some $\alpha,\beta \in S$ with $\varphi(\alpha) = \xi$ $\varphi(\beta) = \zeta$, we have $G(\chi_\xi,\varepsilon,\alpha,\beta)$.

We now verify for this φ and $\{f_\varepsilon : \varepsilon \in \mathcal{N}\}$ the conditions (i)-(iii) in the definition of a good coding.

Condition (iii) is a direct consequence of the Uniform Coding Lemma, applied to

$$f^*(\xi) = \{\beta : \beta \in S \ \& \ f(\xi) = \varphi(\beta)\} .$$

To prove (i) and (ii) notice first that for each fixed $\xi < \kappa$, the relation

$$\chi_\xi(\alpha,\beta) \simeq m$$

is in $\underset{\sim}{\Delta}$, and in fact the following stronger assertion is true: there is a Γ-recursive partial function $F(\alpha,\beta,m,\gamma)$ such that

$$\gamma \in S \Rightarrow F(\alpha,\beta,m,\gamma)\downarrow \ \& \ [F(\alpha,\beta,m,\gamma) \simeq 1 \Leftrightarrow \chi_{\varphi(\gamma)}(\alpha,\beta) \simeq m] ;$$

this follows directly from the fact that φ is a Γ-norm. Now using the hypothesis that Γ is closed under ${}^3E$, it follows immediately that there are relations $P(\varepsilon,\alpha,\beta,\gamma)$ and $\check{P}(\varepsilon,\alpha,\beta,\gamma)$ in Γ and $\check{\Gamma}$ respectively, such that

$$\gamma \in S \Rightarrow [P(\varepsilon,\alpha,\beta,\gamma) \Leftrightarrow \check{P}(\varepsilon,\alpha,\beta,\gamma) \Leftrightarrow G(\chi_{\varphi(\gamma)},\varepsilon,\alpha,\beta)]$$

i.e. the key relation $G(\chi_{\xi},\varepsilon,\alpha,\beta)$ is <u>in</u> $\underset{\sim}{\Delta}$, <u>for each</u> ξ, <u>uniformly in</u> ξ.

We can establish (ii) immediately by setting

$$V(\varepsilon,\alpha,\beta) \Leftrightarrow \check{P}(\varepsilon,\alpha,\beta,\alpha) .$$

To check (i), notice that the relation

$$f_\varepsilon(\xi)\downarrow \wedge f_\varepsilon(\xi) \leq \theta$$

is defined by replacing in (*) above the last clause

$$\beta \in S \ \& \ \beta' \in S \ \& \ \varphi(\beta) = \varphi(\beta')$$

by

$$\beta \in S \ \& \ \beta' \in S \ \& \ \varphi(\beta) = \varphi(\beta') \leq \theta ;$$

for this too we can find some Q, $\check{Q}$ in Γ, $\check{\Gamma}$ respectively such that

$$\gamma \in S \Rightarrow [Q(\beta,\beta',\gamma) \Leftrightarrow \check{Q}(\beta,\beta',\gamma) \Leftrightarrow \beta \in S \ \& \ \beta' \in S \ \& \ \varphi(\beta) = \varphi(\beta') \leq \varphi(\gamma)] .$$

Putting these equivalences together then and using the closure properties of Γ, we find some R, $\check{R}$ in Γ, $\check{\Gamma}$ respectively so that

$$\gamma,\gamma' \in S \Rightarrow [R(\varepsilon,\gamma,\gamma') \Leftrightarrow \check{R}(\varepsilon,\gamma,\gamma') \Leftrightarrow f_\varepsilon(\varphi(\gamma))\downarrow \ \& \ f_\varepsilon(\varphi(\gamma)) \leq \varphi(\gamma')] ,$$

from which (i) follows directly. ⊣

§2. Partition properties imply determinacy. Our main goal now will be to show that partition properties of cardinals imply the dtterminacy of Suslin sets. Combining this and the result of §1 we obtain an elegant set theoretical equivalent of AD within the smallest admissible set above the continuum.

What actually comes up in the proof is a relatively weak consequence of the strong partition property which we establish first.

2.1. Lemma. If $\kappa \to (\kappa)^{\mu}$ for each $\mu < \kappa$, then we can associate with each wellordering W of rank $\leq \kappa$ a countably additive measure μ_W on the set $[\kappa]^W$ of increasing W-term sequences in κ, so that the following coherence property holds: if $W' \subseteq W$ is a subordering of W and $A \subseteq [\kappa]^W$, then

$$\mu_W(A) = 1 \Rightarrow \mu_{W'}\{f \restriction W' : f \in A\} = 1 .$$

Proof. As usual, let $A \not\models^W$ $(A \subseteq \kappa)$ be the set of all increasing maps h from W into A with the property that $h(x) > \sup\{h(y) : y < x\}$ and there is $\{\xi_n^x\}_{x\in W, n\in\omega}$ such that for each $x \in W$, $\xi_0^x < \xi_1^x < \cdots < \xi_n^x < \cdots \to h(x)$. Then the partition property $\kappa \to (\kappa)^{\mu}$, $\forall \mu < \kappa$ easily implies that the following is a (countably additive) measure on $[\kappa]^W$:

$$\mu_W(A) = 1 \text{ iff there is a closed unbounded } C \subseteq \kappa \text{ with } C \not\models^W \subseteq A .$$

The coherence property for these measures follows immediately. ⊣

Recall that a set $A \subseteq \mathfrak{N}$ is called λ-Suslin if there is a tree T on $\omega \times \lambda$ such that

$$A = p[T] = \{\alpha \in \mathfrak{N} : \exists f \in \lambda^{\omega} (\alpha, f) \in [T]\} .$$

2.2. Theorem. Let λ be an ordinal and assume that there is a cardinal $\kappa > \lambda$ with the property that

$$\kappa \to (\kappa)^{\mu}, \qquad \forall \mu < \kappa .$$

Then every λ-Suslin subset of $\mathfrak{N}$ is determined.

Proof. Let $A = p[T]$, where T is a tree on $\omega \times \lambda$. For each $\emptyset \neq \sigma \in \omega^{<\omega} =$ the set of all finite sequences from ω, let $T'(\sigma) = \{u : \text{length}(u) (= m) \leq \text{length}(\sigma) \wedge (\sigma \restriction m, u) \in T\}$. Let $<_{KB}$ be the Kleene-Brouwer ordering on $\lambda^{<\omega}$ and let W_σ be $<_{KB}$ restricted to $T'(\sigma)$, so that W_σ is a wellordering. For convenience we shall identify many times W_σ with its domain $T'(\sigma)$. If

$T(\alpha) = \bigcup_{n\geq 1} T'(\alpha \restriction n)$ is wellfounded, let W_α be again $<_{KB}$ restricted on $T(\alpha)$, so that $W_\alpha = \bigcup_{n\geq 1} W_{\alpha\restriction n}$. Note of course that

$$\emptyset \neq \sigma \subseteq \sigma' \Rightarrow W_\sigma \subseteq W_{\sigma'} .$$

Using Lemma 2.1, let $\{\mu_\sigma : \emptyset \neq \sigma \in \omega^{<\omega}\}$ be an assignment of countably additive measures to the function sets $[\kappa]^{W_\sigma}$ so that the coherence property of the lemma holds and let μ_α be the analogous measure on $[\kappa]^{W_\alpha}$, whenever W_α is a wellordering. We will use these measures to reduce the game (associated with) A to a closed game A^*.

The game A is played as follows:

A	I	$\alpha(0)$		$\alpha(2)$	...
	II		$\alpha(1)$		$\alpha(3)$

$$\text{II wins} \Leftrightarrow \alpha \notin A$$
$$\Leftrightarrow T(\alpha) \text{ is wellfounded} .$$

In the auxiliary game A^*, II makes additional moves as follows;

A^*	I	$\alpha(0)$		$\alpha(2)$	...
	II		$\alpha(1), f_1$		$\alpha(3), f_3$

to win, II must insure that for each n and for each $i \leq n$, f_{2i+1} is an order preserving map from $W_{\alpha\restriction 2i+1}$ into κ, and $f_1 \subseteq f_3 \subseteq \cdots \subseteq f_{2n+1}$.

Clearly this is a closed game for player II, so it is determined. Actually, since the auxiliary moves by player II come from a set which is not necessarily wellorderable, without the axiom of choice one can only assert that either I has a winning strategy or else II has a "multiple-valued" winning strategy or <u>quasistrategy</u>. The easiest way to visualize a winning quasistrategy for player II is as a set Q of sequences $(\alpha(0);\alpha(1),f_1;\ldots;\alpha(2n+1),f_{2n+1})$ closed under subsequences such that the winning conditions for II are satisfied and such that (a) $\forall\alpha(0)\exists\alpha(1),f_1(\alpha(0);\alpha(1),f_1) \in Q)$, and (b) for every $(\alpha(0);\alpha(1),f_1;\ldots;\alpha(2n+1),f_{2n+1}) \in Q$ and every $\alpha(2n+2)$ there is some $\alpha(2n+3)$, f_{2n+3} such that $(\alpha(0);\alpha(1),f_1;\ldots;\alpha(2n+1),f_{2n+1};\alpha(2n+2);\alpha(2n+3),f_{2n+3}) \in Q$. Since however I plays only natural numbers an easy application of DC shows that if II has a winning quasistrategy he actually has a winning strategy.

If now indeed II has a winning strategy in the auxiliary game A^*, he clearly has a winning strategy in the original game. So assume I has a winning strategy τ^* in the auxiliary game. We will define a strategy τ for I in A by using the measures μ_σ to "integrate out" II's auxiliary moves in the usual way.

For any

$$\sigma = (\alpha(0),\alpha(1),\ldots,\alpha(2n + 1))$$

and any order preserving

$$f : T'(\sigma) \to \kappa ,$$

let

$$f_1, f_3, \ldots, f_{2n+1}$$

be the auxiliary moves induced by f, i.e.

$$f_{2i+1} = f \restriction T'(\alpha(0),\ldots,\alpha(2i + 1)) \qquad (i \leq n) .$$

At position σ of the game A then, have I play by

$$\tau(\sigma) = m \Leftrightarrow \text{for } \mu_\sigma\text{-almost all } f,$$
$$\tau^*(\alpha(0),\alpha(1),f_1,\ldots,\alpha(2n + 1),f_{2n+1}) = m .$$

Assume towards a contradiction that I follows τ but loses, in a run of A which produces the play

$$\alpha = \alpha(0),\alpha(1),\alpha(2),\ldots ,$$

so that $T(\alpha)$ is wellfounded, i.e. W_α is a wellordering. By the construction, for each n we have a set

$$B_n \subseteq [\kappa]^{W_\alpha \restriction (2n+1)}$$

of measure 1 in the canonical measure on this space, so that I's play by τ in A together with any member of B_n is a partial play by I in A^* which follows τ^*. Now by the coherence property of the measures, for each n, the set

$$\{f \in [\kappa]^{W_\alpha} : f \restriction T'(\alpha(0),\ldots,\alpha(2n + 1)) \in B_n\}$$

has measure 1, so the intersection of all these sets has measure 1 and is not empty; if f is in this set, then

$$\alpha(0),\alpha(1),f \restriction T'(\alpha(0),\alpha(1)),\ldots,\alpha(2n + 1),f \restriction T'(\alpha(0),\ldots,\alpha(2n + 1)) \ldots$$

is a play in A^* where I plays by τ^* and loses, contrary to our assumptions. ⊣

Let

$$\mathfrak{R}^+ = \text{smallest admissible set containing } \mathcal{R} .$$

By combining Theorems 1.1 and 2.2 we have now

2.3. Theorem. The following are equivalent
(i) $\mathfrak{R}^+ \models AD$
(ii) $\mathfrak{R}^+ \models \forall\kappa$ (power(κ) exists) & $\forall\lambda\exists\kappa > \lambda$($\kappa$ has the strong partition property).

Proof. Assume $\mathfrak{R}^+ \models AD$. That $\mathfrak{R}^+ \models \forall\kappa$ (power(κ) exists), follows immediately from the Coding Lemma 7D.5 of DST. That $\mathfrak{R}^+ \models \forall\lambda\exists\kappa > \lambda$ (κ has the strong partition property), follows from 1.1.

Conversely assume (ii). By the proof of 2.2 we have that

$$(*) \qquad \mathfrak{R}^+ \models \forall A \subseteq \mathcal{R}[A \text{ is } \lambda\text{-Suslin for some } \lambda \Rightarrow A \text{ is determined}]$$

(To adapt the proof of 2.2 in this context one needs to recall that for any admissible set $\mathfrak{a}$ and any open game on a set $x \in \mathfrak{a}$, if the player that tries to win the open side has a winning quasistrategy, he has one which is in $\mathfrak{a}$.) Recall now (DST, 7C and Moschovakis [1974], Ch. 9) that the pointsets $A \subseteq \mathcal{R}$ in $\mathfrak{R}^+$ are exactly the hyperprojective ones. So it is enough to show that every hyperprojective pointset A carries a hyperprojective scale. As a warmup let us see how to do this for the projective sets.

Since every $\underset{\sim}{\Sigma}^1_2$ set carries a $\underset{\sim}{\Sigma}^1_2$-scale we immediately have that it is determined by (*). So we have Determinacy ($\underset{\sim}{\Sigma}^1_2$), and therefore by the Second Periodicity Theorem (DST 6C.3 and 6C.1) every $\underset{\sim}{\Sigma}^1_4$ set carries a $\underset{\sim}{\Sigma}^1_4$-scale. Again by (*) we have Determinacy ($\underset{\sim}{\Sigma}^1_4$) thus by another use of the Second Periodicity Theorem, every $\underset{\sim}{\Sigma}^1_6$ set carries a $\underset{\sim}{\Sigma}^1_6$-scale etc.

We will prove the general result about hyperprojective sets now, by defining an appropriate hierarchy on these sets and extending the above argument through the transfinite.

In the rest of this proof we will follow the terminology and notation of 7C of DST. Recall that $\underset{\sim}{\text{IND}}$ is the pointclass of all inductive pointsets. For each ordinal ξ and each positive analytical operator $\varphi(x,A)$, let φ^ξ be its ξth iterate, i.e.

$$\varphi^\xi(x) \Leftrightarrow \varphi(x, \varphi^{<\xi}) ,$$

where $\varphi^{<\xi}(x) \Leftrightarrow \exists\eta < \xi\varphi^\eta(x)$. For each limit ordinal $\lambda < \kappa^{\mathfrak{R}}$ let

$$\underset{\sim}{\text{IND}}_\lambda = \{A : A \text{ is the continuous preimage of } \varphi^{\lhd\lambda}, \text{ for some positive analytical operator } \varphi\} .$$

We summarize some basic structural properties of $\underset{\sim}{\mathrm{IND}}_\lambda$.

Lemma. For each limit $\lambda < \kappa^{\mathbb{R}}$ the pointclass $\underset{\sim}{\mathrm{IND}}_\lambda$ is closed under continuous substitutions, &, $\vee$, $\cup^{\omega}$, $\exists^{\mathcal{N}}$, has the prewellordering property and is $\mathcal{N}$-parametrized. In fact there is a single "universal" positive analytical operator $\varphi_0(\varepsilon,\alpha,A)$, A varying over subsets of $\mathcal{N} \times \mathcal{N}$ such that for each limit $\lambda < \kappa^{\mathbb{R}}$, $\varphi_0^{<\lambda}$ is universal for the $\underset{\sim}{\mathrm{IND}}_\lambda$ subsets of $\mathcal{N}$, and moreover for each analytical $\varphi(\alpha,B)$ (with B varying over subsets of $\mathcal{N}$) there is a recursive ε such that for all limit $\lambda < \kappa^{\mathbb{R}}$, $\varphi^{<\lambda}(\alpha) \Leftrightarrow \varphi_0^{<\lambda}(\varepsilon,\alpha)$.

For a proof see for example Kechris [A].

We define also the pointclass $\mathrm{IND}_{\lambda+2n}$ for each $n \geq 0$ by the induction

$$\underset{\sim}{\mathrm{IND}}_{\lambda+2n+2} = \exists^{\mathcal{N}}\forall^{\mathcal{N}}\underset{\sim}{\mathrm{IND}}_{\lambda+2n} \,.$$

Again these are closed under &, $\vee$, $\cup^{\omega}$, $\cap^{\omega}$, $\exists^{\mathcal{N}}$ and are $\mathcal{N}$-parametrized. Let us in fact fix canonical universal sets $U_{\lambda+2n}$ for each $\underset{\sim}{\mathrm{IND}}_{\lambda+2n}$ as follows:

$$U_\lambda = \varphi_0^{<\lambda}$$

$$U_{\lambda+2n+2} = \{(\varepsilon,\alpha) : \exists\beta\,\forall\gamma U_{\lambda+2n}(\varepsilon,\langle\alpha,\beta,\gamma\rangle)\} \,.$$

These are universal sets in $\underset{\sim}{\mathrm{IND}}_{\lambda+2n}$ for the $\mathrm{IND}_{\lambda+2n}$ subsets of $\mathcal{N}$, but as in DST 3H.1 one can easily build upon them a good universal system $\{U^{\mathcal{X}}_{\lambda+2n}\}$ where $U^{\mathcal{X}}_{\lambda+2n}$ is universal for the subsets of a product space $\mathcal{X}$ in $\underset{\sim}{\mathrm{IND}}_{\lambda+2n}$. If $A \subseteq \mathcal{X}$ and $x \in A \Leftrightarrow (\varepsilon,x) \in U^{\mathcal{X}}_{\lambda+2n}$, we call ε an $\underset{\sim}{\mathrm{IND}}_{\lambda+2n}$-code of A. One of the basic properties of a good universal system is that the closure properties of $\underset{\sim}{\mathrm{IND}}_{\lambda+2n}$ are uniform in the $\underset{\sim}{\mathrm{IND}}_{\lambda+2n}$-codes; see DST 3H.2.

In order to show that every hyperprojective set carries a hyperprojective scale, we will prove that if $\varphi(\alpha,A)$ is a positive Σ^1_2 operator, then for each $\lambda < \kappa^{\mathbb{R}}$ limit, $\varphi^{<\lambda}$ carries an $\underset{\sim}{\mathrm{IND}}_\lambda$-scale and $\varphi^{\lambda+n}$ carries a $\underset{\sim}{\mathrm{IND}}_{\lambda+2n+2}$-scale for each $n \geq 0$. Since every hyperprojective pointset is the continuous preimage of φ^{ξ} for some such φ and some $\xi < \kappa^{\mathbb{R}}$, this will complete our proof.

The proof will be by effective transfinite induction, on a suitable coding system for ordinals $< \kappa^{\mathbb{R}}$, using the Recursion Theorem.

Let first $\psi(w,A)$, $w \in \mathcal{N}$, $A \subseteq \mathcal{N}$, be positive analytical such that if

$$|w| = \text{least } \xi \text{ such that } \psi^{\xi}(w) \,,$$

and $W = \psi^{\infty}$, then

$$|\;| : W \twoheadrightarrow \kappa^{\mathbb{R}}$$

and there are recursive pointsets $\mathrm{Lim} \subseteq \mathcal{N}$, $\mathrm{Succ} \subseteq \mathcal{N}$ and a total recursive function pd such that

$$w \in W \Rightarrow [\mathrm{Lim}(w) \Leftrightarrow |w| \text{ is limit}] \ \& \ [\mathrm{Succ}(w) \Leftrightarrow |w| \text{ is successor}] \ \& \ [\mathrm{Succ}(w) \Rightarrow |\mathrm{pd}(w)| = |w| - 1] .$$

The following lemma will complete now the proof

Lemma. Given $\varphi(\alpha,A)$ a Σ^1_2 positive operator, there are scales $\{\sigma_i^{\triangleleft\lambda}\}$, $\{\sigma_i^{\lambda+n}\}$ on $\varphi^{\triangleleft\lambda}$, $\varphi^{\lambda+n}$ respectively, for each limit $\lambda < \kappa^{\mathbb{R}}$, and partial recursive functions f_0, g_0, f_1, g_1 such that, letting for each scale $\{\sigma_i\}$ on a pointset P

$$S(i,x,y) \Leftrightarrow x \leq^*_{\sigma_i} y ,$$
$$T(i,x,y) \Leftrightarrow x <^*_{\sigma_i} y ,$$

be its associated relations, we have

$w \in W \Rightarrow$ [if $|w| = \lambda + n$, λ limit, $n \geq 0$, then for $n = 0$ we have that $f_0(w)$, $g_0(w)$ are $\underset{\sim}{\mathrm{IND}}_\lambda$-codes of the relations associated with $\{\sigma_i^{\triangleleft\lambda}\}$, while for all n, $f_1(w)$, $g_1(w)$ are $\underset{\sim}{\mathrm{IND}}_{\lambda+2n+2}$-codes of the relations associated with $\{\sigma_i^{\lambda+n}\}$.

Proof. The scales are defined as in Moschovakis [1978] inductively:

$$\sigma_0^{\triangleleft\lambda}(\alpha) = |\alpha|_\varphi = \text{least } \xi \text{ such that } \alpha \in \varphi^\xi ,$$
$$\sigma_{i+1}^{\triangleleft\lambda}(\alpha) = \langle |\alpha|_\varphi, \sigma_i^{|\alpha|_\varphi}(\alpha)\rangle ;$$

while $\{\sigma_i^{\lambda+n}\}$ is defined by the Second Periodicity Theorem inductively on n starting from $\{\sigma_i^{\triangleleft\lambda}\}$. For this to be a legitimate scale it is sufficient to know that $\underset{\sim}{\mathrm{IND}}_\lambda$ has the scale property by an argument similar to that given for the projective sets in the early stages of the present proof. In any case, however, the associated relations of $\{\sigma_i^{\triangleleft\lambda}\}$, $\{\sigma_i^{\lambda+n}\}$ are defined independently of this, so if we can prove the second assertion of the lemma about these associated relations then by induction on λ we have immediately that $\underset{\sim}{\mathrm{IND}}_\lambda$ has the scale property (recall that every $A \in \underset{\sim}{\mathrm{IND}}_\lambda$ is the continuous preimage of $\varphi^{\triangleleft\lambda}$, where φ can be taken to be Σ^1_2 positive analytical) and thus the proof is complete.

In order to construct f_0, f_1, g_0, g_1 we shall use effective transfinite induction. It is rather routine to define $f_1(w)$, $g_1(w)$ once all $f_0(v)$, $g_0(v)$, $f_1(v)$, $g_1(v)$ are known for $|v| < |w|$, thus we can concentrate on $f_0(w)$, $g_0(w)$. So assume $|w| = \lambda$ and that all $f_0(v)$, $g_0(v)$, $f_1(v)$, $g_1(v)$ are known for

$|v| < \lambda$. We write the first relation associated with $\{\sigma_i^{<\lambda}\}$ (the calculation for the other one is similar):

$$\begin{aligned} S(i,\alpha,\beta) \Leftrightarrow {} & [i = 0 \ \& \ |\alpha|_\varphi \leq |\beta|_\varphi < \lambda] \vee \\ & \{i > 0 \ \& \ [(|\alpha|_\varphi < |\beta|_\varphi < \lambda) \vee \\ & [|\alpha|_\varphi = |\beta|_\varphi < \lambda \ \& \\ & \sigma_i^{|\alpha|_\varphi}(\alpha) \leq \sigma_i^{|\alpha|_\varphi}(\beta)]\} . \end{aligned}$$

By the Stage Comparison Theorem (see Moschovakis [1974]) there are positive analytical $\varphi_1, \varphi_2, \varphi_3, \varphi_4$ such that

$$\begin{aligned} |\alpha|_\varphi \leq |\beta|_\varphi < \lambda & \Leftrightarrow \varphi_1^{<\lambda}(\alpha,\beta) \ \& \ \varphi^{<\lambda}(\beta) \\ |\alpha|_\varphi < |\beta|_\varphi < \lambda & \Leftrightarrow \varphi_2^{<\lambda}(\alpha,\beta) \ \& \ \varphi^{<\lambda}(\beta) \\ |\alpha|_\varphi = |\beta|_\varphi = |v|^{<\lambda} & \Leftrightarrow \varphi_3^{<\lambda}(\alpha,v) \ \& \ \varphi_3^{<\lambda}(\beta,v) \ \& \\ & \quad \varphi_4^{<\lambda}(v,\alpha) \ \& \ \varphi_4^{<\lambda}(v,\beta) \ \& \ \psi^{<\lambda}(v) . \end{aligned}$$

Thus

$$\begin{aligned} S(i,\alpha,\beta) \Leftrightarrow {} & [i = 0 \ \& \ \varphi_1^{<\lambda}(\alpha,\beta) \ \& \ \varphi^{<\lambda}(\beta)] \vee \\ & \{i > 0 \ \& \ [\varphi_2^{<\lambda}(\alpha,\beta) \ \& \ \varphi^{<\lambda}(\beta)] \vee \\ & (\exists v)[\varphi_3^{<\lambda}(\alpha,v) \ \& \ \varphi_3^{<\lambda}(\beta,v) \ \& \ \varphi_4^{<\lambda}(v,\alpha) \\ & \varphi_4^{<\lambda}(v,\beta) \ \& \ \psi^{<\lambda}(v) \ \& \ \sigma_i^{|v|}(\alpha) \leq \sigma_i^{|v|}(\beta)]\} . \end{aligned}$$

By induction hypothesis if $|v| < \lambda$, say $|v| = \lambda' + n'$, we have that $f_1(v)$ is a $\underset{\sim}{\mathrm{IND}}_{\lambda'+2n'+2}$ code of the first relation associated with $\{\sigma_i^{\lambda'+n'}\} = \{\sigma_i^{|v|}\}$ It is therefore enough to establish the following fact:

There is a positive analytical $\theta(\varepsilon,\alpha,w,A)$, $A \subseteq \mathfrak{n}^3$, and ε_0 recursive, such that for each ξ,

$$|w| \leq \xi \ \& \ (\varepsilon,\alpha) \in U_{|w|'} \Leftrightarrow (\varepsilon_0,\varepsilon,\alpha,w) \in \theta^\xi$$

where $|w|' = \lambda + 2n + 2$, if $|w| = \lambda + n$.

By the Simultaneous Induction Lemma (Moschovakis [1974] 1C.1 or DST 7C.11) it is actually enough to construct a system $\theta_1, \theta_2, \theta_3$, such that

$$\theta_3^\xi(\varepsilon,\alpha,w) \Leftrightarrow |w| \leq \xi \ \& \ (\varepsilon,\alpha) \in U_{|w'|} .$$

Let first, by the Stage Comparison Theorem, $X(\varepsilon,\alpha,w,A)$ be a positive analytical operator such that letting

$$|\varepsilon,\alpha|_{\varphi_0} = \text{least } \xi \text{ such that } (\varepsilon,\alpha) \in \varphi_0^{\xi} ,$$

we have

$$|\varepsilon,\alpha|_{\varphi_0} \leq \xi \ \& \ (w \notin W \vee |\varepsilon,\alpha|_{\varphi_0} < |w|) \Leftrightarrow \chi^{\xi}(\varepsilon,\alpha,w) .$$

The system is now as follows:

$$\theta_1^{\xi}(w) \Leftrightarrow \psi(w,\theta_1^{<\xi})$$

$$\theta_2^{\xi}(\varepsilon,\alpha,w) \Leftrightarrow \chi(\varepsilon,\alpha,w,\theta_2^{<\xi})$$

$$\theta_3^{\xi}(\varepsilon,\alpha,w) \Leftrightarrow \psi(w,\theta_1^{<\xi}) \ \& \ \{[\mathrm{Lim}(w) \ \& \ \exists\beta\forall\gamma\theta_2^{<\xi}(\varepsilon,\langle\alpha,\beta,\gamma\rangle,w)] \vee$$
$$[\mathrm{Succ}(w) \ \& \ \exists\beta\forall\gamma\theta_3^{<\xi}(\varepsilon,\langle\alpha,\beta,\gamma\rangle,\mathrm{pd}(w))]\} . \qquad \dashv$$

In view of the preceding result we are led naturally to pose the following

Open problem. Is it true that in $L[\mathbb{R}]$, AD is equivalent to the existence of arbitrarily large below Θ cardinals with the strong partition property?

And we conclude this section by proving an extension of 2.2 for games on ordinals. If $A \subseteq \lambda^{\omega}$ is a pointset on λ, we say that A is ν-Suslin if there is a tree T on $\lambda \times \nu$ such that

$$f \in A \Leftrightarrow \exists g \in \nu^{\omega}(f,g) \in [T].$$

We have now

2.4. Theorem. Let λ, ν be ordinals and assume that there is a cardinal $\kappa > \lambda, \nu$ such that

$$\kappa \to (\kappa)^{\mu}_{\xi}, \qquad \forall\mu,\xi < \kappa .$$

If $A \subseteq \lambda^{\omega}$ is such that both A and $\sim A$ are ν-Suslin, then A is determined.

Proof. We try to imitate the argument in the proof of 2.2. Consider the game $G(A)$ associated with A, and, picking a tree T with $p[T] = A$, consider the auxiliary game $G^*(A,T)$ defined there. If I has a winning strategy in $G^*(A,T)$ we are done. So assume II has a winning quasistrategy in $G^*(A,T)$. This we cannot necessarily convert into a winning strategy however, since λ will be in general uncountable and we do not have AC_{λ}. So we cannot immediately conclude that I will have a winning strategy for $G(A)$.

Here we have to use the fact that $\sim A$ is also ν-Suslin. We fix a tree S so that $p[S] = \sim A$ and we consider an auxiliary game $G'(A,S)$, which is as before except that it is now I's responsibility to make the extra moves.

If II has a winning strategy in $G'(A,S)$ we easily again conclude that II has a winning strategy in $G(A)$. Else I has a winning quasistrategy in $G'(A,S)$. Using DC now it is easy to see that there are runs in $G^*(A,T)$ and $G'(A,S)$ with the same real part $\alpha(0),\alpha(1),\ldots$ in which II (resp. I) has followed his winning quasistrategy in $G^*(A,T)$ (resp. $G'(A,S)$). This is clearly a contradiction. ⊣

From this result and 1.1 we immediately obtain the following, where we call $A \subseteq \lambda^\omega$ Suslin if it is μ-Suslin for some μ.

2.5. Theorem. Assume AD. Let $\lambda < \Theta$ and let $A \subseteq \lambda^\omega$ be such that both A and $\sim A$ are Suslin. Then A is determined.

This strengthens the last conclusion of Theorem 2.2 of Moschovakis [A], by removing the restriction that $\lambda \leq \kappa^{\mathbb{R}}$. It also provides an alternative proof of that result, which however does not give the key definability estimates of the original argument that are needed in the rest of that paper. (Note that a set $A \subseteq \lambda^\omega$ is Suslin iff it admits uniform semiscales in the terminology of Moschovakis [A].)

§3. Mahlo cardinals from determinacy. In this section we prove a strengthening of Theorem 1.1 for Spector pointclasses Γ which are additionally closed under both $\exists^{\mathbb{R}}$ and $\forall^{\mathbb{R}}$. The result is as follows

3.1. Theorem. Assume AD, let Γ be a Spector pointclass closed under both $\exists^{\mathbb{R}}$ and $\forall^{\mathbb{R}}$ and let $\kappa = o(\underset{\sim}{\Delta})$ be the ordinal associated with Γ. Then κ has the strong partition property, and moveover $\{\lambda : \lambda < \kappa \ \&\ \lambda$ has the strong partition property$\}$ is stationary in κ. In particular, κ is Mahlo.

Among other things, this implies that there are arbitrarily large cardinals κ with the above properties below Θ. This is because for each pointset A the pointclass

$$\mathrm{IND}(A) = \text{all pointsets which are inductive in } A\ ,$$

is a Spector pointclass containing both A and its complement and is closed under $\exists^{\mathbb{R}}$ and $\forall^{\mathbb{R}}$.

Also it follows that $\kappa^{\mathbb{R}}$ is Mahlo and in fact $\{\lambda : \lambda < \kappa^{\mathbb{R}} \ \&\ \lambda$ has the strong partition property$\}$ is stationary in $\kappa^{\mathbb{R}}$ (since $\kappa^{\mathbb{R}} = o(\underset{\sim}{\Delta})$, where $\Gamma = \mathrm{IND} = \mathrm{IND}(\emptyset)$). Note here that by Steel [A], $\kappa({}^3E)$ is not Mahlo. In fact Moschovakis has conjectured that $\kappa({}^3E)$ is the first (weakly) inaccessible cardinal (granting AD of course). It can be seen that $\kappa^{\mathbb{R}}$ is not the first Mahlo cardinal. It is conjectured that the first Mahlo cardinal is the ordinal of ${}^4S$ (= the type 4 superjump), granting AD again (see Harrington [1973] for results about the superjump).

We give now the proof of 3.1.

Proof. It is enough by 1.1 to show that

$$\{o(\underset{\sim}{\Delta}^*) : \Gamma^* \text{ is a Spector pointclass closed under } {}^3E \text{ and contained in } \underset{\sim}{\Delta}\}$$

is stationary in $\kappa = o(\underset{\sim}{\Delta})$.

Let $f : \kappa \to \kappa$. We shall find Γ^* as above, such that $o(\underset{\sim}{\Delta}^*)$ is closed under f i.e.

$$\xi < o(\underset{\sim}{\Delta}^*) \Rightarrow f(\xi) < o(\underset{\sim}{\Delta}^*) .$$

Fix $S \subseteq \mathcal{N}$ and a Γ-norm

$$\varphi : S \twoheadrightarrow \kappa$$

Then in the notation of the proof of 1.6 find ε such that $f = f_\varepsilon$. From the definition of f_ε it is obvious that there are two relations $R, \check{R}$ in $\underset{\sim}{\Gamma}, \check{\underset{\sim}{\Gamma}}$ respectively such that

$$(*) \qquad \alpha \in S \Rightarrow [R(\alpha,\beta) \Leftrightarrow \check{R}(\alpha,\beta) \Leftrightarrow \beta \in S \;\&\; f(\varphi(\alpha)) = \varphi(\beta)]$$

To simplify the notation assume that actually $R, \check{R}$ are in $\Gamma, \check{\Gamma}$. (Otherwise replace everywhere below Γ by $\Gamma(\alpha_0)$ for some appropriate parameter α_0.)

A type 3 object is a function

$${}^3F : \omega^{\mathcal{Z}} \times \mathcal{Y} \to \omega ,$$

where $\mathcal{Z}, \mathcal{Y}$ are product spaces. For example Kleene's ${}^3E$ is the type 3 object

$${}^3E : \omega^{\mathcal{N}} \to \omega$$

given by

$${}^3E(h) = \begin{cases} 0, & \text{if } \exists\alpha[h(\alpha) = 0] \\ 1, & \text{if } \forall\alpha[h(\alpha) \neq 0] . \end{cases}$$

We say that a Spector pointclass Γ is closed under ${}^3F$ if for each $h : \mathcal{X} \times \mathcal{Z} \to \omega$, a Γ-recursive partial function, the relation

$$P(i,x,y) \Leftrightarrow \forall z[h(x,z)\downarrow] \;\&\; {}^3F(\lambda z h(x,z),y) = i$$

is in Γ. To each such ${}^3F$ we associate the envelope

$$\mathrm{Env}({}^3E, {}^3F) = \text{all pointsets which are Kleene-semirecursive in } {}^3E, {}^3F .$$

This is a Spector pointclass closed under ${}^3E$, ${}^3F$ and in fact by Moschovakis [1974] it is the smallest one with these properties, thus if Γ is a Spector pointclass closed under ${}^3E$, ${}^3F$ then $\mathrm{Env}({}^3E,{}^3F) \subseteq \Gamma$. Moreover by Moschovakis [1967] if $A \in \mathrm{Env}({}^3E,{}^3F)$, there is $B \in \mathrm{Env}({}^3E,{}^3F)$ with

$$x \notin A \Leftrightarrow \exists\alpha(x,\alpha) \in B .$$

This immediately implies that if Γ is a Spector pointclass closed under both $\exists^{\mathcal{N}}$, $\forall^{\mathcal{N}}$ and if ${}^3F$ is a type 3 object, so that Γ is closed under ${}^3F$, then

$$\mathrm{Env}({}^3E,{}^3F) \subseteq \Delta .$$

This is the key fact that we will need below.

Consider now the following type 3 object ${}^3F$ associated with $f : \kappa \to \kappa$:

$${}^3F : \omega^{\mathcal{N}\times\mathcal{N}} \times \mathcal{N}^2 \to \omega$$

and for each $h \in \omega^{\mathcal{N}\times\mathcal{N}}$, $\alpha \in \mathcal{N}$

$${}^3F(h,\alpha,\beta) = \begin{cases} 0, & \text{if } h \text{ is the characteristic function of a prewellordering} \\ & \leq \text{ on } \mathcal{N} \text{ such that } |\leq| < \kappa,\ \alpha,\beta \in S\ \& \\ & \varphi(\alpha) < \varphi(\beta) < f(|\leq|); \\ 1, & \text{otherwise} . \end{cases}$$

<u>Lemma.</u> Γ is closed under ${}^3F$.

Granting the lemma consider

$$\Gamma^* = \mathrm{Env}({}^3E,{}^3F) .$$

By our preceding remarks, it is enough to show that $\kappa^* = o(\underset{\sim}{\Delta}^*)$ is closed under f. For that let $S^* \in \Gamma^*$ and $\varphi^* : S^* \twoheadrightarrow \kappa^*$ be a Γ^*-norm. If $\xi < \kappa^*$ let $h : \mathcal{N} \times \mathcal{N} \to \omega$ be the characteristic function of $\leq\, = \{(\alpha,\beta) : \varphi^*(\alpha) \leq \varphi^*(\beta) < \xi\}$. Clearly h is in $\underset{\sim}{\Delta}^*$. Since Γ^* is closed under ${}^3F$ this implies that

$$\prec\; = \{(\alpha,\beta) : {}^3F(h,\alpha,\beta) = 0\}$$

is also in $\underset{\sim}{\Delta}^*$. But

$$\prec\; = \{(\alpha,\beta) : \varphi(\alpha) < \varphi(\beta) < f(\xi)\} .$$

Thus $f(\xi)$ is the length of a $\underset{\sim}{\Delta}^*$ prewellordering and so $f(\xi) < \kappa^* = o(\underset{\sim}{\Delta}^*)$, which is what we wanted to prove.

To verify the lemma note that if $h : \mathcal{X} \times \mathcal{N} \times \mathcal{N} \to \omega$ has graph in Γ and for some $x \in \mathcal{X}$, $\forall z[h(x,z)\downarrow]$ then one can check in a $\Delta(x)$ way, uniformly in x, whether or not $\lambda z h(x,z)$ is a characteristic function of a prewellordering of $\mathcal{N}$. If this is the case and we denote by $\leq_x = \leq$ this prewellordering, then by 4C.14 of DST we can find, effectively in x, a $\gamma_x = \gamma \in S$ with $\varphi(\gamma) > |\leq|$, and thus using the Coding Lemma 7D.5 of DST we can see that $\{\delta : \varphi(\delta) = |\leq|\}$ is in $\Delta(x)$, uniformly in x again. From this and (*) it is immediate that we can check in a $\Delta(x)$ way, uniformly in x, whether $\varphi(\alpha) < \varphi(\beta) < f(|\leq|)$ is true or not, thus completing the proof that Γ is closed under ${}^3F$. ⊣

§4. Nonsingular measures from AD. It has been known for quite some time that AD implies the existence of many measurable cardinals below Θ (see for example Kleinberg [1977] or Kechris [1978]). All the normal measures on cardinals κ that were produced however, were of the "singular" type, i.e. they concentrated on the ordinals of cofinality λ, for some regular $\lambda < \kappa$. It has thus been open whether one could obtain from AD measurable cardinals below Θ that carry measures which concentrate on regular cardinals. Our result below provides (when combined with 3.1) many such examples below Θ.

4.1. Theorem. Assume that κ is a Mahlo cardinal satisfying the strong partition property. Then there exists a nontrivial κ-additive normal measure on κ, giving the regular cardinals less than κ measure 1.

Proof. Let Q be any stationary set of regular cardinals less than κ. We will construct a normal measure giving Q measure 1.

First let us define $\hat{Q}$ to be the set of those ξ in Q such that some closed unbounded subset of ξ is disjoint from Q. Note that $\hat{Q}$ consists of the difference between Q and the result of applying the Mahlo operation to Q.

We can now define a function $\mu_Q : 2^\kappa \to 2$ by

$$\mu_Q(X) = 1 \text{ iff for some closed unbounded subset } C \text{ of } \kappa,\ X \supseteq C \cap \hat{Q}.$$

Lemma 1. μ_Q is a nontrivial κ-additive measure on κ.

Proof of Lemma 1. Let $\mathcal{U}_Q = \{A \subseteq \kappa | \mu_Q(A) = 1\}$. We will show that $\mathcal{U}_Q$ is a nonprincipal κ-additive ultrafilter on κ.

It is first important to note that $\hat{Q}$ is stationary. For given any closed unbounded set C, the least limit point of C in Q is a member of $\hat{Q} \cap C$.

Since, now, $\hat{Q}$ is stationary, $\mathcal{U}_Q$ must be nonprincipal. Also, clearly, $\mathcal{U}_Q$ is a filter.

Suppose $X \subseteq \kappa$ is given. We must show that $X \in \mathcal{U}_Q$ or $\sim X \in \mathcal{U}_Q$, and so let us define a partition $F : [\kappa]^\kappa \to 2$

$F(Y) = 1$ iff the least limit point of Y which is a member of Q is also a member of X.

(Identify here $[\kappa]^\kappa$ with the set of subsets of κ of cardinality κ.) Let D be a set of cardinality κ homogeneous for F and let us suppose that $F''[D]^\kappa = \{0\}$.

Claim. $X \supseteq D_{\text{l.p.}} \cap \hat{Q}$, where $D_{\text{l.p.}}$ denotes the set of limit points of D.

(Proof of Claim. Suppose ξ is a limit point of D in $\hat{Q}$. Let C_ξ be a closed unbounded subset of ξ disjoint from Q. By an interlocking argument we can thin out $D \cap \xi$ and C_ξ simultaneously to $(D \cap \xi)'$ and C'_ξ respectively such that $(D \cap \xi)'$ and C'_ξ have the same limit points and such that each is unbounded in ξ. Since $C_\xi \cap Q = \emptyset$, there is no limit point of $(D \cap \xi)'$ less than ξ in Q, and so ξ is the least limit point of $(D \cap \xi)'$ in Q. Since $F((D \cap \xi)' \cup (D - \xi)) = 0$, $\xi \in X$.)

By a similar claim, if we had that $F''[D]^\kappa = \{1\}$, then we would have that $\sim X \supseteq D_{\text{l.p.}} \cap \hat{Q}$. Thus either X or $\sim X$ is in $\mathcal{U}_Q$, and so $\mathcal{U}_Q$ is an ultrafilter.

Suppose now that $X_\xi \in \mathcal{U}_Q$ for each $\xi < \theta < \kappa$. We must show that $\bigcap_{\xi<\theta} X_\xi \in \mathcal{U}_Q$. Let us define a partition $G : [\kappa]^\kappa \to \theta + 1$ as follows: given $Y \in [\kappa]^\kappa$, let η be the least limit point of Y which is a member of Q. Then if $\eta \in \bigcap_{\xi<\theta} X_\xi$ we define $G(Y)$ to be θ. Otherwise, $G(Y)$ is the least $\xi < \theta$ such that $\eta \notin X_\xi$.

Let, now, E be a set of cardinality κ homogeneous for G. Then if $G''[E]^\kappa = \{\theta\}$, an argument similar to one used above would show that $\bigcap_{\xi<\theta} X_\xi \supseteq E_{\text{l.p.}} \cap \hat{Q}$, and hence that $\bigcap_{\xi<\theta} X_\xi \in \mathcal{U}_Q$. Otherwise, our argument above would show that $\sim X_\xi \supseteq E_{\text{l.p.}} \cap \hat{Q}$ for some $\xi < \theta$. Since for some closed unbounded C, $X_\xi \supseteq C \cap \hat{Q}$, we would have $E_{\text{l.p.}} \cap C \cap \hat{Q} \subseteq X_\xi \cap \sim X_\xi = \emptyset$. Since $E_{\text{l.p.}} \cap C \cap \hat{Q}$ is stationary, this is impossible. Thus $\bigcap_{\xi<\theta} X_\xi \in \mathcal{U}_Q$. $\dashv$

Lemma 2. μ_Q is normal.

Proof of Lemma 2. Suppose $f : \kappa \to \kappa$ and $\mu_Q(\{\xi | f(\xi) < \xi\}) = 1$. Let C be a closed unbounded set such that $\{\xi | f(\xi) < \xi\} \supseteq C \cap \hat{Q}$, and let us define a partition $F : [C \cap \hat{Q}]^\kappa \to 2$ by

$F(Y) = 0$ iff the value of f on the least limit point of Y which is a member of Q is less than the least member of Y.

Let D be homogeneous for F and of cardinality κ.

<u>Claim</u>. $F''[D]^\kappa = \{0\}$.

(<u>Proof of claim</u>. Since $D \subseteq C \cap \hat{Q}$, the least limit point of D which is a member of Q, η, is sent by f to some ordinal less than η. Clearly η is still the least limit point of $D - (f(\eta) + 1)$ which is a member of Q, and so $F(D - (f(\eta) + 1)) = 0$. Thus $F''[D]^\kappa = \{0\}$.)

By an argument similar to one used earlier, the above claim yields that $f''(D_{1.p.} \cap \hat{Q}) \subseteq \cap D$. Since μ_Q is κ-additive, $\mu_Q(f^{-1}\{\xi_0\}) = 1$ for some $\xi_0 < \cap D$. $^2\dashv$

By Lemmas 1 and 2, μ_Q is a κ-additive nontrivial normal measure on κ. Since $Q \supseteq \kappa \cap \hat{Q}$, $\mu_Q(Q) = 1$, and our proof is complete. $\dashv$

<u>Remark</u>. With a bit of extra effort, we could carry out the above proof starting with any stationary set Q. Thus for uncountable κ satisfying the strong partition property, each stationary set gets measure 1 under some normal measure. For these and further results we refer to Kleinberg [A].

<u>References</u>

L. A. Harrington [1973], Contributions to recursion theory in higher types, Ph.D. Thesis, M.I.T., 1973.

A. S. Kechris [1978], AD and projective ordinals, Cabal Seminar 76-77, Lecture Notes in Mathematics, Springer-Verlag, Vol. 689, 1978, 91-132.

A. S. Kechris [A], Souslin cardinals, κ-Souslin sets and the scale property in the hyperprojective hierarchy, this volume.

E. M. Kleinberg [1977], Infinitary combinatorics and the Axiom of Determinacy, Lecture Notes in Mathematics, Springer-Verlag, Vol. 612, 1977.

E. M. Kleinberg [A], A measure representation theorem for strong partition cardinals, Jour. Symb. Logic, to appear.

Y. N. Moschovakis [1967], Hyperanalytic predicates, Trans. Am. Math. Soc. **129**, 1967, 249-282.

Y. N. Moschovakis [1970], Determinacy and prewellorderings of the continuum, in Math. Logic and Found. of Set Theory, ed. by Y. Bar-Hillel, North Holland, 1970, 24-62.

Y. N. Moschovakis [1974], Structural characterizations of classes of relations, Generalized Recursion Theory, ed. by J. E. Fenstad and P. G. Hinman, North Holland, 1974, 53-79.

Y. N. Moschovakis [1974], Elementary Induction on Abstract Structures, North Holland, 1974.

Y. N. Moschovakis [1978], Inductive scales on inductive sets, Cabal Seminar 76-77, Lecture Notes in Mathematics, Springer-Verlag, Vol. 689, 1978, 185-192.

Y. N. Moschovakis [1980], Descriptive Set Theory, North Holland, 1980.

Y. N. Moschovakis [A], Ordinal games and playful models, this volume.

J. R. Steel [A], Closure properties of pointclasses, this volume.

THE AXIOM OF DETERMINACY AND THE PREWELLORDERING PROPERTY

Alexander S. Kechris[1]
Department of Mathematics
California Institute of Technology
Pasadena, California 91125

Robert M. Solovay[2]
Department of Mathematics
University of California
Berkeley, California 94720

John R. Steel[3]
Department of Mathematics
University of California
Los Angeles, California 90024

§1. Introduction. Let $\omega = \{0,1,2,\ldots\}$ be the set of natural numbers and $\mathcal{R} = \omega^\omega$ the set of all functions from ω into ω, or for simplicity reals. A product space is of the form

$$\mathfrak{X} = X_1 \times X_2 \times \cdots \times X_k ,$$

where $X_i = \omega$ or $\mathcal{R}$. Subsets of these product spaces are called pointsets. A boldface pointclass is a class of pointsets closed under continuous preimages and containing all clopen pointsets (in all product spaces).

The following results have been proved in Steel [1980]:

If Γ is a boldface pointclass, $\check{\Gamma}$ its dual, i.e. $\check{\Gamma} = \{\sim A : A \in \Gamma\}$, and Δ its ambiguous part, i.e. $\Delta = \Gamma \cap \check{\Gamma}$, then assuming ZF + DC + AD,

(1) Either Γ or $\check{\Gamma}$ has the separation property.

(2) If Δ is closed under (finite) intersections and unions, then either Γ or $\check{\Gamma}$ has the reduction property.

Note that by a result of van Wesep [1978b], if Γ is not closed under complements, then it is impossible for both Γ and $\check{\Gamma}$ to have the separation property (assuming again ZF + DC + AD).

---

(1) Partially supported by NSF Grant MCS79-20465. The author is an A.P. Sloan Foundation Fellow.

(2) Partially supported by NSF Grant MCS77-01640.

(3) Partially supported by NSF Grant MCS78-02989.

Our purpose here is to investigate the situation concerning a stronger structural property of pointclasses, namely the prewellordering property. We establish in §2 the following criterion:

Theorem (ZF + DC + AD). Let Γ be a boldface pointclass, closed under countable intersections and unions and either existential or universal quantification over $\mathbb{R}$, but not complements. Then the following are equivalent,

(i) Γ or $\check{\Gamma}$ has the prewellordering property.

(ii) Δ is not closed under wellordered unions (of arbitrary length).

By an application of this criterion and some further analysis done in §§3, 4, 5 we obtain for example the following,

Theorem (ZF + DC + AD). Let Γ be a boldface pointclass, closed under countable intersections and unions and either existential or universal quantification over $\mathbb{R}$, but not complements. Then

$$\Gamma \subseteq L[\mathbb{R}] \Rightarrow \Gamma \text{ or } \check{\Gamma} \text{ has the prewellordering property.}$$

Here $L[\mathbb{R}]$ is the smallest inner model of ZF containing all the reals.

Finally we address ourselves to the following problem: If a pointclass Γ as above is closed under only one kind of quantification over $\mathbb{R}$, i.e. either only existential or only universal (in which case it is reasonable to call such a pointclass projective-like), then in view of the inherent asymmetry in the closure properties between Γ and $\check{\Gamma}$, is it possible to determine in which side of the pair $(\Gamma,\check{\Gamma})$ we have the prewellordering property, as it is done by the work of Martin [1968] and Moschovakis (see Addison-Moschovakis [1968]) for the projective pointclasses $\utilde{\Sigma}^1_n$, $\utilde{\Pi}^1_n$? We provide an affirmative answer in §§4, 5: We first embed each projective-like pointclass Γ in a uniquely determined projective-like hierarchy. We then classify all possible projective-like hierarchies into four types, I-IV and demonstrate that (at least within $L[\mathbb{R}]$) each of these types exhibits a unique prewellordering pattern, identical to that of the classical projective hierarchy for the types I and III, but dual to that for the other two types II and IV. Of course ZF + DC + AD is assumed throughout.

The concept of and some results about Wadge degrees will be of the essence in this paper. The papers Van Wesep [1978a], Van Wesep [1978b] and Steel [1980] provide all the necessary information. For general results in descriptive set theory needed below we refer to Moschovakis [1980]. Finally, it is convenient to assume ZF + DC throughout this paper and explicitly indicate only any further hypotheses as they are needed.

§2. A criterion for prewellordering (Γ). Let Λ be a pointclass (i.e. an arbitrary collection of pointsets). We say that Λ is closed under wellordered unions if for any sequence $\{A_\eta\}_{\eta<\xi}$ of members of Λ, $\bigcup_{\eta<\xi} A_\eta \in \Lambda$. (Here ξ is an arbitrary ordinal, and it is understood that all A_η are subsets of some arbitrary product space $\mathfrak{X}$.) Recall also that a pointclass Γ has the prewellordering property if for every $A \in \Gamma$ there is a norm $\varphi : A \to \kappa$ such that the associated relations

$$x \leq^*_\varphi y \Leftrightarrow x \in A \wedge [y \notin A \vee \varphi(x) \leq \varphi(y)],$$

$$x <^*_\varphi y \Leftrightarrow x \in A \wedge [y \notin A \vee \varphi(x) < \varphi(y)]$$

are in Γ. Such a norm Γ is called a Γ-norm.

We have now

2.1 Theorem (AD). Let Γ be a boldface pointclass, closed under countable unions and intersections and either $\exists^{\mathbb{R}}$ or $\forall^{\mathbb{R}}$, but not complements. Then the following are equivalent:

(i) Prewellordering (Γ),

(ii) Reduction (Γ) and Δ is not closed under wellordered unions.

In view of the result of Steel mentioned in the introduction, we have

2.2 Corollary (AD). Let Γ be a boldface pointclass closed under countable unions and intersections, and either $\exists^{\mathbb{R}}$ or $\forall^{\mathbb{R}}$, but not complements. Then the following are equivalent

(i) Prewellordering $(\Gamma) \vee$ Prewellordering $(\check{\Gamma})$,

(ii) Δ is not closed under wellordered unions.

Proof of 2.1. If Γ has the prewellordering property, let $A \in \Gamma - \Delta$ and let φ be a regular Γ-norm on A, of length κ. (Regular means that $\varphi : A \twoheadrightarrow \kappa$). For $\xi < \kappa$, let $A_\xi = \varphi^{-1}[\{\xi\}]$. Then each $A_\xi \in \Delta$ but $\bigcup_{\xi<\kappa} A_\xi = A \notin \Delta$, so Δ is not closed under wellordered unions. Finally we clearly have Reduction (Γ).

So assume now that Δ is not closed under wellordered unions, and Reduction (Γ) holds. We shall distinguish three cases, depending on the closure properties of Γ.

2.3. Γ is closed under both $\exists^{\mathbb{R}}$ and $\forall^{\mathbb{R}}$.

We shall start in this case with a lemma which will be also useful later on.

Call a boldface pointclass Λ strongly closed if it is closed under (finite) unions and intersections, complements and quantification (of both types) over $\mathbb{R}$. For example $\Lambda = \bigcup_n \Delta^1_n$ is strongly closed. Note that Λ need not be closed under countable unions.

2.3.1. Lemma. Let Λ be strongly closed. Then the following three ordinals associated with Λ are equal:

(i) $\sup\{\xi : \xi$ is the length of a Λ prewellordering of $\mathbb{R}\}$,

(ii) $\sup\{\xi : \xi$ is the rank of a Λ wellfounded relation on $\mathbb{R}\}$,

(iii) $\sup\{w(A) : A \in \Lambda\}$.

Here $w(A)$ is the Wadge ordinal of $A \subseteq \mathbb{R}$. (Note that $w(A) = w(\sim A)$.)

Proof. Clearly (i) $\leq$ (ii) and (iii) $\leq$ (i). So enough to show that (ii) $\leq$ (iii).

For each list of pointsets $A_1, A_2, \ldots, A_n$ let $\underset{\sim}{\Sigma}^1_1(A_1, A_2, \ldots, A_n)$ be the smallest boldface pointclass closed under unions and intersections, $\exists^\omega$, $\forall^\omega$ and $\exists^{\mathbb{R}}$, which contains $A_1, \ldots, A_n$. By Moschovakis [1970] §3, $\underset{\sim}{\Sigma}^1_1(A_1, A_2, \ldots, A_n)$ is $\mathbb{R}$-parametrized i.e. has universal sets. Let $\underset{\sim}{\Pi}^1_1(A_1, \ldots, A_n)$ be the dual of $\underset{\sim}{\Sigma}^1_1(A_1, \ldots, A_n)$. Note that $\underset{\sim}{\Pi}^1_1(A, \sim A)$ cannot be closed under $\exists^{\mathbb{R}}$, so that we can build up $\underset{\sim}{\Sigma}^1_2(A, \sim A)$, $\underset{\sim}{\Pi}^1_2(A, \sim A), \ldots$ in the usual way.

Recall now the 0th Periodicity Theorem (see Kechris [1977]), which asserts that if Γ is a boldface pointclass closed under countable intersections and unions, then if $\forall^{\mathbb{R}}\Gamma = \{\forall\alpha(x,\alpha) \in A : A \in \Gamma\}$ and similarly for $\exists^{\mathbb{R}}\Gamma$, we have

(i) $\exists^{\mathbb{R}}\Gamma \subseteq \Gamma \wedge \text{Reduction}(\Gamma) \Rightarrow \text{Reduction}(\forall^{\mathbb{R}}\Gamma)$,

(ii) $\forall^{\mathbb{R}}\Gamma \subseteq \Gamma \wedge \text{Reduction}(\Gamma) \Rightarrow \text{Reduction}(\exists^{\mathbb{R}}\Gamma)$.

(For the convenience of the reader we repeat this proof in an appendix.) From this and Steel's Theorem (mentioned in the introduction), we can find for each pointset A an integer $N > 1$ so that $\underset{\sim}{\Pi}^1_N(A, \sim A)$ has the reduction property. ($N = 2$ or $N = 3$ will of course suffice.)

From these remarks it follows that if $<$ is a wellfounded relation in Λ, there is a boldface pointclass $\Gamma \subsetneq \Lambda$ such that Γ is $\mathbb{R}$-parametrized, closed under countable intersections and unions and $\forall^{\mathbb{R}}$, such that Γ has the reduction property and moreover $\Delta \supseteq \underset{\sim}{\Sigma}^1_1(<)$. (Note that if Λ is strongly closed and $A_1, \ldots, A_n \in \Lambda$, $\underset{\sim}{\Pi}^1_n(A_1, \ldots, A_n) \subsetneq \Lambda$).

For a Γ with these structural properties we can introduce the following Γ coding of the Δ sets of reals:

Let W^0, W^1 be a universal pair for Γ (i.e. for each $A, B \subseteq \mathbb{R}$ in Γ there is $\varepsilon \in \mathbb{R}$ with $A = W^0_\varepsilon = \{\alpha : (\varepsilon,\alpha) \in W^0\}$, $B = W^1_\varepsilon$). Let $\overline{W}^0$, $\overline{W}^1$ in Γ reduce W^0, W^1. Put

$$\varepsilon \in C \Leftrightarrow \overline{W}^0_\varepsilon \cup \overline{W}^1_\varepsilon = \mathbb{R}$$

and for $\varepsilon \in C$

$$H_\varepsilon = \overline{W}^0_\varepsilon \quad (= \sim\overline{W}^1_\varepsilon) .$$

Then $C \in \Gamma$ and

$$\{H_\varepsilon : \varepsilon \in C\} = \{A \subseteq \mathbb{R} : A \in \Delta\} .$$

We can finish the proof now as follows: By the Recursion Theorem we can find a partial continuous function f such that

(1) $\alpha \in \mathrm{Field}(\prec) \Rightarrow f(\alpha) \in C$,

(2) $\alpha \prec \beta \Rightarrow w(H_{f(\alpha)}) < w(H_{f(\beta)})$.

We use here the fact that $\underset{\sim}{\Sigma}^1_1(\prec) \subseteq \Delta$. Thus

$$\mathrm{rank}(\prec) \leq \sup\{w(A) : A \in \Delta\} \leq \text{(iii)} . \qquad 2.3.1.\dashv$$

We proceed now to complete the proof of case 2.3. Let Γ satisfy its hypotheses. Let δ be the ordinal associated to $\Lambda \equiv \Delta$ by the preceding lemma. Let also θ be the least ordinal such that there is a θ-sequence $\{A_\xi\}_{\xi<\theta}$ of members of Δ with $\bigcup_{\xi<\theta} A_\xi \notin \Delta$. We claim that $\theta = \delta$.

To see that $\delta \leq \theta$, notice first that we can code Δ sets using Reduction (Γ) as in the proof of 2.3.1. (That Γ is $\mathbb{R}$-parametrized follows from the nonclosure of Γ under complements - see Van Wesep [1978].) Let $C, \varepsilon \mapsto H_\varepsilon$ denote the set of codes and the coding map respectively. If $\xi < \delta$, let $\varphi : \mathbb{R} \to \xi$ be a Δ-norm and then for each ξ-sequence $\{B_\eta\}_{\eta<\xi}$ of Δ sets let, by Moschovakis [1970], $R(w,\varepsilon)$ be a Δ relation such that

(i) $\varphi(w) = \varphi(v) \Rightarrow [R(w,\varepsilon) \Leftrightarrow R(v,\varepsilon)]$

(ii) $R(w,\varepsilon) \Rightarrow \varepsilon \in C$

(iii) $\forall w \exists \varepsilon[R(w,\varepsilon) \wedge H_\varepsilon = B_{\varphi(w)}]$.

Then

$$\alpha \in \bigcup_{\eta<\xi} B_\eta \Leftrightarrow \exists w \exists \varepsilon[R(w,\varepsilon) \wedge \alpha \in H_\varepsilon] ,$$

thus $\bigcup_{\eta<\xi} B_\eta \in \Delta$.

To prove that conversely $\theta \leq \delta$, we use the minimality of θ. It easily implies that there is a θ-sequence of Δ sets $\{A_\xi\}_{\xi<\theta}$ such that $A_\xi \subsetneqq A_\eta$, if $\xi < \eta < \theta$, $A_\lambda = \bigcup_{\xi<\lambda} A_\xi$, if $\lambda < \theta$ is limit, and $A = \bigcup_{\xi<\theta} A_\xi \notin \Delta$. For each $x \in A$, let

$$\varphi(x) = \text{least } \xi,\ x \in A_{\xi+1} - A_\xi .$$

Then φ is a regular norm on A of length θ. But if $x \in A$ and $\varphi(x) = \xi < \theta$, then the prewellordering

$$\{(y,z) : \varphi(y) \leq \varphi(z) < \varphi(x)\} ,$$

of length ξ, is in Δ, since it is equal to

$$\bigcup_{\eta<\xi}(A_\eta \times \bigcap_{\eta'<\eta} \sim A_{\eta'}) ,$$

while $\xi < \theta$. So $\xi < \delta$, thus $\theta \leq \delta$.

Let now

$$\Gamma^* = \{\bigcup_{\xi<\theta} A_\xi : \forall \xi < \theta (A_\xi \in \Delta)\} .$$

First notice that, by an argument of Martin [198?]), Γ^* has the prewellordering property. Indeed, if $A = \bigcup_{\xi<\theta} A_\xi$ is in Γ^* and we let

$$\psi(x) = \text{least } \xi,\ x \in A_\xi ,$$

be the norm associated to this wellordered union, then ψ is a Γ^*-norm. This is because if $\leq^*_\psi$, $<^*_\psi$ are its two associated relations, then

$$x \leq^*_\psi y \Leftrightarrow \exists \xi < \theta[x \in A_\xi \wedge \forall \xi' < \xi (y \notin A_{\xi'})] ,$$

$$x <^*_\psi y \Leftrightarrow \exists \xi < \theta[x \in A_\xi \wedge \forall \xi' \leq \xi (y \notin A_{\xi'})] ,$$

so these are in Γ^*, by the minimality of θ again.

So it is enough to show that $\Gamma^* = \Gamma$. Since $\Gamma^* \supsetneqq \Delta$, by Wadge's Lemma it is enough to show that $\Gamma^* \subseteq \Gamma$. For that let

$$\varepsilon \in W \Leftrightarrow \varepsilon \in C \wedge H_\varepsilon \text{ is wellfounded}$$

(view H_ε as a subset of $\mathbb{R}^2$ here), and for $\varepsilon \in W$, let

$$|\varepsilon| = \text{rank}(H_\varepsilon) .$$

Then $W \in \Gamma$ and $\{|\varepsilon| : \varepsilon \in C\} = \delta = \theta$. Given now $\{B_\xi\}_{\xi<\theta}$ a θ-sequence of Δ sets, consider the following Solovay-type game:

I	II	II wins iff $\varepsilon \in W \Rightarrow \alpha \in C \wedge \exists \eta[|\varepsilon| < \eta < \delta$
ε	α	$\wedge H_\alpha = \bigcup_{\xi<\eta} B_\xi]$.

If I has a winning strategy f then the relation

$$(\varepsilon,x) \prec (\varepsilon',y) \Leftrightarrow \varepsilon = \varepsilon' \in f[\mathbb{R}] \wedge (x,y) \in H_\varepsilon$$

is wellfounded and in Δ, so if II plays any $\alpha \in C$ with $H_\alpha = \bigcup_{\xi<\eta} B_\xi$ for any $\delta > \eta > \text{rank}(\prec)$, he beats f. So II must have a winning strategy g. Then

$$x \in \bigcup_{\xi<\theta} B_\xi \Leftrightarrow \exists \varepsilon[\varepsilon \in W \wedge x \in H_{g(\varepsilon)}] ,$$

so $\bigcup_{\xi<\theta} B_\xi \in \Gamma$, thus $\Gamma^* \subseteq \Gamma$, and we are done. 2.3.⊣

2.4. Γ <u>is closed under</u> $\exists^{\mathbb{R}}$ <u>but not</u> $\forall^{\mathbb{R}}$.

In this case the result follows from the following lemma which will be also useful later on.

2.4.1 <u>Lemma.</u> Assume Γ is a boldface pointclass, closed under countable unions and intersections and $\exists^{\mathbb{R}}$ but not $\forall^{\mathbb{R}}$ (thus Γ is not closed under complements). Then if Reduction (Γ) holds, Γ is closed under wellordered unions.

<u>Proof.</u> Let θ be least such that for some $\{A_\xi\}_{\xi<\theta}$ with all A_ξ in Γ, $\bigcup_{\xi<\theta} A_\xi \notin \Gamma$, towards a contradiction. It is easy to check that

$$\theta \text{ is a regular uncountable cardinal .}$$

Put

$$\Gamma' = \{\bigcup_{\xi<\theta} A_\xi : \forall \xi < \theta (A_\xi \in \Gamma)\} .$$

Since $\Gamma' \supsetneqq \Gamma$, by Wadge's Lemma $\Gamma' \supseteq \check{\Gamma}$, and since clearly Γ' is closed under $\exists^{\mathbb{R}}$, $\Gamma' \supseteq \exists^{\mathbb{R}}\check{\Gamma}$. Put

$$\Gamma_+ = \forall^{\mathbb{R}}\Gamma, \quad \text{so that } \check{\Gamma}_+ = \exists^{\mathbb{R}}\check{\Gamma} ,$$

therefore $\Gamma' \supseteq \check{\Gamma}_+$. Let also

$$\delta_+ = \sup\{\xi : \xi \text{ is the rank of a } \check{\Gamma}_+ \text{ wellfounded relation}\} .$$

Since $\Gamma' \supsetneqq \Delta_+$ $(= \Gamma_+ \cap \check{\Gamma}_+)$ and $\Gamma \subseteq \Delta_+$ (because $\forall^{\mathbb{R}}\Gamma = \Gamma^+ \not\subseteq \Gamma$, thus $\Gamma \subseteq \check{\Gamma}_+$ by Wadge), Δ_+ is not closed under wellordered unions, so let θ_+ be the least ordinal such that some union of a θ_+-sequence of Δ_+ sets is not in Δ_+. Clearly $\theta_+ \leq \theta$.

We have now that $\theta_+ \geq \delta_+$, thus

$$\theta \geq \delta_+ .$$

(This is essentially an argument of Martin [198?]: If $\theta_+ < \delta_+$, then as in the proof of 2.3 an application of the Moschovakis Coding Lemma shows that

$$\{\bigcup_{\xi<\theta_+} A_\xi : \forall\xi < \theta_+ (A_\xi \in \Delta_+)\} = \check{\Gamma}_+$$

and thus again as in 2.3, $\check{\Gamma}_+$ has the prewellordering property, contradicting the fact that Γ_+ has the reduction property by the 0th Periodicity Theorem).

Let

$$\delta = \sup\{\xi : \xi \text{ is the rank of a } \Gamma \text{ wellfounded relation}\} .$$

By the standard argument (see for example Kechris [1974]) $\delta_+ > \delta$, thus

$$\theta > \delta .$$

Let now $\prec$ be a $\check{\Gamma}_+$ wellfounded relation. Then, since $\check{\Gamma}_+ \subseteq \Gamma'$,

$$\prec = \bigcup_{\xi<\theta} \prec_\xi ,$$

where each $\prec_\xi$ is a Γ wellfounded relation, and by the minimality of θ we can assume that $\xi \leq \eta < \theta \Rightarrow \prec_\xi \subseteq \prec_\eta$. For each $x \in \text{Field}(\prec)$ put

$$f_x(\xi) = \begin{cases} 0, & \text{if } x \notin \text{Field}(\prec_\xi) , \\ |x|_{\prec_\xi} \equiv \text{rank of } x \text{ in } \prec_\xi, & \text{otherwise} . \end{cases}$$

Then $f_x : \theta \to \delta$ is nondecreasing, so since $\delta < \theta$ and θ is regular, we have

$$f_x(\xi) = \text{constant} \equiv \xi(x) < \delta$$

for all large enough $\xi < \theta$. Then

$$x \prec y \Rightarrow \xi(x) < \xi(y) ,$$

thus $\text{rank}(\prec) \leq \delta$, so $\delta_+ \leq \delta$, a contradiction. 2.4.1.⊣
2.4.

2.5. Γ <u>is closed under</u> $\forall^{\mathbb{R}}$ <u>but not</u> $\exists^{\mathbb{R}}$.

Let again θ be least such that some union of a θ-sequence of Δ sets is not in Δ. Put

$$\Gamma' = \{\bigcup_{\xi<\theta} A_\xi : \forall \xi < \theta(A_\xi \in \Delta)\} .$$

Since Γ' has the prewellordering property $\Gamma' \neq \check{\Gamma}$ so, since $\Gamma' \supsetneq \Delta$, $\Gamma' \supseteq \Gamma$.

Call now a sequence $\{A_\xi\}_{\xi<\theta}$ Λ-<u>bounded</u> (for any pointclass Λ), if

$$\forall X \in \Lambda[X \subseteq \bigcup_{\xi<\theta} A_\xi \Rightarrow \exists \xi < \theta(X \subseteq A_\xi)] .$$

Put

$$\Gamma^* = \{\bigcup_{\xi<\theta} A_\xi : \forall \xi < \theta(A_\xi \in \Delta) \wedge \{A_\xi\}_{\xi<\theta} \text{ is } \check{\Gamma}\text{-bounded}\} .$$

The result will follow from the following three lemmas.

2.5.1. <u>Lemma.</u> Γ^* is a boldface pointclass with the prewellordering property.

2.5.2. <u>Lemma.</u> If $\Gamma^* \supsetneq \Delta$, then $\Gamma^* = \Gamma$.

2.5.3. <u>Lemma.</u> $\Gamma^* \not\supseteq \Delta$.

<u>Proof of 2.5.1.</u> Since $\check{\Gamma}$ is closed under $\exists^{\mathbb{R}}$ it is trivial to check that Γ^* is closed under continuous preimages. Let now $\{A_\xi\}_{\xi<\theta}$ be a $\check{\Gamma}$-bounded sequence of Δ sets. Let $A = \bigcup_{\xi<\theta} A_\xi$ and let

$$\varphi(x) = \text{least } \xi,\ x \in A_\xi$$

the associated norm on A. Then

$$x \leq_\varphi^* y \Leftrightarrow \exists\xi < \theta[\overbrace{\exists\xi' \leq \xi(x \in A_{\xi'} \wedge \forall\eta' < \xi'(y \notin A_{\eta'}))}^{B_\xi(x,y)}]$$

$$x <_\varphi^* y \Leftrightarrow \exists\xi < \theta[\overbrace{\exists\xi' \leq \xi(x \in A_{\xi'} \wedge \forall\eta' \leq \xi'(y \notin A_{\eta'}))}^{C_\xi(x,y)}] .$$

Let $B_\xi(x,y)$, $C_\xi(x,y)$ the two pointsets indicated above. It is enough to show that $\{B_\xi\}_{\xi<\theta}$, $\{C_\xi\}_{\xi<\theta}$ are $\check{\Gamma}$-bounded. Take for example $\{B_\xi\}_{\xi<\theta}$. Let $Y \in \check{\Gamma}$, $Y \subseteq \bigcup_{\xi<\theta} B_\xi$. If $X = \{x : \exists y(x,y) \in Y\}$, then $X \in \check{\Gamma}$ and $X \subseteq \bigcup_{\xi<\theta} A_\xi$ so for some $\xi < \theta$, $X \subseteq A_\xi$. Then $Y \subseteq B_\xi$. 2.5.1.⊣

<u>Proof of 2.5.2.</u> Assume $\Gamma^* \not\supseteq \Delta$. If $\Gamma^* \neq \Gamma$, then by Wadge's Lemma, $\Gamma^* \supseteq \check{\Gamma}$. Since, by the 0th Periodicity Theorem, Reduction $(\Gamma) \Rightarrow$ Reduction $(\exists^{\mathbb{R}}\Gamma)$, we have by Lemma 2.4.1 that $\Gamma^* \subseteq \exists^{\mathbb{R}}\Gamma$, so

$$\check{\Gamma} \subseteq \Gamma^* \subseteq \exists^{\mathbb{R}}\Gamma .$$

Let now $A \in \check{\Gamma}$ be such that

$$x \in B \Leftrightarrow \forall\alpha(x,\alpha) \in A$$

is not in $\exists^{\mathbb{R}}\Gamma$. Write

$$A = \bigcup_{\xi<\theta} A_\xi ,$$

where each $A_\xi \in \Delta$ and $\{A_\xi\}_{\xi<\theta}$ is $\check{\Gamma}$-bounded. Then

$$x \in B \Leftrightarrow \forall\alpha\exists\xi < \theta(x,\alpha) \in A_\xi .$$

So if $x \in B$, then $\{x\} \times \mathbb{R} \subseteq \bigcup_{\xi<\theta} A_\xi$, thus by $\check{\Gamma}$-boundedness, $\{x\} \times \mathbb{R} \subseteq A_\xi$ for some $\xi < \theta$. Thus

$$x \in B \Leftrightarrow \exists\xi < \theta\ \forall\alpha(x,\alpha) \in A_\xi .$$

If

$$x \in B_\xi \Leftrightarrow \forall\alpha(x,\alpha) \in A_\xi ,$$

then $B_\xi \in \Gamma \subseteq \exists^{\mathbb{R}}\Gamma$, so by 2.4.1 again $B = \bigcup_{\xi<\theta} B_\xi \in \exists^{\mathbb{R}}\Gamma$, a contradiction. 2.5.2. $\dashv$

Proof of 2.5.3. Let $A \in \Gamma - \Delta$. Let $\mathcal{S}$ be universal for $\check{\Gamma}$, and let

$$B = \{\varepsilon : \mathcal{S}_\varepsilon \subseteq A\} .$$

Thus $B \in \Gamma$. As $\Gamma' \supseteq \Gamma$ we can write

$$B = \bigcup_{\xi<\theta} B_\xi ,$$

where each B_ξ is in Δ. Put

$$A_\xi = \bigcup \{\mathcal{S}_\varepsilon : \varepsilon \in B_\xi\} .$$

Then each $A_\xi \in \check{\Gamma}$, $\bigcup_{\xi<\theta} A_\xi = A$ and $\{A_\xi\}_{\xi<\theta}$ is $\check{\Gamma}$-bounded. Since, by the minimality of θ, we can choose the sequence $\{B_\xi\}_{\xi<\theta}$ to be increasing and continuous, the same can be assumed to be true for the sequence $\{A_\xi\}_{\xi<\theta}$. We claim now that $E = \{\xi < \theta : A_\xi \in \Delta\}$ is ω-closed unbounded in θ, which of course completes the proof since if $\{\rho_\xi\}_{\xi<\theta}$ is its increasing enumeration and $A'_\xi = A_{\rho_\xi}$, then $\{A'_\xi\}_{\xi<\theta}$ is a $\check{\Gamma}$-bounded sequence of Δ sets with union $A \notin \Delta$.

Clearly E is ω-closed. To see that it is unbounded let $\xi < \theta$ be given. As $A_\xi \in \check{\Gamma}$, $A \in \Gamma$ and $A_\xi \subseteq A$ we can find, using Separation $(\check{\Gamma})$, a set $X_0 \in \Delta$, with $A_\xi \subseteq X_0 \subseteq A$. By $\check{\Gamma}$-boundedness let $\xi_0 > \xi$ be such that $X_0 \subseteq A_{\xi_0}$. Repeat now this process with ξ_0 to define X_1, ξ_1 etc. If $\xi' = \lim_{i<\omega} \xi_i < \theta$ (recall that θ must be a regular uncountable cardinal), then $A_{\xi'} = \bigcup_{\eta<\xi'} A_\eta = \bigcup_{i<\omega} A_{\xi_i} = \bigcup_{i<\omega} X_i \in \Delta$. 2.5.3.
2.5.
2.1. $\dashv$

§3. Inductive-like pointclasses and projective algebras. We shall examine now the extent to which the condition (ii) of the criterion 2.2 is satisfied. We consider first the case when Γ is closed under both types of real quantification. Let us give the following definitions.

Definition. A boldface pointclass Γ is called inductive-like if it is closed under countable intersections and unions, $\exists^{\mathbb{R}}$ and $\forall^{\mathbb{R}}$, but not complements.

The typical inductive-like pointclasses are $\underset{\sim}{\text{IND}}(\mathbb{R})$, the class of all inductive over the structure of analysis $\mathbb{R}$ pointsets, and its dual $\check{\underset{\sim}{\text{IND}}}(\mathbb{R})$.

Definition. A boldface pointclass Λ is called a projective algebra iff it is closed under complements, wellordered unions, $\exists^{\mathbb{R}}$ and $\forall^{\mathbb{R}}$.

Then 2.2 implies immediately

3.1 Theorem (AD). Let Γ be an inductive-like pointclass. Then the following are equivalent:

(i) Prewellordering $(\Gamma) \vee$ Prewellordering $(\check{\Gamma})$,

(ii) Δ is not a projective algebra.

The following result now provides a good first impression about the concept of projective algebra.

3.2 Theorem (AD). Let Λ be a projective algebra and let $L[\Lambda]$ be the smallest inner model of ZF containing $\mathcal{R} \cup \Lambda$. Then $\text{power}(\mathcal{R}) \cap L[\Lambda] = \Lambda$.

Proof. Let Λ be a projective algebra. Let $\langle \xi, \eta \rangle$ be the bijection between ORD $\times$ ORD and ORD corresponding to the Gödel wellordering of pairs of ordinals. Let $\langle \xi, \eta, \zeta \rangle = \langle \xi, \langle \eta, \zeta \rangle \rangle$ and $\xi = \langle I_\xi, J_\xi, K_\xi \rangle$. Let $\mathfrak{F}_1, \mathfrak{F}_2, \ldots, \mathfrak{F}_9$ be the (binary) Gödel operations as given in Shoenfield [1967]. Put for $\alpha \in \mathcal{R}$, $A \in \Lambda \cap \text{power}(\mathcal{R})$:

$$G_i(x,y;\alpha,A) = \mathfrak{F}_i(x,y), \qquad 1 \leq i \leq 9,$$

$$G_{10}(x,y;\alpha,A) = x \cap \alpha,$$

$$G_{11}(x,y;\alpha,A) = x \cap A.$$

Then define $F : \text{ORD} \times \mathcal{R} \times \Lambda \to V$ by

$$\begin{aligned} F(\xi;\alpha,A) &= \{F(\eta;\beta,B) : \eta < \xi,\ \beta \in \mathcal{R},\ B \in \Lambda\}, \quad \text{if } I_\xi = 0 \\ &= G_{I_\xi}(F(J_\xi;(\alpha)_0,(A)_0), F(K_\xi;(\alpha)_1,(A)_1);(\alpha)_2,(A)_2), \quad \text{if } 0 < I_\xi \leq 11, \\ &= \{F(J_\xi;(\alpha)_0,(A)_0), F(K_\xi,(\alpha)_1,(A)_1)\}, \quad \text{if } I_\xi \geq 12, \end{aligned}$$

where $(\alpha)_i(n) = \alpha(2^i \cdot 3^n)$, $(A)_i = \{\alpha : i^\frown\alpha \in A\}$. We have then

$$L[\Lambda] = \{F(\xi;\alpha,A) : \xi \in \text{ORD},\ \alpha \in \mathcal{R},\ A \in \Lambda\}.$$

(Notice here that if M is an inner model of ZF containing $\mathcal{R}$ and $\Lambda \subseteq M$ then actually $\Lambda \in M$, since either $\Lambda = M \cap \text{power}(\mathcal{R})$ or else if $A \notin \Lambda$ has least Wadge ordinal, $A \in M$, so $\Lambda \in M$).

Let us say now that a relation $\Phi(x, A_1, \ldots, A_n)$, where $x \in \mathfrak{X}$, $A_i \subseteq \mathcal{R}$ and $A_i \in \Lambda$ is in the class $\overline{\Lambda}$, if for each fixed $P_1, \ldots, P_n \in \Lambda$ the pointset $\Phi^* \equiv \Phi^*_{P_1,\ldots,P_n}$ given by

$$\Phi^*(x,\varepsilon_1,\ldots,\varepsilon_n) \Leftrightarrow \varepsilon_1,\ldots,\varepsilon_n \text{ code continuous functions } f_{\varepsilon_1},\ldots,f_{\varepsilon_n}$$
$$\text{and } \Phi(x, f^{-1}_{\varepsilon_1}[P_1], \ldots, f^{-1}_{\varepsilon_n}[P_n]),$$

is in Λ.

Lemma 1. $\bar{\Lambda}$ contains $\Phi(\alpha,A) \Leftrightarrow \alpha \in A$, and is closed under continuous substitutions, complements, $\exists^{\mathbb{R}}$, $\forall^{\mathbb{R}}$, $\exists^{\Lambda}$, $\forall^{\Lambda}$ and wellordered unions.

Proof. Everything is obvious except perhaps $\exists^{\Lambda}$. Let $\Phi(x,A,B)$ be in $\bar{\Lambda}$ and consider $\Psi(x,A) \Leftrightarrow \exists B \in \Lambda\Phi(x,A,B)$. Fix $P \in \Lambda$. Then

$$\Psi^*_P(x,\varepsilon) \Leftrightarrow \varepsilon \text{ codes a continuous function } f_\varepsilon \wedge \Psi(x,f_\varepsilon^{-1}[P])$$
$$\Leftrightarrow \varepsilon \text{ codes a continuous function } f_\varepsilon \wedge \exists B \in \Lambda\Phi(x,f_\varepsilon^{-1}[P],B)$$
$$\Leftrightarrow \varepsilon \text{ codes a continuous function}$$
$$f_\varepsilon \wedge \exists\xi < \lambda\exists B[w(B) \leq \xi \wedge \Phi(x,f_\varepsilon^{-1}[P],B)] ,$$

where $\lambda = \sup\{w(B) : B \in \Lambda\}$. So enough to show that if

$$S_\xi(x,\varepsilon) \Leftrightarrow \varepsilon \text{ codes a continuous function}$$
$$f_\varepsilon \wedge \exists B[w(B) \leq \xi \wedge \Phi(x,f_\varepsilon^{-1}[P],B)] ,$$

then $S_\xi \in \Lambda$, as $\Psi^*_P = \bigcup_{\xi<\lambda} S_\xi$. For that assume without loss of generality that ξ is such that $w(Q) = \xi \Rightarrow Q$ is self-dual, and pick such a Q. Then $Q \in \Lambda$ and

$$S_\xi(x,\varepsilon) \Leftrightarrow \varepsilon \text{ codes a continuous function } f_\varepsilon \wedge \exists B \leq_w Q\Phi(x,f_\varepsilon^{-1}[P],B)$$
$$\Leftrightarrow \varepsilon \text{ codes a continuous function } f_\varepsilon \wedge \exists\varepsilon' [\varepsilon' \text{ codes a continuous}$$
$$\text{function } f_{\varepsilon'} \wedge \Phi(x,f_\varepsilon^{-1}[P],f_{\varepsilon'}^{-1}[Q])]$$
$$\Leftrightarrow \exists\varepsilon' \Phi^*_{P,Q}(x,\varepsilon,\varepsilon') ,$$

so $S_\xi \in \Lambda$. ⊣

Lemma 2. For each fixed ξ, η the following relations are in $\bar{\Lambda}$:

(i) $\Phi^{\in}_{\xi,\eta}(\alpha,A;\beta,B) \Leftrightarrow F(\xi;\alpha,A) \in F(\eta;\beta,B)$,

(ii) $\Phi^{=}_{\xi,\eta}(\alpha,A;\beta,B) \Leftrightarrow F(\xi;\alpha,A) = F(\eta;\beta,B)$,

(iii) $\Psi^1_{\xi,\eta}(\alpha;\beta,B) \Leftrightarrow \alpha \in F(\max\{\xi,\eta\};\beta,B)$

$\Psi^2_{\xi,\eta}(n;\beta,B) \Leftrightarrow n \in \omega \wedge n \in F(\max\{\xi,\eta\};\beta,B)$.

Proof. Routine induction on $\langle\xi,\eta\rangle$. ⊣

To complete the proof, let now $S \subseteq \mathbb{R}$ be in $L[\Lambda]$ and find $\xi_0 \in \mathrm{ORD}$, $\alpha_0 \in \mathbb{R}$, $A_0 \in \Lambda$ with $S = F(\xi_0;\alpha_0,A_0)$. Then

$$\alpha \in S \Leftrightarrow \alpha \in F(\xi_0;\alpha_0,A_0)$$
$$\Leftrightarrow \Psi^1_{\xi_0,\xi_0}(\alpha;\alpha_0,A_0) .$$

So if ε_0 codes the identity function and $(\Psi^1_{\xi_0,\xi_0})^*_{A_0} \equiv R$, we have

$$\alpha \in S \Leftrightarrow (\alpha,\alpha_0,\varepsilon_0) \in R$$

thus $S \in \Lambda$. ⊣

In particular for any projective algebra Λ,

$$\text{power}(\mathbb{R}) \cap L[\mathbb{R}] \subseteq \Lambda ,$$

and if $A \in \Lambda$ and $\exists B \subseteq \mathbb{R}(B \notin L[A])$, or equivalently by Steel-Van Wesep [1981], $A^{\#}$ exists, then $A^{\#} \in \Lambda$. Thus any projective algebra is quite big. As a simple consequence we have

3.3 <u>Corollary</u> (AD). Let Γ be any inductive-like pointclass contained in $L[\mathbb{R}]$. Then Prewellordering $(\Gamma) \vee$ Prewellordering $(\check{\Gamma})$.

Also we immediately obtain,

3.4 <u>Corollary</u> (AD). If Γ is any inductive-like pointclass then $\Gamma \supseteq \underset{\sim}{\mathrm{IND}}(\mathbb{R})$ or $\Gamma \supseteq \check{\underset{\sim}{\mathrm{IND}}}(\mathbb{R})$. In particular, $\underset{\sim}{\mathrm{IND}}(\mathbb{R})$ is the smallest inductive-like pointclass satisfying reduction (or even not satisfying separation).

<u>Proof</u>. By Moschovakis [1974], $\underset{\sim}{\mathrm{IND}}(\mathbb{R})$ is the smallest inductive-like pointclass satisfying prewellordering. ⊣

We conclude this section by offering some speculations on the extent of projective algebras.

Let $\underset{\sim}{P}_\infty$ denote the smallest projective algebra, $\underset{\sim}{B}_\infty$ the smallest boldface pointclass closed under complements and wellordered unions (the class of ∞-Borel sets) and $\underset{\sim}{S}_\infty = \bigcup_\kappa \underset{\sim}{S}_\kappa$, where $\underset{\sim}{S}_\kappa$ is the class of κ-Souslin sets. Clearly

$$\underset{\sim}{S}_\infty \subseteq \underset{\sim}{B}_\infty \subseteq \underset{\sim}{P}_\infty .$$

By an (unpublished) result of Martin and Steel

$$AD + V = L[\mathbb{R}] \Rightarrow \underset{\sim}{S}_\infty = \underset{\sim}{\Sigma}^2_1 .$$

Using this it can be shown that

$$AD + V = L[\mathbb{R}] \Rightarrow \underset{\sim}{B}_\infty = \text{power}(\mathbb{R}) .$$

A proof of this is given in an appendix. On the other hand, if $AD_{\mathbb{R}}$ denotes the Axiom of Determinacy for games on reals and Θ is the sup of the lengths of prewellorderings on $\mathbb{R}$, then the following question has been raised in Solovay [1978]:

$$AD_{\mathbb{R}} + \Theta \text{ is regular} \Rightarrow \underset{\sim}{S}_{\infty} = \text{power}(\mathbb{R}) \ ?$$

An affirmative answer would also imply that $\underset{\sim}{P}_{\infty} = \text{power}(\mathbb{R})$ and thus extend 3.3 and further results of the present paper to all appropriate Γ's. Even if the above question admits a negative answer it is still conceivable that some reasonable hypothsis extending AD (not necessarily properly) might still imply that $\underset{\sim}{P}_{\infty} = \text{power}(\mathbb{R})$.

§4. Projective-like pointclasses and hierarchies. We proceed now to discuss the case of pointclasses closed under only one kind of real quantification.

Definition. A boldface pointclass is projective-like if it is closed under countable intersections and unions, and either $\exists^{\mathbb{R}}$ or $\forall^{\mathbb{R}}$ but not both, and is not closed under complements.

The prototypes of such pointclasses are of course $\underset{\sim}{\Sigma}^1_n$, $\underset{\sim}{\Pi}^1_n$.

It is convenient for the work below to embed every porjective-like pointclass in a (unique) projective-like hierarchy, where this concept is defined as follows

Definition. A projective-like hierarchy is a sequence $\Gamma_1, \Gamma_2, \Gamma_3, \ldots$ of projective-like pointclasses such that

(i) Γ_1 is closed under $\forall^{\mathbb{R}}$,

(ii) $\Gamma_{i+1} = \exists^{\mathbb{R}}\Gamma_i$, if Γ_i is closed under $\forall^{\mathbb{R}}$, and $\Gamma_{i+1} = \forall^{\mathbb{R}}\Gamma_i$, if Γ_i is closed under $\exists^{\mathbb{R}}$,

(iii) $\{\Gamma_i\}$ is maximal, i.e. there is no projective-like pointclass Γ_0 closed under $\exists^{\mathbb{R}}$ such that $\Gamma_1 = \forall^{\mathbb{R}}\Gamma_0$.

We say that a projective-like pointclass Γ belongs to the hierarchy $\{\Gamma_i\}$ if $\Gamma = \Gamma_i$ or $\Gamma = \check{\Gamma}_i$, for some i.

Again $\underset{\sim}{\Pi}^1_1, \underset{\sim}{\Sigma}^1_2, \underset{\sim}{\Pi}^1_3, \underset{\sim}{\Sigma}^1_4, \ldots$ is the prototype of this notion.

Note first that each projective-like Γ belongs to exactly one projective-like hierarchy $\{\Gamma_i\}$. This is easy to see from the following two facts, where we note that $\Game\Gamma = \exists^{\mathbb{R}}\Gamma$, if $\forall^{\mathbb{R}}\Gamma \subseteq \Gamma$, and $\Game\Gamma = \forall^{\mathbb{R}}\Gamma$, if $\exists^{\mathbb{R}}\Gamma \subseteq \Gamma$ ($\Game$ is the game quantifier) and we define $w(\Gamma) \equiv w(A)$, for each $A \in \Gamma - \Delta$:

(i) $w(\Gamma) < w(\Game\Gamma)$

(ii) If Γ', Γ'' are projective-like and $\Game\Gamma' = \Game\Gamma''$, then $\Gamma' = \Gamma''$ (otherwise, by Wadge, $\Gamma' \subseteq \Delta''$ or $\Gamma'' \subseteq \Delta'$, so let us say in the first case, $\Game\Gamma'' = \Game\Gamma' \subseteq \Game\check{\Gamma}'' = $ (by AD) $\widecheck{(\Game\Gamma'')}$, a contradiction).

We shall classify now all the projective-like hierarchies into four types I-IV. Let us recall a basic fact about Wadge degrees first.

Call $A \subseteq \mathbb{R}$ <u>self-dual</u> iff $A \leq_w \sim A$. (The Wadge reducibility $\leq_w$ is defined by $A \leq_w B \Leftrightarrow \exists f : \mathbb{R} \to \mathbb{R}(f \text{ continuous} \wedge f^{-1}[B] = A)$). Then Steel-Van Wesep (see Van Wesep [1978]) prove: If $w(A)$ is limit, then

$$A \text{ is self-dual iff } w(A) \text{ has cofinality } \omega .$$

We describe now the classification. (AD is assumed throughout.)

<u>TYPE I.</u> $\{\Gamma_i\}$ is such that $\Gamma_1 = \boldsymbol{\Pi}^1_1(A)$, for some A with $\mathrm{cof}(w(A)) = \omega$ and with the property that $\Lambda = \{B : w(B) < w(A)\}$ is strongly closed (recall the beginning of 2.3 here).

We visualize this by the following picture in the Wadge hierarchy:

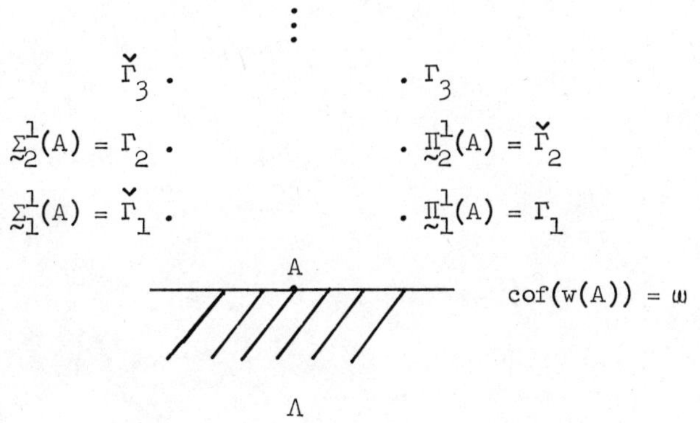

By convention we accept the case $w(A) = 0$ as being within this type, so that the classical projective hierarchy is of Type I. The next example is generated by taking A to be a set of least Wadge ordinal above the projective sets, so that $\Lambda = \{B : w(B) < w(A)\} = \bigcup_n \boldsymbol{\Delta}^1_n$.

<u>TYPE II.</u> $\{\Gamma_i\}$ is such that $\Gamma_1 = \boldsymbol{\Pi}^1_1(A)$ for some A with $\mathrm{cof}(w(A)) > \omega$, with $\sim A \in \boldsymbol{\Pi}^1_1(A)$, and $\Lambda = \{B : w(B) < w(A)\}$ strongly closed.

The relevant picture looks now as follows:

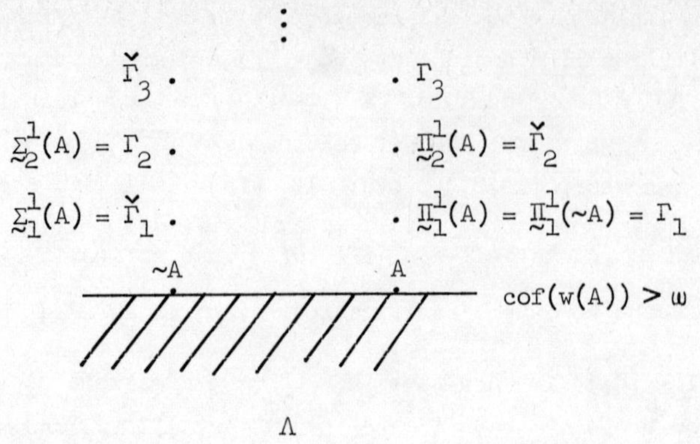

The smallest example of such a hierarchy is constructed as follows: Let Γ_η, $1 \leq \eta < \omega_1$ be defined by

$$\Gamma_1 = \Pi^1_1 ,$$
$$\Gamma_{\eta+1} = \Game\Gamma_\eta ,$$
$$\Gamma_\lambda = \text{all countable unions of sets in } \bigcup_{\eta<\lambda} \Gamma_\eta, \text{ if } \lambda = \bigcup\lambda > 0 .$$

Let then A be such that

$$\Lambda = \{B : w(B) < w(A)\} = \bigcup_{\eta<\omega_1} \Gamma_\eta .$$

Since $\mathrm{cof}(w(A)) = \omega_1$ it follows that $\{B : B \leq_w A\}$ is not closed under both intersections and unions (see for example Steel [1980]), so $\Pi^1_1(A) = \Pi^1_1(\sim A)$, thus $\sim A \in \Pi^1_1(A)$.

<u>TYPE III</u>. $\{\Gamma_i\}$ is such that $\Delta_1 = \Gamma_1 \cap \check{\Gamma}_1$ is strongly closed.

Picture:

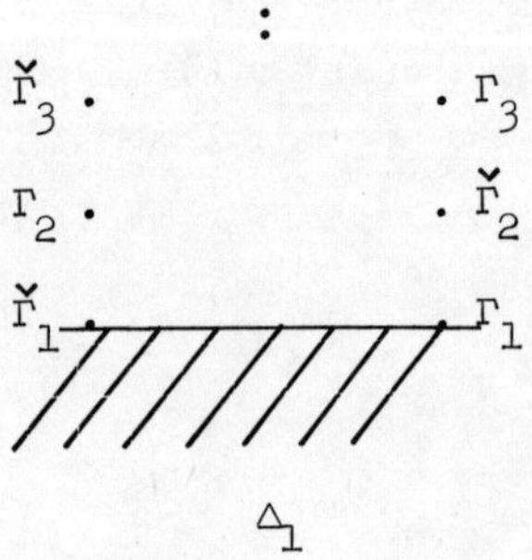

A typical example (the smallest as it follows from the results in §5)

is $\Gamma_1 =_2 \underset{\sim}{\text{ENV}}({}^3E)$ ≡ the pointclass of sets Kleene semirecursive in ${}^3E$ and a real.

<u>TYPE IV.</u> $\{\Gamma_i\}$ is such that $\Gamma_1 = \forall^{\mathbb{R}}(\Gamma \vee \check{\Gamma})$, where Γ is inductive-like, and $\Gamma \vee \check{\Gamma} = \{A \cup B : A \in \Gamma \wedge B \in \check{\Gamma}\}$.

Picture:

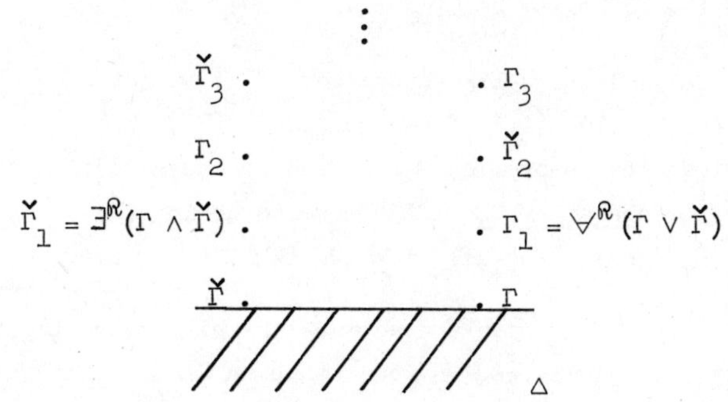

Again the smallest example is constructed by taking $\Gamma = \underset{\sim}{\text{IND}}(\mathbb{R})$.

We have now the following

4.1 <u>Theorem</u> (AD). Every projective-like hierarchy is of exactly one of the types I-IV.

<u>Proof.</u> As no projective-like hierarchy can be of two different types, it is enough to prove that every projective-like hierarchy $\{\Gamma_i\}$ is of one of the types I-IV.

Let Λ be the largest strongly closed pointclass contained in Δ_1. Let $\lambda = \sup\{w(A) : A \in \Lambda\}$. Then exactly one of the four possibilities below must hold:

I. $\text{cof}(\lambda) = \omega$

II. $\text{cof}(\lambda) > \omega$ and if A is such that $w(A) = w(\sim A) = \lambda$ then $\{B : B \leq_w A\}$ (and thus $\{B : B \leq_w \sim A\}$) is neither projective-like nor inductive-like).

III. $\text{cof}(\lambda) > \omega$ and if $A, \sim A$ are as above, then $\{B : B \leq_w A\}$ (and thus $\{B : B \leq_w \sim A\}$) is projective-like.

IV. $\text{cof}(\lambda) > \omega$ and if $A, \sim A$ are as above, then $\{B : B \leq_w A\}$ (and thus $\{B : B \leq_w \sim A\}$) is inductive-like.

If I holds, let A be such that $w(A) = \lambda$. By the maximality of Λ we must have $\underset{\sim}{\Pi}^1_1(A) = \Gamma_1$, so that $\{\Gamma_i\}$ is of type I. If II holds then (by Wadge) $\sim A \in \underset{\sim}{\Pi}^1_1(A)$, and $\Gamma_1 = \underset{\sim}{\Pi}^1_1(A)$, so that $\{\Gamma_i\}$ is of type II. If III holds and without loss of generality $\{B : B \leq_w A\}$ is closed under $\forall^{\mathbb{R}}$ but not $\exists^{\mathbb{R}}$, then $\Gamma_1 = \{B : B \leq_w A\}$ and $\Delta_1 = \Lambda$, so that $\{\Gamma_i\}$ is of type III. Finally, if IV holds and $\Gamma = \{B : B \leq_w A\}$, we have $\Gamma_1 = \forall^{\mathbb{R}}(\Gamma \vee \check{\Gamma})$ so that $\{\Gamma_i\}$ is of type IV. ⊣

§5. The prewellordering pattern in projective-like hierarchies. We conclude now our analysis by establishing (under certain assumptions) that each type of projective-like hierarchy has only one prewellordering pattern. For convenience let us introduce the following terminology.

Definition. A projective-like hierarchy $\{\Gamma_i\}$ is of character Π iff Prewellordering (Γ_1) holds (iff Prewellordering (Γ_i) holds for each $i \geq 1$, by the First Periodicity Theorem of Martin and Moschovakis). A projective-like hierarchy is of character Σ iff Prewellordering $(\check{\Gamma}_1)$ holds (iff Prewellordering $(\check{\Gamma}_i)$ holds for all $i \geq 1$).

Definition. The ground of a projective-like hierarchy $\{\Gamma_i\}$ is defined to be the largest strongly closed Λ such that $\Lambda \subseteq \Delta_1$. Thus the ground Λ coincides with the strongly closed Λ occurring in the definition of $\{\Gamma_i\}$ being of type I or II, with Δ_1 if $\{\Gamma_i\}$ is of type III, and with $\Delta = \Gamma \cap \check{\Gamma}$ if $\{\Gamma_i\}$ is of type IV and Γ is given in the definition of this type.

We have now

5.1 Theorem (AD). Let $\{\Gamma_i\}$ be a projective-like hierarchy. Then

(i) If $\{\Gamma_i\}$ is of type I, then $\{\Gamma_i\}$ has character Π.

(ii) If $\{\Gamma_i\}$ is of type II, then $\{\Gamma_i\}$ has character Σ, provided its ground is not a projective algebra.

(iii) If $\{\Gamma_i\}$ is of type III, then $\{\Gamma_i\}$ has character Π iff its ground is not a projective algebra.

(iv) If $\{\Gamma_i\}$ is of type IV, then $\{\Gamma_i\}$ has character Σ, provided its ground is not a projective algebra.

5.2 Corollary (AD). Let $\{\Gamma_i\}$ be a projective-like hierarchy contained in $\underset{\sim}{P}_\infty$. Then

(i) If $\{\Gamma_i\}$ is of type I, III, then $\{\Gamma_i\}$ has character Π.

(ii) If $\{\Gamma_i\}$ is of type II, IV, then $\{\Gamma_i\}$ has character Σ.

In particular, this holds if $\{\Gamma_i\}$ is contained in $L[\mathbb{R}]$.

5.3 Corollary (AD). Let Γ be a boldface pointclass closed under countable intersections and unions and $\exists^{\mathbb{R}}$ or $\forall^{\mathbb{R}}$, but not complements. If $\Gamma \subseteq \underset{\sim}{P}_\infty$, in particular if $\Gamma \subseteq L[\mathbb{R}]$, then Prewellordering (Γ) or Prewellordering $(\check{\Gamma})$.

Proof of 5.1. (i) Let A be as in the definition of a type I hierarchy so that in particular $\Gamma_1 = \underset{\sim}{\Pi}^1_1(A)$. Let $\Lambda = \{B : w(B) < w(A)\}$. Since $\mathrm{cof}(w(A)) = \omega$, Λ is not closed under countable unions so if

$$\Gamma_0 = \{\bigcup_n A_n : \forall n\,(A_n \in \Lambda)\},$$

then Γ_0 is a boldface pointclass, closed under intersections and unions, countable unions and $\exists^{\mathbb{R}}$. Moreover we have Prewellordering (Γ_0). Since $\Gamma_1 = \forall^{\mathbb{R}}\Gamma_0$, we have Prewellordering (Γ_1) by the First Periodicity Theorem. (i)⊣

(iii) By 2.1 and the fact that Δ_1 is not a projective algebra it is enough to show Reduction (Γ_1) and by Steel's Theorem it is enough to show $\neg$ Separation (Γ_1). This is immediate from the following

Lemma. If Γ is a boldface pointclass not closed under complements, and Δ is closed under $\exists^{\mathbb{R}}, \forall^{\mathbb{R}}$ then

$$\text{Separation}\,(\Gamma) \Rightarrow \exists^{\mathbb{R}}\Gamma \subseteq \Gamma .$$

Proof. Let P, Q be disjoint sets in $\exists^{\mathbb{R}}\Gamma$. Let

$$P(x) \Leftrightarrow \exists\alpha(x,\alpha) \in A, \quad Q(x) \Leftrightarrow \exists\beta(x,\beta) \in B ,$$

where $A, B \in \Gamma$ are also disjoint. Put

$$A'(x,\alpha,\beta) \Leftrightarrow A(x,\alpha)$$

$$B'(x,\alpha,\beta) \Leftrightarrow B(x,\beta) .$$

Then $A', B' \in \Gamma$ are disjoint so find $C \in \Delta$ separating them. Then

$$S(x) \Leftrightarrow \exists\alpha\forall\beta(x,\alpha,\beta) \in C$$

is in Δ and separates P, Q (this is a variant of an argument of Addison).

So if $\exists^{\mathbb{R}}\Gamma \not\subseteq \Gamma$, then by Wadge $\check{\Gamma} \subseteq \exists^{\mathbb{R}}\Gamma$, thus if $A \in \Gamma$ then $P = A$, $Q = {\sim}A$ are disjoint in $\exists^{\mathbb{R}}\Gamma$ and thus can be separated by $S \in \Delta$. So $A = S \in \Delta$, a contradiction. (iii)⊣

(iv) Let Γ be the inductive-like pointclass appearing in the definition of a projective-like hierarchy of type IV, so that $\check{\Gamma}_1 = \exists^{\mathbb{R}}(\Gamma \wedge \check{\Gamma})$. Assume without loss of generality that Reduction (Γ) holds. Then by 2.1 Prewellordering (Γ) holds. Then $\Pi = \Gamma \wedge \check{\Gamma}$ is a boldface pointclass closed under intersections and $\forall^{\mathbb{R}}$ and has the prewellordering property. (If $A \in \Gamma$, $B \in \check{\Gamma}$ and φ is a Γ-norm on A, then $\varphi \restriction A \cap B$ is a $\Gamma \wedge \check{\Gamma}$-norm on $A \cap B$.) So by the usual proof $\check{\Gamma}_1 = \exists^{\mathbb{R}}(\Gamma \wedge \check{\Gamma})$ has the prewellordering property. (iv)⊣

(ii) Let A be as in the definition of type II hierarchy so that $\Gamma_1 = \underset{\sim}{\Pi}^1_1(A)$ and A is chosen (from $A, \sim A$) so that

$$\Pi = \{B : B \leq_w A\}$$

has the reduction property. Put $\Sigma = \check{\Pi}$. Then by the lemma in (iii) Π is closed under $\forall^{\mathcal{R}}$ and hence, since Π is $\mathcal{R}$-parametrized, under countable intersections. Thus $\underset{\sim}{\Sigma}^1_1(A) = \exists^{\mathcal{R}}\Pi$.

Since $\Lambda = \{B : w(B) < w(A)\}$ is not a projective algebra, it is not closed under wellordered unions. Let θ be least such that some θ-sequence of elements of Λ has union outside Λ. (Again θ is a regular cardinal $> \omega$, since Λ is closed under countable unions as $w(A)$ has cofinality $> \omega$.) Let $\lambda = w(A) = \sup\{w(B) : B \in \Lambda\}$. As in 2.3 $\theta \leq \lambda$. Put

$$\Gamma' = \{\cup_{\xi<\theta} A_\xi : \forall\xi < \theta(A_\xi \in \Lambda)\} .$$

Then Γ' is a boldface pointclass closed under countable intersections and unions and $\exists^{\mathcal{R}}$. So $\Gamma' \supseteq \underset{\sim}{\Sigma}^1_1(A)$. If $\Gamma' = \underset{\sim}{\Sigma}^1_1(A) = \check{\Gamma}_1$ we are done, since clearly Prewellordering (Γ').

Otherwise, $\Gamma' \supsetneq \underset{\sim}{\Sigma}^1_1(A)$, so by Wadge, $\Gamma' \supseteq \underset{\sim}{\Pi}^1_1(A) = \Gamma_1$, thus $\Gamma' \supseteq \Gamma_2 = \exists^{\mathcal{R}}\Gamma_1$. Now notice that there is a wellfounded relation $<$ of rank $\geq \lambda$ ($\geq \theta$) in Γ_2. Indeed, let

$$(\sigma,\alpha) < (\sigma',\beta) \Leftrightarrow \sigma = \sigma' \text{ codes a continuous function } f_\sigma \ \& \ f_\sigma^{-1}[A] \text{ is wellfounded } \& \ (\alpha,\beta) \in f_\sigma^{-1}[A] .$$

So by Moschovakis [1980], Ch. 7, Γ_2 is closed under wellordered unions of length rank($<$), thus $\Gamma' \subseteq \Gamma_2$. So

$$\Gamma' = \Gamma_2 .$$

Put now

$$\Gamma^* = \{\cup_{\xi<\theta} A_\xi : \forall\xi < \theta(A_\xi \in \Lambda) \wedge \{A_\xi\}_{\xi<\theta} \text{ is } \underset{\sim}{\Sigma}^1_1\text{-bounded}\} .$$

Note first that Γ^* is closed under $\forall^{\mathcal{R}}$. The argument is similar to that in 2.5.2. Indeed let $(x,\alpha) \in B \Leftrightarrow \exists\xi < \theta(x,\alpha) \in B_\xi$, where $\{B_\xi\}_{\xi<\theta}$ is $\underset{\sim}{\Sigma}^1_1$-bounded. Let $x \in C \Leftrightarrow \forall\alpha\,(x,\alpha) \in B$. Then as in 2.5.2 $x \in C \Leftrightarrow \exists\xi < \theta\,\forall\alpha\,(x,\alpha) \in B_\xi$. Let $C_\xi = \{x : \forall\alpha(x,\alpha) \in B_\xi\}$. It is enough to show that $\{C_\xi\}_{\xi<\theta}$ is $\underset{\sim}{\Sigma}^1_1$-bounded. Let $X \in \underset{\sim}{\Sigma}^1_1$, $X \subseteq \cup_{\xi<\theta} C_\xi$. Then $\forall x \in X\,\exists\xi\,\forall\alpha\,(x,\alpha) \in A_\xi$, thus $\forall x \in X\,\forall\alpha\,\exists\xi(x,\alpha) \in B_\xi$, so $X \times \mathcal{R} \subseteq B$ thus for some $\xi < \theta$, $X \times \mathcal{R} \subseteq B_\xi$, therefore $\forall x \in X\,\forall\alpha\,(x,\alpha) \in B_\xi$ i.e. $X \subseteq C_\xi$, and we are done.

So if $\Gamma^* \nsubseteq \Lambda$ then either $\Gamma^* = \Pi$ in which case we have Prewellordering (Π) (as in 2.5.1) thus Prewellordering ($\check{\Gamma}_1$) since $\check{\Gamma}_1 = \exists^{\mathcal{R}}\Pi$, and we are done, or else $\Gamma^* \supseteq \Sigma$ so $\Gamma^* \supseteq \forall^{\mathbb{R}}\Sigma = \check{\Gamma}_1$. Then we must have $\Gamma^* = \Gamma_1$ (since otherwise $\Gamma^* \supseteq \check{\Gamma}_1$, so $\Gamma^* \supseteq \forall^{\mathbb{R}}\check{\Gamma}_1 = \check{\Gamma}_2$, contradicting the fact that $\Gamma^* \subseteq \Gamma' = \Gamma_2$). So if $W \in \Gamma_1 - \Delta_1$ we can write $W = \bigcup_{\xi<\theta} W_\xi$, with $W_\xi \in \Lambda$ and $\{W_\xi\}_{\xi<\theta}$ a $\underset{\sim}{\Sigma}{}^1_1$-bounded union. Let

$$\varphi(x) = \text{least } \xi,\ x \in W_\xi$$

be the associated norm, which by the argument in 2.5.1 is a $(\Gamma^* =)\ \Gamma_1$-norm. But $A \in \Delta_1$, so let f continuous be such that $x \in A \Leftrightarrow f(x) \in W$. By the usual boundedness argument (recall that Γ_1 is closed under $\forall^{\mathbb{R}}$) there is $\xi < \theta$, with $x \in A \Leftrightarrow f(x) \in W_\xi$, so $A \in \Lambda$, a contradiction.

So we can complete the proof by showing that $\Gamma^* \not\subseteq \Lambda$. For that we use the same argument as in 2.5.3. Let $\mathcal{S}$ be universal $\underset{\sim}{\Sigma}{}^1_1$ and put

$$C = \{\varepsilon : \mathcal{S}_\varepsilon \subseteq A\}\ .$$

Then $C \in \underset{\sim}{\Pi}{}^1_1(A) = \Gamma_1 \subseteq \Gamma'$, so $C = \bigcup_{\xi<\theta} C_\xi$, where each $C_\xi \in \Lambda$. Put

$$A_\xi = \bigcup\{\mathcal{S}_\varepsilon : \varepsilon \in C_\xi\}\ .$$

Then $A_\xi \in \Lambda$, $\{A_\xi\}_{\xi<\theta}$ is $\underset{\sim}{\Sigma}{}^1_1$-bounded, and $A = \bigcup_{\xi<\theta} A_\xi \notin \Lambda$, so we are done.$^{(iii)}\dashv$

5.1.$\dashv$

An alternative approach to the proof of 5.1 is given in §3 of Steel [A].

§6. Problems and conjectures. Assume AD in this section. Let $\Lambda \subsetneq \underset{\sim}{P}_\infty$ be a strongly closed boldface pointclass and $\{\Gamma_i\}$ the first projective-like hierarchy not contained in Λ or in other words the projective-like hierarchy with ground Λ. Let λ be the ordinal associated with Λ as in 2.3.1, i.e. the supremum of the ordinals $w(B)$ for $B \in \Lambda$. The question is whether λ determines the type of $\{\Gamma_i\}$ (λ is always a limit cardinal).

Clearly if $\text{cof}(\lambda) = \omega$, $\{\Gamma_i\}$ is of type I. It is easy also to see that if $\text{cof}(\lambda) > \omega$ and λ is singular then $\{\Gamma_i\}$ is of type II (otherwise $\Lambda = \Delta_1$ and Prewellordering (Γ_1) for a type III $\{\Gamma_i\}$, so since $\forall^{\mathbb{R}}\Gamma_1 = \Gamma_1$, we have by Moschovakis [1970] that $\lambda = \delta_1 \equiv \sup\{\xi : \xi$ is the rank of a Δ_1 prewellordering$\}$, thus λ is regular). We offer now the following conjecture:

If λ is regular, then $\{\Gamma_i\}$ is of type III or IV.

(As pointed out above the converse is true.) As positive evidence we consider the fact that this conjecture is true for at least $\Lambda \subseteq \underset{\sim}{\text{IND}}(\mathbb{R})$, as shown by Steel [A].

It is not clear what property of a regular λ would distinguish between types III and IV. However the results in §5(A) of Kechris [A] suggest that some combination of indescribability and Mahlo properties of λ could guarantee that

$\{\Gamma_i\}$ is of type IV.

And we conclude with one more conjecture and two problems concerning closure of classes ander wellordered unions:

(a) Conjecture: If Γ is inductive-like and has the prewellordering property, then Γ is closed under wellordered unions (compare with 2.4.1).

(b) If Γ is projective-like, can Δ be closed under wellordered unions?

(c) Can the ground of type II or III hierarchies be a projective algebra?

Appendix A

We give here a proof of the 0th Periodicity Theorem. Assume below AD. Let Γ be a boldface pointclass closed under countable unions and intersections, which has the reduction property. We show that

(i) $\exists^{\mathbb{R}}\Gamma \subseteq \Gamma \Rightarrow \text{Reduction}(\forall^{\mathbb{R}}\Gamma)$,

(ii) $\forall^{\mathbb{R}}\Gamma \subseteq \Gamma \Rightarrow \text{Reduction}(\exists^{\mathbb{R}}\Gamma)$.

Take first (i). Assume $\exists^{\mathbb{R}}\Gamma \subseteq \Gamma$. Then note that $\forall^{\mathbb{R}}\Gamma = \{\supset\!\!\!\!\partial\alpha A(x,\alpha) : A \in \Gamma$ and A is Turing invariant on $\alpha\}$, where $A(x,\alpha)$ is Turing invariant on α iff $\alpha \equiv_T \beta \wedge A(x,\alpha) \Rightarrow A(x,\beta)$. This is because $\forall\alpha B(x,\alpha) \Leftrightarrow \supset\!\!\!\!\partial\alpha\forall\beta \leq_T \alpha B(x,\beta)$. So let P, Q be in $\forall^{\mathbb{R}}\Gamma$ and say $x \in P \Leftrightarrow \supset\!\!\!\!\partial\alpha(x,\alpha) \in A$, $x \in Q \Leftrightarrow \supset\!\!\!\!\partial\alpha(x,\alpha) \in B$, with $A,B \in \Gamma$ and Turing invariant on α. By Burgess and Miller [1975], we can find $A_1,B_1 \in \Gamma$ Turing invariant reducing A, B. [Indeed let $A',B' \in \Gamma$ reduce A, B and then put

$$A_1(x,\alpha) \Leftrightarrow \exists\alpha' \equiv_T \alpha A'(x,\alpha')$$

$$B_1(x,\alpha) \Leftrightarrow \forall\alpha' \equiv_T \alpha B'(x,\alpha).]$$

Now let $x \in P_1 \Leftrightarrow \supset\!\!\!\!\partial\alpha(x,\alpha) \in A_1$ and $x \in Q_1 \Leftrightarrow \supset\!\!\!\!\partial\alpha(x,\alpha) \in B_1$. Then $P_1,Q_1 \in \forall^{\mathbb{R}}\Gamma$ and they reduce P, Q.

The proof of (ii) is similar utilizing the equivalence $\exists\alpha B(x,\alpha) \Leftrightarrow \supset\!\!\!\!\partial\alpha\exists\beta \leq_T \alpha$ $B(x,\beta)$.

Appendix B

We prove here the fact that

$$AD + V = L[\mathbb{R}] \Rightarrow \underset{\sim}{B}_\infty = \text{power}(\mathbb{R}) .$$

Assume $AD + V = L[\mathbb{R}]$. Assume also that $\underset{\sim}{B}_\infty \neq \text{power}(\mathbb{R})$ towards a contradiction. Pick then $A \subseteq \mathbb{R}$ such that $A \notin \underset{\sim}{B}_\infty$. As every set of reals is ordinal definable from a real, let $\varphi(x,\xi,\alpha)$ be a formula and $\alpha \in \mathbb{R}$, $\xi_0 \in \text{ORD}$ be such that

$$x \in A \Leftrightarrow \varphi(x,\xi_0,\alpha_0) .$$

Fix now α_0 and pick the least ordinal ξ such that $\{x \in \mathbb{R} : \varphi(x,\xi,\alpha_0)\} \notin \underset{\sim}{B}_\infty$. Clearly this ξ is definable from α_0, so we conclude that there is a set of reals B definable from α_0, say

$$x \in B \Leftrightarrow \psi(x,\alpha_0) ,$$

such that $B \notin \underset{\sim}{B}_\infty$. By Skolem-Lowenheim let $\lambda < \Theta$ be least such that

$$L_\lambda[\mathbb{R}] \models ZF_N + DC + AD + \{x : \psi(x,\alpha_0)\} \notin \underset{\sim}{B}_\infty .$$

Here ZF_N is a large enough finite fragment of ZF. Let

$$C = \{x \in \mathbb{R} : L_\lambda[\mathbb{R}] \models \psi(x,\alpha_0)\} .$$

We have that

$$L_\lambda[\mathbb{R}] \models C \notin \underset{\sim}{B}_\infty .$$

But then we claim that actually $C \notin \underset{\sim}{B}_\infty$. Indeed, if $\Delta = \{D \subseteq \mathbb{R} : L_\lambda[\mathbb{R}] \models D \in \underset{\sim}{B}_\infty\}$ and

$$\delta = \sup\{\eta : \eta \text{ is the length of a } \Delta \text{ prewellordering of } \mathbb{R}\} ,$$

then Δ is closed under wellordered unions of length δ. To see this let $\{A_\xi\}_{\xi<\delta}$ be a sequence of Δ sets. Let $S \in L_\lambda[\mathbb{R}]$ be such that all $D \in \Delta$ are Wadge reducible to S (S exists since $L_\lambda[\mathbb{R}] \models \underset{\sim}{B}_\infty \neq \text{power}(\mathbb{R})$). Let $C_\xi = \{\varepsilon : \varepsilon$ codes a continuous function f_ε and $f_\varepsilon^{-1}[S] = A_\xi\}$. Notice now that there is a norm $\chi : \mathbb{R} \twoheadrightarrow \delta$ in $L_\lambda[\mathbb{R}]$, thus by the Moschovakis Coding Lemma there is a function h in $L_\lambda[\mathbb{R}]$ such that $\forall \xi < \delta(h(\xi) \neq \emptyset \wedge h(\xi) \subseteq C_\xi)$. Consequently, $\{A_\xi\}_{\xi<\delta} = \{\bigcap_{\varepsilon \in h(\xi)} f_\varepsilon^{-1}[S]\}_{\xi<\delta} \in L_\lambda[\mathbb{R}]$, so $\bigcup_{\xi<\delta} A_\xi \in \Delta$.

Since Δ is closed under δ unions, clearly Δ is closed under arbitrary wellordered unions, thus $\Delta \supseteq \underset{\sim}{B}_\infty$, so $C \notin \underset{\sim}{B}_\infty$.

We shall complete the proof by showing that $C \in \underset{\sim}{\Delta}{}^2_1$. This leads immediately to a contradiction since $\underset{\sim}{\Sigma}^2_1 = \underset{\sim}{S}_\infty$ and so $C \in \underset{\sim}{B}_\infty$. To see that $C \in \underset{\sim}{\Delta}^2_1$ notice that

$$x \in C \Leftrightarrow \exists\lambda < \Theta[L_\lambda[\mathbb{R}] \models \text{"}ZF_N + DC + AD + \{x : \psi(x,\alpha_0)\} \notin \underset{\sim}{B}_\infty\text{"} \wedge \lambda$$
$$\text{is least with that property} \wedge L_\lambda[\mathbb{R}] \models \psi(x,\alpha_0)] .$$

As structures $L_\lambda[\mathbb{R}]$, for $\lambda < \Theta$, can be coded in a straightforward fashion by sets of reals, this shows that $C \in \underset{\sim}{\Sigma}^2_1$ and a similar computation shows that $C \in \underset{\sim}{\Pi}^2_1$, so we are done.

References

J. W. Addison and Y. N. Moschovakis [1968], Some consequences of the axiom of definable determinateness, Proc. Nat. Acad. Sci. USA, 59 (1968), 708-712.

J. Burgess and D. Miller [1975], Remarks on invariant descriptive set theory, Fund. Math. XC (1975), 53-75.

A. S. Kechris [1974], On projective ordinals, J. Symb. Logic, 39 (1974), 269-282.

A. S. Kechris [1977], Classifying projective-like hierarchies, Bull. Greek Math. Soc., 18 (1977), 254-275.

A. S. Kechris [A], Souslin cardinals, κ-Souslin sets and the scale property in the hyperprojective hierarchy, this volume.

D. A. Martin [1968], The axiom of determinateness and reduction principles in the analytical hierarchy, Bull. Amer. Math. Soc. 74 (1968), 687-689.

D. A. Martin [198?], Projective sets and cardinal numbers: Some questions related to the continuum problem, J. Symb. Logic, to appear.

Y. N. Moschovakis [1970], Determinacy and prewellorderings of the continuum, Math. Logic and Found. of Set Theory, Y. Bar-Hillel Ed., North Holland, (1970), 24-62.

Y. N. Moschovakis [1974], Elementary Induction on Abstract Structures, North Holland, (1974).

Y. N. Moschovakis [1980], Descriptive Set Theory, North Holland, (1980).

J. R. Shoenfield [1967], Mathematical Logic, Addison-Wesley, (1967).

R. M. Solovay [1978], The independence of DC from AD, Cabal Seminar 76-77, Lecture Notes in Mathematics, Springer-Verlag, Vol. 689 (1978), 171-184.

J. R. Steel [1980], Determinateness and the separation property, J. Symb. Logic, to appear.

J. R. Steel [A], Closure properties of pointclass, this volume.

J. R. Steel and R. Van Wesep [1981], Two consequences of determinacy consistent with choice, Trans. Amer. Math. Soc., to appear.

R. Van Wesep [1978a], Wadge degrees and descriptive set theory, Cabal Seminar 76-77, Lecture Notes in Mathematics, Springer-Verlag, Vol. 689 (1978), 171-184.

R. Van Wesep [1978b], Separation principles and the axiom of determinateness, Jour. Symb. Logic, 43 (1978), 77-81.

SOUSLIN CARDINALS, κ-SOUSLIN SETS AND THE SCALE PROPERTY IN THE HYPERPROJECTIVE HIERARCHY

Alexander S. Kechris[1]
Department of Mathematics
California Institute of Technology
Pasadena, California 91125

Let $\Theta = \sup\{\xi : \xi$ is the length of a prewellordering of the set of reals $\mathbb{R}(= \omega^\omega)\}$. Let $\kappa < \Theta$ be an infinite cardinal The class $\underset{\sim}{S}(\kappa)$ of κ-Souslin sets has some well-known closure properties, i.e. it is closed under continuous substitutions, countable intersections and unions, and existential quantification over $\mathbb{R}$. We investigate in §1-2 the question whether $\underset{\sim}{S}(\kappa)$ is $\mathbb{R}$-parametrized or not (i.e. whether $\underset{\sim}{S}(\kappa)$ admits universal sets), assuming AD. Let us call a cardinal κ <u>Souslin</u> iff there is a new κ-<u>Souslin</u> set i.e. $\underset{\sim}{S}(\kappa) - \bigcup_{\lambda<\kappa} \underset{\sim}{S}(\lambda) \neq \emptyset$. Let $\kappa_0, \kappa_1, \kappa_2, \ldots, \kappa_\xi, \ldots$ be the increasing enumeration of the Souslin cardinals (the first few of them are $\kappa_0 = \omega$, $\kappa_1 = \omega_1$, $\kappa_2 = \omega_\omega$, $\kappa_3 = \omega_{\omega+1}, \ldots$). We show in §1 that $\underset{\sim}{S}(\kappa)$ is $\mathbb{R}$-parametrized iff there is a largest Souslin cardinal $\leq \kappa$. Thus $\underset{\sim}{S}(\kappa)$ is $\mathbb{R}$-parametrized for all $\kappa < \Theta$ iff the sequence $\{\kappa_\xi\}$ is normal (and has a largest element if bounded below Θ). We conjecture that this is indeed the case, and in §2 we verify this conjecture at least below $\kappa^{\mathbb{R}}$ = the first non-hyperprojective ordinal, so that $\underset{\sim}{S}(\kappa)$ is always $\mathbb{R}$-parametrized for $\kappa \leq \kappa^{\mathbb{R}}$.

In §3 we study the hyperprojective hierarchy $\{\underset{\sim}{\Sigma}^1_\xi, \underset{\sim}{\Pi}^1_\xi, \underset{\sim}{\Delta}^1_\xi\}_{\xi<\kappa^{\mathbb{R}}}$ and, using AD again, we establish that the scale property propagates throughout this hierarchy following the pattern established for the prewellordering property in Kechris-Solovay-Steel [A]. In particular, it follows that if Γ is a projective-like pointclass, contained in the inductive sets, then either Γ or $\check{\Gamma}$ has the scale property. This smooth propagation of scales breaks down immediately past the inductive sets and this "gap phenomenon" is discussed briefly in §4. Finally §5 contains a number of open problems, conjectures and remarks related to the preceding work.

This paper draws heavily on the terminology, notation and results established in Kechris-Solovay-Steel [A]. We refer also to Moschovakis [1980] for the basic results from descriptive set theory that we use.

Our underlying theory is ZF + DC, and any further assumptions (mainly AD)

---

[1] Research partially supported by NSF Grant MCS79-20465. The author is an A. P. Sloan Foundation Fellow.

are explicitly indicated. Of course those of our results wiich are also absolute for the inner model $L[\mathbb{R}]$, although stated as proved from AD, need only $AD^{L[\mathbb{R}]}$ and in fact (in most cases) much weaker forms of definable determinacy, like Hyperprojective Determinacy, etc. An interested and patient reader should have no trouble figuring out how much determinacy is needed in each case.

§1. Souslin cardinals and κ-Souslin sets. Let κ be an infinite ordinal. A set $A \subseteq \mathbb{R}$ is κ-Souslin if there is a tree T on $\omega \times \kappa$ such that $A = p[T] \equiv \{\alpha \in \mathbb{R} : \exists f \in \kappa^{\omega}(\alpha, f) \in [T]\}$. Similar definitions apply to pointsets $A \subseteq \mathcal{X}$ contained in arbitrary product spaces. Denote by $\underset{\sim}{S}(\kappa)$ the pointclass of κ-Souslin pointsets. It is immediate that $\underset{\sim}{S}(\kappa) = \underset{\sim}{S}(\kappa')$, where κ' is the cardinality of κ, so that it is enough to consider only $\underset{\sim}{S}(\kappa)$, when κ is a cardinal.

The class $\underset{\sim}{S}(\kappa)$ has some well-known closure properties which we summarize below.

1.1 Proposition. The pointclass $\underset{\sim}{S}(\kappa)$ is closed under continuous substitutions, countable unions and intersections and $\exists^{\mathbb{R}}$.

The obvious question now is whether $\underset{\sim}{S}(\kappa)$ is $\mathbb{R}$-parametrized i.e. has universal sets. We provide below a sufficient condition for this to be true. For this we need to introduce first the following notion.

Definition. A cardinal is called Souslin iff $\underset{\sim}{S}(\kappa) - \bigcup_{\lambda<\kappa} \underset{\sim}{S}(\lambda) \neq \emptyset$ i.e. there is a κ-Souslin set which is not λ-Souslin for any $\lambda < \kappa$.

Assuming AD, let $\kappa_0, \kappa_1, \kappa_2, \ldots, \kappa_\xi, \ldots$, $\xi < \Xi$ enumerate the Souslin cardinals. Thus

$$\kappa_0 = \omega,\ \kappa_1 = \omega_1 = \underset{\sim}{\delta}^1_1,\ \kappa_2 = \omega_\omega,\ \kappa_3 = \omega_{\omega+1} = \underset{\sim}{\delta}^1_3,\ \ldots .$$

The only Souslin cardinals below $\sup_{n<\omega} \underset{\sim}{\delta}^1_n$ are λ_{2n+1}, $\underset{\sim}{\delta}^1_{2n+1}$, where $(\lambda_{2n+1})^+ = \underset{\sim}{\delta}^1_{2n+1}$. We have now the following main result.

1.2 Theorem (AD). Let κ be a Souslin cardinal. Then $\underset{\sim}{S}(\kappa)$ is $\mathbb{R}$-parametrized.

Proof. Assume not towards a contradiction. Then, by Wadge, $\underset{\sim}{S}(\kappa) \equiv \Lambda$ is closed under complements, so also under $\forall^{\mathbb{R}}$ i.e. it is strongly closed in the sense of Kechris-Solovay-Steel [A], §2.3. By the results of that paper (especially 2.4.1) if $M \subsetneq \Lambda$ then there is a pointclass Γ, with $M \subsetneq \Gamma \subsetneq \Lambda$ such that Γ is closed under wellordered unions. We shall use this fact repeatedly below.

Note first that it is sufficient to show that cofinality$(\kappa) > \omega$. Because then letting $M = \bigcup_{\xi<\kappa} \underset{\sim}{S}(\xi)$, if $A \in \underset{\sim}{S}(\kappa)$, then A is a wellordered union of sets in M (if T is a tree on $\omega \times \kappa$, then $p[T] = \bigcup_{\xi<\kappa} p[T \upharpoonright \xi]$). Let $M \subsetneq \Gamma \subsetneq \Lambda$ be closed under wellordered unions. Then $\underset{\sim}{S}(\kappa) = \Lambda \subseteq \Gamma$, a contradiction.

Note now that if $\lambda = \sup\{\xi : \xi$ is the length of a Λ prewellordering of $\mathbb{R}\}$, then λ has cofinality $> \omega$, since Λ is closed under countable unions. Thus it is enough to prove the lemma below to obtain the desired contradiction.

<u>Lemma.</u> $\lambda = \kappa$.

<u>Proof of Lemma.</u> If $\kappa < \lambda$, then there is a Λ-norm $\varphi : \mathbb{R} \twoheadrightarrow \kappa$. By the Moschovakis Coding Lemma we can code trees on $\omega \times \kappa$ within $\underset{\sim}{\Sigma}^1_1(\leq_\varphi, <_\varphi)$, where $x \leq_\varphi y \Leftrightarrow \varphi(x) \leq \varphi(y)$ and similarly for $<_\varphi$. Thus $\underset{\sim}{S}(\kappa) \subseteq \underset{\sim}{\Sigma}^1_1(\leq_\varphi, <_\varphi) \subsetneq \Lambda$, a contradiction.

So $\kappa \geq \lambda$. Pick now $A \in \underset{\sim}{S}(\kappa) - \bigcup_{\xi<\kappa} \underset{\sim}{S}(\xi)$. Since A is κ-Souslin, A carries a κ'-scale $\{\varphi_n\}_{n\in\omega}$, where κ' has the same cardinality as κ. We can also assume that $\{\varphi_n\}$ is <u>regular</u> i.e. each φ_n maps A onto an initial segment of the ordinals. For $n \in \omega$, let μ_n = length of φ_n. Clearly $\kappa \leq \sup_{n\in\omega} \mu_n$, so it will be enough to show that for each n,

$$\operatorname{card}(\mu_n) \leq \lambda .$$

So fix $n \in \omega$. For any $\xi < \mu_n$, let

$$A_\xi = \{\alpha \in A : \varphi_n(\alpha) = \xi\} .$$

Then if T is the tree associated with the scale $\{\varphi_n\}$, i.e.

$$T = \{(\alpha(0), \varphi_0(\alpha), \ldots, \alpha(m-1), \varphi_{m-1}(\alpha)) : m \in \omega,\ \alpha \in A\}$$

and for $\xi \leq \mu_n$ we let

$$T_{\leq\xi} = \{(a_0, \xi_0, \ldots, a_{m-1}, \xi_{m-1}) \in T : \xi_n \leq \xi\}$$

and similarly for $T_{<\xi}$, we have by the semicontinuity property of scales that

$$A_\xi = p[T_{\leq\xi}] - p[T_{<\xi}] .$$

Thus $A_\xi \in \Lambda$, since $p[T_{\leq\xi}]$, $p[T_{<\xi}]$ are κ'-Souslin and Λ is closed under complements. Moreover

$$\xi \neq \eta < \mu_n \Rightarrow A_\xi \cap A_\eta = \emptyset .$$

Put for $\xi < \mu_n$,

$$g(\xi) = w(A_\xi) \equiv \text{the Wadge ordinal of } A_\xi .$$

Since by Lemma 2.3.1 of Kechris-Solovay-Steel [A], we have $\lambda = \sup\{w(B) : B \in \Lambda\}$, it follows that $g : \mu_n \to \lambda$.

We claim now that for each $\rho < \lambda$, the order type of

$$\{\xi < \lambda : g(\xi) < \rho\} = g^{-1}[\rho]$$

is less than λ. Granting this, let for $\rho < \lambda$, $f_\rho : \lambda \twoheadrightarrow \varphi^{-1}[\rho]$. Then if $f(\rho,\xi) = f_\rho(\xi)$, $f : \lambda \times \lambda \twoheadrightarrow \bigcup_{\rho<\lambda} g^{-1}[\rho] = \mu_n$, thus $\mathrm{card}(\mu_n) \leq \lambda$ and we are done.

We prove the claim now: Let for each $\rho < \lambda$,

$$N = \{A : w(A) < \rho\} .$$

Then $N \subsetneqq \Lambda$, so find Γ closed under wellordered unions and such that $N \subsetneqq \Gamma \subsetneqq \Lambda$. Define

$$x \prec y \Leftrightarrow \exists\xi\exists\xi' [\xi,\xi' \in g^{-1}[\rho] \wedge \xi < \xi' \wedge x \in A_\xi \wedge y \in A_{\xi'}] .$$

Clearly $\prec$ is wellfounded, has rank the order type of $g^{-1}[\rho]$ and belongs to Γ, thus the order type of $g^{-1}[\rho]$ is less than λ, and we are done. $\dashv$

By the preceding result the only case when $\underset{\sim}{S}(\kappa)$ is not $\mathbb{R}$-parametrized is when

$$\underset{\sim}{S}(\kappa) = \underset{\sim}{S}(\sigma) ,$$

where σ is a limit of Souslin cardinals but is itself not Souslin. In this case $\underset{\sim}{S}(\sigma) = \bigcup_{\substack{\kappa<\sigma \\ \kappa \text{ Souslin}}} \underset{\sim}{S}(\kappa)$ is closed under complements. We conjecture that this never happens i.e.

<u>Conjecture</u> (AD + anything reasonable). If $\kappa < \Theta$ is a limit of Souslin cardinals, then κ is a Souslin cardinal.

Thus if $\{\kappa_\xi\}_{\xi<\Xi}$ is the enumeration of the Souslin cardinals, our conjecture means that for $\nu < \Xi$ limit

$$\kappa_\nu = \sup_{\xi<\nu} \kappa_\xi ,$$

and moreover that either $\sup_{\xi<\Xi} \kappa_\xi = \Theta$ i.e. if $\underset{\sim}{S}_\infty = \bigcup_\lambda \underset{\sim}{S}(\lambda)$, then

$$\underset{\sim}{S}_\infty = \mathrm{power}(\mathbb{R})$$

or else $\Xi = \xi + 1$ is successor and $\underset{\sim}{S}_\infty = \underset{\sim}{S}(\kappa_\xi)$. Thus in particular we have the following

<u>Conjecture</u> (AD + anything reasonable). The pointclass $\underset{\sim}{S}_\infty$ is $\mathbb{R}$-parametrized,

unless $\underset{\sim}{S}_\infty = \text{power}(\mathbb{R})$.

It has been proved by Martin and Steel that

$$AD + V = L[\mathbb{R}] \Rightarrow \underset{\sim}{S}_\infty = \underset{\sim}{\Sigma}_1^2 \wedge (\Xi = \underset{\sim}{\delta}_1^2 + 1, \kappa_{\underset{\sim}{\delta}_1^2} = \underset{\sim}{\delta}_1^2) .$$

On the other hand Solovay [1978] raises the question whether

$$AD_{\mathbb{R}} + \Theta \text{ is regular} \Rightarrow \underset{\sim}{S}_\infty = \text{power}(\mathbb{R}) \ ?$$

We shall see now in the next section that our first conjecture is verified for $\kappa \leq \kappa^{\mathbb{R}}$ (and in fact for a while beyond that, by §4).

§2. Souslin cardinals below $\kappa^{\mathbb{R}}$. Our main purpose here is to prove the following.

2.1 Theorem (AD). The class of Souslin cardinals is closed unbounded below $\kappa^{\mathbb{R}}$.

As an immediate corollary we have

2.2 Theorem (AD). For each $\kappa \leq \kappa^{\mathbb{R}}$, the pointclass $\underset{\sim}{S}(\kappa)$ is $\mathbb{R}$-parametrized.

It will be convenient at this stage to introduce the hyperprojective hierarchy to facilitate the presentation of the proof of 2.1. In the next section we shall take up a detailed study of the structural properties of the hyperprojective hierarchy itself.

First let $\underset{\sim}{IND}$ denote the class of inductive over the structure of analysis $\mathbb{R}$ pointsets and $\underset{\sim}{HYP} \equiv \underset{\sim}{IND} \cap \underset{\sim}{\check{IND}}$ the pointclass of hyperprojective sets. We define below a hierarchy on $\underset{\sim}{HYP}$.

Definition (AD). For each $1 \leq \xi < \kappa^{\mathbb{R}}$ let the pointclasses $\underset{\sim}{\Sigma}_\xi^1$, $\underset{\sim}{\Pi}_\xi^1$, $\underset{\sim}{\Delta}_\xi^1$ be defined as follows:

(i) For $\xi < \omega$, these are the usual projective pointclasses.

(ii) $\underset{\sim}{\Sigma}_{\xi+1}^1 = \exists^{\mathbb{R}} \underset{\sim}{\Pi}_\xi^1, \underset{\sim}{\Pi}_{\xi+1}^1 = \underset{\sim}{\check{\Sigma}}_{\xi+1}^1$.

(iii) For limit λ,

$\underset{\sim}{\Sigma}_\lambda^1$ = the smallest projective-like pointclass closed under $\exists^{\mathbb{R}}$, which contains $\bigcup_{\xi<\lambda} \underset{\sim}{\Sigma}_\xi^1$,

$\underset{\sim}{\Pi}_\lambda^1 = \underset{\sim}{\check{\Sigma}}_\lambda^1$.

(iv) $\underset{\sim}{\Delta}_\xi^1 = \underset{\sim}{\Sigma}_\xi^1 \cap \underset{\sim}{\Pi}_\xi^1$.

Put also for limit λ,

$$\Lambda^1_\lambda = \bigcup_{\xi<\lambda} \underset{\sim}{\Delta}^1_\xi .$$

We have of course

$$\underset{\sim}{\mathrm{HYP}} = \bigcup_{\xi<\kappa^{\mathbb{R}}} \underset{\sim}{\Sigma}^1_\xi \; (= \bigcup_{\xi<\kappa^{\mathbb{R}}} \underset{\sim}{\Delta}^1_\xi) .$$

Finally define

$$\underset{\sim}{\delta}^1_\xi = \sup\{\eta : \eta \text{ is the rank of a } \underset{\sim}{\Sigma}^1_\xi \text{ wellfounded relation}\}$$

and for limit ν

$$\lambda^1_\nu = \sup_{\xi<\nu} \underset{\sim}{\delta}^1_\xi .$$

(It follows from the results in §3 that also $\underset{\sim}{\delta}^1_\xi = \sup\{\eta : \eta$ is the rank of a $\underset{\sim}{\Delta}^1_\xi$ prewellordering$\}$, so that this notation is justified.)

We shall need of course crucially below some facts about positive elementary inductive definability on the structure of analysis $\mathbb{R}$; see Moschovakis [1974] here.

For each monotone operator $\Phi(x,A)$, where x varies over a product space $\mathcal{X}$ and A over power$(\mathcal{X})$, let Φ^ξ be its ξth iterate, defined by $\Phi^\xi(x) \Leftrightarrow \Phi(x,\Phi^{<\xi})$, $\Phi^{<\xi} = \bigcup_{\eta<\xi} \Phi^\eta$.

<u>Definition</u>. For each <u>limit</u> ordinal $\lambda \leq \kappa^{\mathbb{R}}$ define the following pointclass

$$\underset{\sim}{\mathrm{IND}}_\lambda = \{A : A \text{ is Wadge reducible to some } \Phi^{<\lambda}, \text{ where } \Phi \text{ is a positive elementary operator on } \mathbb{R}\} .$$

(Recall that A is <u>Wadge reducible</u> to $B (A \leq_W B)$ iff there is continuous f with $A = f^{-1}[B]$). Thus $\underset{\sim}{\mathrm{IND}}_{\kappa^{\mathbb{R}}} = \underset{\sim}{\mathrm{IND}}$.

We will need the following two lemmas of which the first one is implicit in Moschovakis [1974]. We will sketch their proofs after deriving from them a proof of 2.1.

2.3 <u>Lemma</u>. For each limit $\lambda \leq \kappa^{\mathbb{R}}$, $\underset{\sim}{\mathrm{IND}}_\lambda$ is closed under continuous substitutions, $\cap$, $\cup$, $\bigcup^\omega$, $\exists^{\mathbb{R}}$ and is $\mathbb{R}$-parametrized. In fact there is a "universal" positive elementary over $\mathbb{R}$ operator $\Phi(\alpha,A)$ such that $\Phi^{<\lambda}$ is $\underset{\sim}{\mathrm{IND}}_\lambda$-complete for all limit $\lambda \leq \kappa^{\mathbb{R}}$ (i.e. each member of $\underset{\sim}{\mathrm{IND}}_\lambda$ is Wadge reducible to it). If cofinality$(\lambda) > \omega$, then $\underset{\sim}{\mathrm{IND}}_\lambda$ is also closed under $\bigcap^\omega$ as well. Finally $\underset{\sim}{\mathrm{IND}}_\lambda$ has the prewellordering property.

(We have actually put a lot more information in 2.3 than we need for the proof of 2.1. It will be used in later sections.)

2.4 Lemma (AD). For each limit $\lambda < \kappa^{\mathbb{R}}$,

$$\underset{\sim}{\Lambda}^1_\lambda \subsetneqq \underset{\sim}{\mathrm{IND}}_\lambda \subsetneqq \underset{\sim}{\Lambda}^1_{\lambda+\omega}$$

(The class $\underset{\sim}{\mathrm{IND}}_\lambda$ can be actually computed precisely in the hyperprojective hierarchy by the results in the next section.)

Granting these two lemmas let us proceed to give the

Proof of 2.1. It is clear that the sequence $\{\kappa_\xi\}_{\xi<\kappa^{\mathbb{R}}}$ is unbounbed below $\kappa^{\mathbb{R}}$. We have to show that it is closed. So let $\theta < \kappa^{\mathbb{R}}$ be limit. Put $\kappa = \sup_{\eta<\theta} \kappa_\eta$. We have to show that κ is a Souslin cardinal i.e. $\underset{\sim}{S}(\kappa) \not\subseteq \bigcup_{\eta<\theta} \underset{\sim}{S}(\kappa_\eta) \equiv \Lambda$. Since Λ is strongly closed and $\Lambda \subsetneqq \underset{\sim}{\mathrm{HYP}}$, there is a limit ordinal $\sigma < \kappa^{\mathbb{R}}$ with

$$\Lambda = \underset{\sim}{\Lambda}^1_\sigma .$$

Note that $\sigma \leq \lambda^1_\sigma \equiv \sup_{\xi<\sigma} \underset{\sim}{\delta}^1_\xi \leq \kappa$, where the last bound comes from the Kunen-Martin Theorem.

Let now Φ be the universal operator asserted to exist in 2.3. For each ordinal $\xi \leq \kappa^{\mathbb{R}}$ let $\{\varphi^\xi_n\}_{n\in\omega}$ be the canonical scale on Φ^ξ defined by the procedure of Moschovakis [1978]. Let $\rho_\xi = \sup_n(\mathrm{length}(\varphi^\xi_n))$. By the definition of $\{\varphi^\xi_n\}$ it is easy to check that for each limit ordinal ν and each $m \in \omega$

$$\rho_{\nu+m} < \lambda^1_{\nu+\omega} .$$

Moreover again by the precedure of Moschovakis [1978], for each limit ordinal θ, $\Phi^{<\theta}$ admits a scale on

$$\max\{\theta, \sup_{\xi<\theta} \rho_\xi\} ,$$

thus $\Phi^{<\sigma}$ admits a scale on

$$\max\{\sigma, \sup_{\xi<\sigma} \rho_\xi\} \leq \lambda^1_\sigma \leq \kappa$$

i.e. a κ-scale. Since $\Phi^{<\sigma} \in \underset{\sim}{\mathrm{IND}}_\sigma - \check{\underset{\sim}{\mathrm{IND}}}_\sigma$, we have by 2.4 that $\Phi^{<\sigma} \notin \underset{\sim}{\Lambda}^1_\sigma = \Lambda$. Thus $\Phi^{<\sigma} \in \underset{\sim}{S}(\kappa) - \Lambda$ and we are done.

We conclude this section by giving the proofs of 2.3 and 2.4.

Proof of 2.3. The proof of the closure properties follows that of the closure properties of $\underset{\sim}{\mathrm{IND}}$ as in Chapter 1 of Moschovakis [1974]. As an example let us prove closure under $\exists^{\mathbb{R}}$: Let $A \in \underset{\sim}{\mathrm{IND}}_\lambda$ and let Ψ be a positive elementary on $\mathbb{R}$ operator such that for some continuous f

$$(x,\alpha) \in A \Leftrightarrow f(x,\alpha) \in \Psi^{<\lambda} .$$

Let

$$x \in B \Leftrightarrow \exists\alpha(x,\alpha) \in A .$$

Then

$$x \in B \Leftrightarrow \exists\alpha\exists\xi < \lambda f(x,\alpha) \in \Psi^{\xi}$$
$$\Leftrightarrow \exists\xi < \lambda[\exists\alpha f(x,\alpha) \in \Psi^{\xi}] .$$

Consider now the following simultaneous induction

$$\Psi_1^{\xi}(\beta) \Leftrightarrow \Psi(\beta,\Psi_1^{<\xi})$$
$$\Psi_2^{\xi}(x) \Leftrightarrow \exists\alpha[f(x,\alpha) \in \Psi_1^{<\xi}] .$$

Then

$$\Psi_1^{\xi} = \Psi^{\xi} \text{ and}$$
$$\Psi_2^{\xi}(x) \Leftrightarrow \exists\alpha[f(x,\alpha) \in \Psi^{<\xi}] ,$$

so for limit λ,

$$\Psi_2^{<\lambda}(x) \Leftrightarrow \exists\xi < \lambda\exists\alpha[f(x,\alpha) \in \Psi^{\xi}]$$
$$\Leftrightarrow x \in B .$$

Now, by the Simultaneous Induction Lemma 1C.1 of Moschovakis [1974], there is a positive elementary on $\mathbb{R}$ operator $\mathfrak{X}$ and constants x_0 such that

$$\Psi_2^{\xi}(x) \Leftrightarrow \mathfrak{X}^{\xi}(x_0,x) ,$$

thus

$$x \in B \Leftrightarrow \mathfrak{X}^{<\lambda}(x_0,x)$$

thus $B \in \underset{\sim}{\text{IND}}_{\lambda}$ and we are done.

For the prewellordering property we just use again the proof that $\underset{\sim}{\text{IND}}$ has the prewellordering property; see Moschovakis [1974], Chapter 2. According to it, to each positive elementary on $\mathbb{R}$ operator Ψ we can assign two other positive elementary on $\mathbb{R}$ operators Ψ_1, Ψ_2 such that if ψ is the canonical norm on Ψ^{∞} associated with the induction, and $\leq_{\psi}^{*}$, $<_{\psi}^{*}$ its corresponding relations, then for each ξ we have

$$\psi(x) \leq \xi \wedge x \leq_{\psi}^{*} y \Leftrightarrow \Psi_1^{\xi}(x,y)$$
$$\psi(x) \leq \xi \wedge x <_{\psi}^{*} y \Leftrightarrow \Psi_2^{\xi}(x,y) ,$$

so that for limit λ,

$$x \in \Psi^{\lambda} \wedge x \leq^*_\psi y \Leftrightarrow \Psi_1^{\lambda}(x,y)$$
$$x \in \Psi^{\lambda} \wedge x <^*_\psi y \Leftrightarrow \Psi_2^{\lambda}(x,y) .$$

thus $\psi \restriction \Psi^{\lambda}$ is a $\underset{\sim}{\mathrm{IND}}_\lambda$-norm and this establishes Prewellordering($\underset{\sim}{\mathrm{IND}}_\lambda$).

Finally we prove the statement about parametrization:

First recall that by Moschovakis [1970] every positive $\underset{\sim}{\Pi}^1_1$ relation $\Psi(x,A)$, where $x \in \mathfrak{X}$, $A \subseteq \mathcal{R}$ can be brought in the form

$$\Psi(x,A) \Leftrightarrow \forall\beta[R(x,\beta) \vee \exists n((\beta)_n \in A)] \wedge \overline{R}(x) ,$$

with $R, \overline{R} \in \underset{\sim}{\Pi}^1_1$. Similarly every positive $\underset{\sim}{\Sigma}^1_1$ relation $\Psi'(x,A)$ is equivalent to one in the form

$$\Psi'(x,A) \Leftrightarrow \exists\beta[S(x,\beta) \wedge \forall n((\beta)_n \in A)] \vee \overline{S}(x) ,$$

with $S, \overline{S} \in \underset{\sim}{\Sigma}^1_1$. Similar normal forms are valid for positive $\underset{\sim}{\Pi}^1_1$ or $\underset{\sim}{\Sigma}^1_1$ relations $\Psi(x,A)$ for $A \subseteq \mathcal{Y}$ any arbitrary product space. As a result, for each fixed $\mathfrak{X}$, $\mathcal{Y}$ there is a positive $\underset{\sim}{\Pi}^1_1$ (resp. $\underset{\sim}{\Sigma}^1_1$) relation of the form $\Psi(\varepsilon,x,A)$ ($\varepsilon \in \mathcal{R}$, $x \in \mathfrak{X}$, $A \subseteq \mathcal{Y}$) which is universal for the positive $\underset{\sim}{\Pi}^1_1$ (resp. $\underset{\sim}{\Sigma}^1_1$) relations of the form $\Psi(x,A)$.

According to Moschovakis [1974] Chapter 1, Ex. 1.15, one can find for each positive elementary on $\mathbb{R}$ operator $\Psi(x,A)$ an operator $\Psi'(y,x,A)$ which is a disjunction of a $\underset{\sim}{\Pi}^1_1$ positive operator and a $\underset{\sim}{\Sigma}^1_1$ positive operator, and constants y_0 such that for all limit λ,

$$\Psi^{\lambda}(x) \Leftrightarrow \Psi'^{\lambda}(y_0,x) .$$

So it is clear that it is enough to find an operator $\Phi(\varepsilon,\alpha,A)$ ($\varepsilon \in \mathcal{R}$, $\alpha \in \mathcal{R}$, $A \subseteq \mathcal{R} \times \mathcal{R}$) which is the disjunction of a positive $\underset{\sim}{\Pi}^1_1$ and a positive $\underset{\sim}{\Sigma}^1_1$ operator such that for all such $\Psi(\alpha,X)$ ($\alpha \in \mathcal{R}$, $X \subseteq \mathcal{R}$) there is $\varepsilon_0 \in \mathcal{R}$ (depending on Ψ) so that:

$$\Psi^{\lambda}(\alpha) \Leftrightarrow \Phi^{\lambda}(\varepsilon_0,\alpha), \quad \text{for all limit } \lambda .$$

Note that to achieve this it is sufficient to find Φ as above such that for all Ψ as above there is ε_0 (depending on Ψ) such that

$$\Phi(\varepsilon_0,\alpha,A) \Leftrightarrow \Psi(\alpha,\{\alpha' : A(\varepsilon_0,\alpha')\})$$

(Then by a simple induction on ξ,

$$\Phi^{\xi}(\varepsilon_0,\alpha) \Leftrightarrow \Psi^{\xi}(\alpha) .)$$

For that, let $\mathfrak{X}(\delta,\alpha,X)$, where δ, $\alpha \in \mathcal{R}$, $X \subseteq \mathcal{R}$, be universal in the class of disjunctions of positive $\underset{\sim}{\Pi}^1_1$ and $\underset{\sim}{\Sigma}^1_1$ formulas. Put then

$$\Phi(\varepsilon,\alpha,A) \Leftrightarrow \chi(\varepsilon,\alpha,\{\alpha' : A(\varepsilon,\alpha')\}) .$$

To verify that this works, fix an appropriate Ψ and let ε_0 be such that

$$\Psi(\alpha,X) \Leftrightarrow \chi(\varepsilon_0,\alpha,X) .$$

Then

$$\Phi(\varepsilon_0,\alpha,A) \Leftrightarrow \chi(\varepsilon_0,\alpha,\{\alpha' : A(\varepsilon_0,\alpha')\}$$
$$\Leftrightarrow \Psi(\alpha,\{\alpha' : A(\varepsilon_0,\alpha')\}) . \qquad 2.3\dashv$$

Proof of 2.4. By induction on λ. Obvious for $\lambda = \omega$. If $\lambda = \lambda' + \omega$ is a successor limit ordinal then by induction hypothesis,

$$\Delta^1_{\lambda'} \subsetneqq \underset{\sim}{\mathrm{IND}}_{\lambda'} \subsetneqq \Delta^1_{\lambda'+\omega} = \Delta^1_\lambda .$$

Since Δ^1_λ is strongly closed, it is easy to check that for each positive elementary on $\mathbb{R}$ operator Ψ one have $\Psi^{\lambda'+n} \in \Delta^1_\lambda$ for each $n \in \omega$, thus $\Psi^{<\lambda} \in \underset{\sim}{\Sigma}^1_{\lambda+1} \subsetneqq \Delta^1_{\lambda+\omega}$. So $\underset{\sim}{\mathrm{IND}}_\lambda \subsetneqq \Delta^1_{\lambda+\omega}$. To see that $\Delta^1_\lambda \subsetneqq \underset{\sim}{\mathrm{IND}}_\lambda$ notice that for any $n \in \omega$, if $A \in \underset{\sim}{\Sigma}^1_{\lambda'+n}$ then A is gotten from some $B \in \underset{\sim}{\mathrm{IND}}_{\lambda'}$ by repeatedly applying a finite number of real quantifications. This is because $\Delta^1_{\lambda'} \subsetneqq \underset{\sim}{\mathrm{IND}}_{\lambda'}$. Then by the Simultaneous Induction Lemma 1C.1 in Moschovakis [1974] it follows that for some positive elementary on $\mathbb{R}$ operator Ψ_1, $A \leq_w \Psi_1^{\lambda'}$. But by the proof of the prewellordering property for $\underset{\sim}{\mathrm{IND}}_\lambda$, clearly $\Psi_1^{\lambda'} \in \underset{\sim}{\mathrm{IND}}_\lambda$, so $A \in \underset{\sim}{\mathrm{IND}}_\lambda$. Thus $\Delta^1_\lambda = \bigcup_n \underset{\sim}{\Sigma}^1_{\lambda'+n} \subseteq \underset{\sim}{\mathrm{IND}}_\lambda$ and since Δ^1_λ is closed under complements $\Delta^1_\lambda \subsetneqq \underset{\sim}{\mathrm{IND}}$.

For the case when λ is the limit of limit ordinals, it immediately follows by induction hypothesis that $\Delta^1_\lambda \subsetneqq \underset{\sim}{\mathrm{IND}}_\lambda$. For the other inclusion, notice that by our induction hypothesis each member of $\underset{\sim}{\mathrm{IND}}_\lambda$ is the wellordered union of sets in Δ^1_λ and thus belongs to $\underset{\sim}{\Sigma}^1_{\lambda+1}$ (since by Kechris-Solovay-Steel [A] either $\underset{\sim}{\Sigma}^1_\lambda$ or $\underset{\sim}{\Sigma}^1_{\lambda+1}$ has the prewellordering property and thus is closed under wellordered unions). So $\underset{\sim}{\mathrm{IND}}_\lambda \subsetneqq \Delta^1_{\lambda+\omega}$. 2.4$\dashv$

§3. The scale property in the hyperprojective hierarchy. As a special case of the results in Kechris-Solovay-Steel [A] one can determine the prewellordering pattern in the hyperprojective hierarchy. We do the same thing for the scale property. Moreover we identify the pointclasses $\underset{\sim}{S}(\kappa)$, for κ a Souslin cardinal $< \kappa^{\mathbb{R}}$, within the hyperprojective hierarchy. It will be convenient to introduce first the following terminology, motivated by the classification of projective-like hierarchies in Kechris-Solovay-Steel [A] §4:

For θ a limit ordinal $< \kappa^{\mathbb{R}}$, we shall say that,

(i) θ is of type I iff cofinality$(\theta) = \omega$, or equivalently iff the projective-like hierarchy $\{\underset{\sim}{\Pi}^1_\theta, \underset{\sim}{}^1_{\theta+1}, \underset{\sim}{\Pi}^1_{\theta+2}, \underset{\sim}{}^1_{\theta+3}, \ldots\}$ is of type I.

(ii) θ is of type II iff $\{\underset{\sim}{\Pi}^1_\theta, \underset{\sim}{}^1_{\theta+1}, \ldots\}$ is of type II,

(iii) θ is of type III iff $\{\underset{\sim}{\Pi}^1_\theta, \underset{\sim}{}^1_{\theta+1}, \ldots\}$ is of type III.

John Steel has proved a beautiful result (see Steel [A]) which allows us to characterize types II and III above in terms of ordinal invariants. The characterization is as follows.

Theorem (AD) (Steel [A]). A limit cardinal $\theta < \kappa^{\mathbb{R}}$ is of type III iff λ^1_θ $(= \kappa_\theta)$ is regular (iff $\theta = \lambda^1_\theta$ $(= \kappa_\theta)$ and θ is regular). (Thus $\theta < \kappa^{\mathbb{R}}$ is of type II iff cofinality$(\theta) > \omega$ and λ^1_θ $(= \kappa_\theta)$ is singular.)

We now have the following

3.1 Theorem (AD). Let θ be a limit ordinal $< \kappa^{\mathbb{R}}$. Then we have,

(i) If θ is of type I, the following classes have the scale property

$$\underset{\sim}{\Pi}^1_\theta, \underset{\sim}{\Sigma}^1_{\theta+1}, \underset{\sim}{\Pi}^1_{\theta+2}, \underset{\sim}{\Sigma}^1_{\theta+3}, \ldots .$$

Moreover, $\underset{\sim}{S}(\kappa_{\theta+m}) = \underset{\sim}{\Sigma}^1_{\theta+m}$, for all $m \geq 0$. Finally, for all $m \geq 0$,

$(\kappa_{\theta+m})^+ = \underset{\sim}{\delta}^1_{\theta+m}$ and is measurable,

$\underset{\sim}{\delta}^1_{\theta+2m+1} = (\underset{\sim}{\delta}^1_{\theta+2m})^+$, so $\kappa_{\theta+2m+1} = \underset{\sim}{\delta}^1_{\theta+2m}$, and

$\kappa_{\theta+2m}$ has cofinality ω.

(ii) If θ is of type II, the following classes have the scale property

$$\underset{\sim}{\Sigma}^1_\theta, \underset{\sim}{\Pi}^1_{\theta+1}, \underset{\sim}{\Sigma}^1_{\theta+2}, \underset{\sim}{\Pi}^1_{\theta+3}, \ldots .$$

Moreover,

$$\underset{\sim}{S}(\kappa_{\theta+m}) = \underset{\sim}{\Sigma}^1_{\theta+m}, \text{ for all } m \geq 0 .$$

Finally, for all $m \geq 0$,

$(\kappa_{\theta+m})^+ = \underset{\sim}{\delta}^1_{\theta+m}$ and is measurable,

$\underset{\sim}{\delta}^1_{\theta+2m+2} = (\underset{\sim}{\delta}^1_{\theta+2m+1})^+$, so $\kappa_{\theta+2m+2} = \delta^1_{\theta+2m+1}$, and

$\kappa_{\theta+2m+1}$ has cofinality ω.

(iii) If θ is of type III, the following classes have the scale property

$$\underset{\sim}{\Pi}^1_\theta, \underset{\sim}{\Sigma}^1_{\theta+1}, \underset{\sim}{\Pi}^1_{\theta+2}, \underset{\sim}{\Sigma}^1_{\theta+3}, \ldots .$$

Moreover

$$\underset{\sim}{S}(\kappa_{\theta+m}) = \underset{\sim}{\Sigma}^1_{\theta+m+1}, \quad \text{for all} \quad m \geq 0 .$$

Finally, for all $m \geq 0$,

$$(\kappa_{\theta+m})^+ = \underset{\sim}{\delta}^1_{\theta+m+1} \quad \text{and is measurable}$$
$$\underset{\sim}{\delta}^1_{\theta+2m+1} = (\underset{\sim}{\delta}^1_{\theta+2m})^+, \quad \text{so} \quad \kappa_{\theta+2m} = \underset{\sim}{\delta}^1_{\theta+2m},$$
$$\kappa_{\theta+2m+1} \quad \text{has cofinality} \quad \omega, \quad \text{and}$$
$$\kappa_\theta = \underset{\sim}{\delta}^1_\theta \quad \text{and is measurable.}$$

Proof. By induction on θ.

(i) By induction hypothesis (or by standard facts about the projective hierarchy when $\theta = \omega$) $\Lambda^1_\theta = \bigcup_{\eta<\theta} \underset{\sim}{S}(\kappa_\eta)$. Moreover $\kappa_\theta = \lim_{\eta<\theta} \kappa\eta = \lambda^1_\theta$ has cofinality ω. Let $\Sigma = \{\bigcup_n A_n : \text{For all } n, A_n \in \Lambda^1_\theta\}$. Then by our induction hypothesis, Σ has the scale property and is closed under $\exists^{\mathcal{R}}$, so since $\underset{\sim}{\Pi}^1_\theta = \forall^{\mathcal{R}}\Sigma$ we have, by the Second Periodicity Theorem of Moschovakis [1980], that $\underset{\sim}{\Pi}^1_\theta, \underset{\sim}{\Sigma}^1_{\theta+1}, \underset{\sim}{\Pi}^1_{\theta+2}, \ldots$ all have the scale property.

Since $\kappa_\theta = \lambda^1_\theta$, there is a $\underset{\sim}{\Delta}^1_\theta$ prewellordering of $\mathcal{R}$ of length κ_θ, thus $\underset{\sim}{S}(\kappa_\theta) \subseteq \underset{\sim}{\Sigma}^1_\theta$. That $\underset{\sim}{\Sigma}^1_\theta \subseteq \underset{\sim}{S}(\kappa_\theta)$ follows from the fact that $\Lambda^1_\theta \subsetneq \underset{\sim}{S}(\kappa_\theta)$ and the closure properties of $\underset{\sim}{S}(\kappa_\theta)$. Thus $\underset{\sim}{\Sigma}^1_\theta = \underset{\sim}{S}(\kappa_\theta)$. The equalities $\underset{\sim}{S}(\kappa_{\theta+m}) = \underset{\sim}{\Sigma}^1_{\theta+m}$ follow now by the usual arguments as for the projective hierarchy. The same holds for the proofs of the fact about $\underset{\sim}{\delta}^1_{\theta+m}$ and $\kappa_{\theta+m}$. (A quicker proof that $\kappa_{\theta+2m+2}$ has cofinality ω makes use of the fact that $\underset{\sim}{S}(\kappa_{\theta+2m+1}) = \underset{\sim}{\Sigma}^1_{\theta+2m+1}$ is closed under wellordered unions.)

(ii) First notice that if the cofinality of θ is bigger than ω then

$$(*) \qquad \underset{\sim}{S}(\kappa_\theta) = \underset{\sim}{\mathrm{IND}}_\theta .$$

To prove (*) first use the fact that each element of $\underset{\sim}{S}(\kappa_\theta)$ is a wellordered union of sets in $\bigcup_{\eta<\theta} \underset{\sim}{S}(\kappa_\eta)$, since cofinality $(\kappa_\theta) > \omega$. But $\bigcup_{\eta<\theta} S(\kappa_\eta) \subseteq \Lambda^1_\theta$ by induction hypothesis and $\Lambda^1_\theta \subsetneq \underset{\sim}{\mathrm{IND}}_\theta$ by 2.4. Since $\underset{\sim}{\mathrm{IND}}_\theta$ is projective-like closed under $\exists^{\mathcal{R}}$ and has the prewellordering property by 2.3, it is closed under wellordered unions, thus $\underset{\sim}{S}(\kappa_\theta) \subseteq \underset{\sim}{\mathrm{IND}}_\theta$. For the inclusion $\underset{\sim}{\mathrm{IND}}_\theta \subseteq \underset{\sim}{S}(\kappa_\theta)$, just look at the proof of 2.1 and note that in the notation there, $\Lambda = \Lambda^1_\theta$ by our induction hypothesis, thus $\theta = \sigma$.

Since in case θ is of type II $\underset{\sim}{\Sigma}^1_\theta$ consists (by the analysis of type II hierarchies in Kechris-Solovay-Steel [A]) of all wellordered unions in Λ^1_θ, we also have in this case that

$$\underset{\sim}{S}(\kappa_\theta) = \underset{\sim}{\mathrm{IND}}_\theta = \underset{\sim}{\Sigma}^1_\theta .$$

The rest of the conclusions of (ii) will follow routinely once we can show that each set in $\underset{\sim}{\text{IND}}_\theta$ admits an $\underset{\sim}{\text{IND}}_\theta$-scale, each norm of which has length $\leq \kappa_\theta$.

For that let Φ be the universal positive elementary on $\mathbb{R}$ operator of 2.3. Let $\{\varphi_n^\xi\}$ be the canonical scale on Φ^ξ as defined in Moschovakis [1978]. Let also $\{\varphi_n\}$ be the scale on $\Phi^{<\theta}$ defined there i.e.

$$\varphi_0(x) = \text{least } \xi < \theta \text{ such that } x \in \Phi^\xi$$
$$\varphi_{n+1}(x) = \langle \varphi_0(x), \varphi_n^{\varphi_0(x)} \rangle .$$

Clearly φ_0 is a $\underset{\sim}{\text{IND}}_\theta$-norm. Now we have

$$x \leq^*_{\varphi_{n+1}} y \text{ iff } x <^*_{\varphi_0} y \vee \exists \xi < \theta[\varphi_0(x) = \varphi_0(y) = \xi \wedge \varphi_n^\xi(x) \leq \varphi_n^\xi(y)] .$$

Now it is easy to verify that for each fixed $\xi < \theta$ the relation

$$\varphi_0(x) = \varphi_0(y) = \xi \wedge \varphi_n^\xi(x) \leq \varphi_n^\xi(y)$$

is in Λ^1_θ, thus $\leq^*_{\varphi_{n+1}}$ is the wellordered union of Λ^1_θ relations, thus belongs to $\underset{\sim}{\text{IND}}_\theta$. Similarly for $<^*_{\varphi_{n+1}}$ and we are done. The fact that all these norms have length $\leq \kappa_\theta$ is included in the proof of 2.1.

(iii) By (*) of case (ii) we know that

$$\underset{\sim}{S}(\kappa_\theta) = \underset{\sim}{\text{IND}}_\theta = \underset{\sim}{\Sigma}^1_{\theta+1} ,$$

so all the assertions of (iii) follow except that $\underset{\sim}{\Pi}^1_\theta$ has the scale property. We prove this now.

First recall from Kechris-Solovay-Steel [A] 2.4 that $\underset{\sim}{\Pi}^1_\theta$ is equal to the class of all $\underset{\sim}{\Delta}^1_\theta$ $(= \Lambda^1_\theta)$-bounded wellordered unions of $\underset{\sim}{\Delta}^1_\theta$ sets.

Fix now $A \in \underset{\sim}{\Pi}^1_\theta$ and let Φ be the universal positive elementary on $\mathbb{R}$ operator considered before. For $x \in \Phi^\infty$, let $|x| \equiv \varphi_0(x) =$ least ξ such that $x \in \Phi^\xi$. Put then

$$(\varepsilon,x) \in C \Leftrightarrow x \in \Phi^{<\theta} \wedge \varepsilon \text{ codes a continuous function } (\equiv f_\varepsilon) ,$$

and for $(\varepsilon,x) \in C$, let

$$Y_{\varepsilon,x} = f_\varepsilon^{-1}[\Phi^{<|x|}] .$$

Then put

$$B = \{(\varepsilon,x) \in C : Y_{\varepsilon,x} \subseteq A\} ,$$

so that $B \in \underset{\sim}{\Sigma}^1_{\theta+1} = \underset{\sim}{\text{IND}}_\theta$. Then there is continuous g such that

$$B = g^{-1}[\Phi^{<\theta}] .$$

Put finally for $\xi < \theta$,

$$A_\xi = \cup\{Y_{\varepsilon,x} : g(\varepsilon,x) \in \Phi^{<\xi} \wedge x \in \Phi^{<\xi}\} .$$

Clearly $A = \cup_{\xi<\theta} A_\xi$, $A_\xi \in \underset{\sim}{\Delta}^1_\theta$ and $\{A_\xi\}_{\xi<\theta}$ is a $\underset{\sim}{\Delta}^1_\theta$-bounded union of sets in $\underset{\sim}{\Delta}^1_\theta$. Now

$$\begin{aligned} \alpha \in A_\xi &\Leftrightarrow \exists\varepsilon\exists x[g(\varepsilon,x) \in \Phi^{<\xi} \wedge x \in \Phi^{<\xi} \wedge f_\varepsilon(\alpha) \in \Phi^{<|x|}] \\ &\Leftrightarrow \exists\varepsilon\exists x[g(\varepsilon,x) \in \Phi^{<\xi} \wedge x \in \Phi^{<\xi} \wedge f_\varepsilon(\alpha) <^*_{\varphi_0} x] . \end{aligned}$$

Let Φ_1 be a positive elementary on $\mathbb{R}$ operator such that

$$z \in \Phi^\eta \wedge z <^*_{\varphi_0} y \Leftrightarrow \Phi_1^\eta(z,y) .$$

Then

$$\alpha \in A_\xi \Leftrightarrow \exists\varepsilon\exists x[g(\varepsilon,x) \in \Phi^{<\xi} \wedge x \in \Phi^{<\xi} \wedge \Phi_1^{<\xi}(f_\varepsilon(\alpha),x)] .$$

From this it is clear that we can define a map $\xi \mapsto \{\psi_n^\xi\}$, where for each $\xi < \theta$, $\{\psi_n^\xi\}$ is a $\underset{\sim}{\Delta}^1_\theta$-scale on A_ξ, making use of the canonical scales being put, as in Moschovakis [1978], on $\Phi^{<\xi}$ and $\Phi_1^{<\xi}$. This in turn defines the following scale on $A = \cup_{\xi<\theta} A_\xi$:

$$\psi_0(\alpha) = \text{least } \xi \text{ such that } \alpha \in A_\xi ,$$

$$\psi_{n+1}(\alpha) = \langle \psi_0(\alpha), \psi_n^{\psi_0(\alpha)}(\alpha) \rangle .$$

We how that this is a $\underset{\sim}{\Pi}^1_\theta$-scale. Clearly $\leq^*_{\psi_0}$, $<^*_{\psi_0}$ are in $\underset{\sim}{\Pi}^1_\theta$ (see Kechris-Solovay-Steel [A], 2.5). Consider now $\leq^*_{\psi_{n+1}}$ (the argument for $<^*_{\psi_{n+1}}$ being similar). We have

$$\begin{aligned} \alpha \leq^*_{\psi_{n+1}} \beta \text{ iff } \alpha <^*_{\psi_0} \beta \vee \exists\xi < \theta[\exists\xi' \leq \xi(\psi_0(\alpha) = \psi_0(\beta) = \xi' \\ \wedge \psi_n^{\xi'}(\alpha) \leq \psi_n^{\xi'}(\beta))] , \end{aligned}$$

thus

$$\leq^*_{\psi_{n+1}} = <^*_{\psi_0} \cup \cup_{\xi<\theta} D_\xi ,$$

where

$$D_\xi = \cup_{\xi'\leq\xi} \{(\alpha,\beta) : \psi_0(\alpha) = \psi_0(\beta) = \xi' \wedge \psi_n^{\xi'}(\alpha) \leq \psi_n^{\xi'}(\beta)\} \in \underset{\sim}{\Delta}^1_\theta .$$

So it is enough to check that $\{D_\xi\}_{\xi<\theta}$ is $\underset{\sim}{\Delta}^1_\theta$-bounded. For that let $X \in \underset{\sim}{\Delta}^1_\theta$, $X \subseteq \cup_{\xi<\theta} D_\xi$ be given. Then

$$\{\alpha : \exists\beta(\alpha,\beta) \in X\} = Y$$

is in $\underset{\sim}{\Delta}^1_\theta$ and $Y \subseteq \bigcup_{\xi<\theta} A_\xi$, so for some $\xi < \theta$, $Y \subseteq A_\xi$. But then $X \subseteq D_\xi$. Indeed, if $(\alpha,\beta) \in X$ then $\alpha \in Y$, thus $\psi_0^{\xi}(\alpha) = \xi' \leq \xi$. So also $\psi_0(\beta) = \xi'$ and since $(\alpha,\beta) \in \bigcup_{\xi<\theta} D_\xi$ we have that $\psi_n^{\xi}(\alpha) \leq \psi_n^{\xi'}(\beta)$, thus $(\alpha,\beta) \in D$. ⊣

We have now the following immediate corollaries

3.2 Corollary (AD). Let Γ be any projective-like pointclass contained in $\underset{\sim}{\mathrm{IND}}$. Then one of Γ or $\check{\Gamma}$ has the scale property.

3.3 Corollary (AD). Let κ be a Souslin cardinal $\leq \kappa^{\mathbb{R}}$. Then

(i) $\underset{\sim}{S}(\kappa)$ has the scale property iff cofinality$(\kappa) > \omega$.

(ii) $\check{\underset{\sim}{S}}(\kappa)$ has the scale property iff cofinality$(\kappa) = \omega$.

(iii) κ^+ is always measurable.

(iv) If $\kappa = \kappa_\eta$ and $\kappa^* = \kappa_{\eta+1}$ is the next Souslin cardinal, then

(a) If κ has cofinality ω, then $\kappa^* = \kappa^+$ is a measurable cardinal, while

(b) If κ has cofinality $> \omega$, then κ^* has cofinality ω.

3.4 Corollary (AD). Let $\kappa < \kappa^{\mathbb{R}}$ be a Souslin cardinal. Then there is a prewellordering of $\mathbb{R}$ in $\underset{\sim}{S}(\kappa) \cap \check{\underset{\sim}{S}}(\kappa)$ of length κ.

Finally, let for each $A \in \underset{\sim}{S}_\infty$, $\kappa(A) \equiv$ Souslin cardinal of $A \equiv$ the smallest κ such that $A \in \underset{\sim}{S}(\kappa)$. Recall that $w(A)$ denotes the Wadge ordinal of A. We have now the following estimate.

3.5 Corollary (AD). For each $A \in \underset{\sim}{\mathrm{IND}}$, $\kappa(A) \leq w(A)$.

Proof. Let $\kappa(A) = \kappa_\eta$, $\eta < \kappa^{\mathbb{R}}$ (the case $\kappa(A) = \kappa^{\mathbb{R}}$ is obvious). If $\eta = \theta$ is limit and $w(A) < \kappa_\theta = \lambda^1_\theta$, then $A \in \Lambda^1_\theta$, so $\kappa(A) < \lambda^1_\theta$, a contradiction. If $\eta = \theta + m$, θ limit, $m > 0$ consider cases according to the type of θ. ⊣

§4. Gaps in the propagation of scales. Assume AD for the discussion in this section.

The uninterrupted propagation of the scale property throughout the hyperprojective hierarchy is disturbed as one goes past the class of inductive sets. If $\{\Gamma_i\}$ is a projective-like hierarchy of type IV (see Kechris-Solovay-Steel [A]), then Kechris has shown that neither Γ_1 nor $\check{\Gamma}_1$ can have the scale property and Martin substantially extended this to show that in fact no Γ_i or $\check{\Gamma}_i$ can have the scale property. In particular, if $\{\underset{\sim}{\Pi}^*_1, \underset{\sim}{\Sigma}^*_2, \underset{\sim}{\Pi}^*_3, \underset{\sim}{\Sigma}^*_4, \ldots\}$ is the least projective-like hierarchy of type IV (so that $\underset{\sim}{\Pi}^*_1 = \forall^{\mathbb{R}}(\underset{\sim}{\mathrm{IND}} \vee \check{\underset{\sim}{\mathrm{IND}}})$), no $\underset{\sim}{\Sigma}^*_n$ or $\underset{\sim}{\Pi}^*_n$ can have the scale property. Thus Corollary 3.2 cannot be extended past $\underset{\sim}{\text{IND}}$. Compare this with the corresponding result about the prewellordering property (see Kechris-Solovay-Steel [A]), which extends to all $\Gamma \subseteq L(\mathbb{R})$ and beyond. (In particular $\underset{\sim}{\Sigma}^*_1, \underset{\sim}{\Pi}^*_2, \underset{\sim}{\Sigma}^*_3, \ldots$ all have the prewellordering property.)

Now if $\underset{\sim}{\Sigma}^*_\omega$ denotes the smallest projective-like pointclass closed under $\exists^{\mathbb{R}}$ and containing $\bigcup_n \underset{\sim}{\Sigma}^*_n$, then Moschovakis [1979] has shown that Π^*_ω has the scale property and thus so do $\underset{\sim}{\Sigma}^*_{\omega+1}, \Pi^*_{\omega+2}, \ldots$, and the familiar pattern for the propagation of scales resumes again for a while, after this gap of length ω. But later on wider and wider gaps occur, relfecting eventually the "unbounded" gap occurring in $L[\mathbb{R}]$ beyond $\underset{\sim}{S}_\infty = \underset{\sim}{\Sigma}^2_1$. We will not however pursue this matter any further here.

Beyond this first occurrence of gaps, other phenomena happen for the first time at the level of $\kappa^{\mathbb{R}}$. For example, $\kappa^{\mathbb{R}}$ is the least Souslin cardinal κ for which the conclusion of Corollary 3.4 fails and also the least ordinal κ for which there is an A with $w(A) = \kappa$ but $\kappa(A) > w(A)$. (Take A to be a complete co-inductive set.)

§5. Miscellaneous remarks, questions, and conjectures. We assume again AD throughout this section.

(A) Closure properties of $\underset{\sim}{S}(\kappa)$. Let κ be a Souslin cardinal. Of course $\underset{\sim}{S}(\kappa)$ is closed under $\exists^{\mathbb{R}}$, but when is it also closed under $\forall^{\mathbb{R}}$? The following result gives a necessary and sufficient condition when cofinality$(\kappa) > \omega$.

We need first to define a notion of indescribability of ordinals. For any limit ordinal λ let $B_\lambda = \{x : \exists \xi < \lambda, x \subseteq L_\xi\}$. Thus $L_\lambda \subseteq B_\lambda$ and B_λ is transitive. Recall that a formula in the language of ZF augmented by extra predicates is Π_2 if it has the form $\forall x \exists y \varphi$, where φ is bounded. We call now an ordinal λ ${}^b\Pi^1_2$-indescribable if for each $X \subseteq L_\lambda$ and each Π_2 formula φ of the appropriate language, with parameters from B_λ, we have

$$\langle B_\lambda, \varepsilon, X\rangle \models \varphi \Rightarrow \exists \theta < \lambda, \ \langle B_\theta, \varepsilon, X \cap L_\theta\rangle \models \varphi .$$

It is easy to check that such λ's are regular cardinals. We have now

5.1 Theorem (AD). Let κ be a Souslin cardinal of cofinality greater than ω. Then the following are equivalent:

(i) $\underset{\sim}{S}(\kappa)$ is closed under $\forall^{\mathbb{R}}$,

(ii) κ is ${}^b\Pi^1_2$-indescribable.

Proof. (i) $\Rightarrow$ (ii) Let $\Gamma = \underset{\sim}{S}(\kappa)$. Then Γ is a Spector class on the structure of analysis $\mathbb{R}$, so if $\delta = \sup\{\xi : \xi$ is the length of a Δ prewellordering of $\mathbb{R}\}$, by the Companion Theorem of Moschovakis [1974], δ is the ordinal of its companion admissible set M above $\mathbb{R}$. As every admissible set is Π_2 reflecting and every set $A \subseteq L_\delta$ is Δ_1 in M, by the Moschovakis Coding Lemma (see his [1970]), it is easy to verify that δ is ${}^b\Pi^1_2$-indescribable (this argument is due to Moschovakis).

So it is enough to show that $\delta = \kappa$. By Kunen-Martin $\delta \leq \kappa^+$ and since δ is a limit cardinal $\delta \leq \kappa$. Let now $A \in \underset{\sim}{S}(\kappa) - \bigcup_{\eta<\kappa} \underset{\sim}{S}(\eta)$. Let $\{\varphi_n\}$ be a regular κ-scale on A. We can assume moreover that $\{\varphi_n\}$ is good i.e.

$$\varphi_i(x) \leq \varphi_i(y) \Rightarrow \forall j \leq i(\varphi_j(x) \leq \varphi_j(y)) .$$

Fix now n. For $\xi < \text{length}(\varphi_n)$, let

$$K_n(\xi) = \sup\{\varphi_j(x) : x \in A \wedge \varphi_n(x) = \xi\} .$$

Then $K_n(\xi) < \kappa$, since if $\varphi_n(x_0) > \xi$ then for any x with $\varphi_n(x) = \xi < \varphi_n(x_0)$ we must have $\varphi_j(x) \leq \varphi_j(x_0)$, $\forall j \leq n$, by goodness, but also $\varphi_j(x) < \varphi_j(x_0)$, $\forall j \geq n$, by goodness again, so $K_n(\xi) \leq \sup \varphi_j(x_0) < \kappa$.

Let now T be the tree on $\omega \times \kappa$ associated with $\{\varphi_i\}$. Let

$$T_n(\xi) = \{(a_0,u_0,\ldots,a_{m-1},u_{m-1}) \in T : u_n \leq \xi \wedge \forall i < m,\ u_i \leq K_n(\xi)\}$$

and similarly for $T'_n(\xi)$, replacing $u_n \leq \xi$ by $u_n < \xi$. Then

$$A^\xi_n = \{x : \varphi_n(x) \leq \xi\} = p[T_n(\xi)]$$
$$B^\xi_n = \{x : \varphi_n(x) < \xi\} = p[T'_n(\xi)] ,$$

so since $T_n(\xi)$, $T'_n(\xi)$ are trees on $K_n(\xi) < \kappa$, clearly $A^\xi_n, B^\xi_\eta \in \Delta$. Thus

$$C^\xi_n = \{x : \varphi_n(x) = \xi\} \in \Delta .$$

The rest of the argument is as in the proof of the lemma within the proof of 1.2.

(ii) $\Rightarrow$ (i) Let T be a tree on $\omega \times \kappa$. Let $(x,\alpha) \in B \Leftrightarrow T(x,\alpha)$ is not wellfounded. Let

$$\begin{aligned} x \in A &\Leftrightarrow \forall\alpha(x,\alpha) \in B \\ &\Leftrightarrow \forall\alpha,\ T(x,\alpha) \text{ is not wellfounded} \\ &\Leftrightarrow \langle B_\kappa, \varepsilon, x, T\rangle \models \forall\alpha\exists\xi(T \upharpoonright \xi(x,\alpha) \text{ is not wellfounded}) \\ &\quad (\text{since cofinality}(\kappa) > \omega), \\ &\Leftrightarrow \langle B_\kappa, \varepsilon, x, T\rangle \models \forall\alpha\exists\xi\exists f \in \xi^\omega\forall n(x \upharpoonright n, \alpha \upharpoonright n, f \upharpoonright n) \in T \\ &\Rightarrow \exists\kappa' < \kappa,\ \langle B_{\kappa'}, \varepsilon, x, T \upharpoonright \kappa'\rangle \models (*) \end{aligned}$$

(where $(*)$ is the Π_2 formula

$$\forall\alpha\exists\xi\exists f \in \xi^{\omega}(x \restriction n,\alpha \restriction n,f \restriction n) \in T))$$

$$\Rightarrow \exists\kappa' < \kappa,\ \forall\alpha[T \restriction \kappa'(x,\alpha) \text{ is not wellfounded}] .$$

Thus

$$x \in A \Leftrightarrow \forall\alpha[S(x,\alpha) \text{ is not wellfounded}] ,$$

where S is a tree on $\omega^2 \times \kappa'$. So

$$A \in \forall^{\mathbb{R}}\underset{\sim}{S}(\kappa') .$$

Since B is an arbitrary element of $\underset{\sim}{S}(\kappa)$ we have that $\underset{\sim}{S}(\kappa) \subseteq \forall^{\mathbb{R}}\underset{\sim}{S}(\kappa')$, so if $\forall^{\mathbb{R}}\underset{\sim}{S}(\kappa') \subseteq \underset{\sim}{S}(\kappa)$ we are done. Otherwise, by Wadge, $\underset{\sim}{S}(\kappa) \subseteq \exists^{\mathbb{R}}\check{\underset{\sim}{S}}(\kappa')$, so since $\underset{\sim}{S}(\kappa) \supseteq \check{\underset{\sim}{S}}(\kappa')$ we have $\underset{\sim}{S}(\kappa) = \exists^{\mathbb{R}}\check{\underset{\sim}{S}}(\kappa')$. Let $\bar{\kappa}$ be the largest Souslin cardinal $\leq \kappa'$ (which must exist since $\underset{\sim}{S}(\kappa)$ is $\mathbb{R}$-parametrized). Then $\underset{\sim}{S}(\kappa) = \exists^{\mathbb{R}}\check{\underset{\sim}{S}}(\bar{\kappa})$ and $\underset{\sim}{S}(\bar{\kappa})$ is a projective-like pointclass closed under $\exists^{\mathbb{R}}$, thus $\exists^{\mathbb{R}}\check{\underset{\sim}{S}}(\bar{\kappa}) = \underset{\sim}{S}(\kappa) \not\subseteq \forall^{\mathbb{R}}\underset{\sim}{S}(\bar{\kappa})$, a contradiction. $\dashv$

Note that from the preceding argument and the fact that for every Spector class Γ on $\mathbb{R}$ its associated ordinal δ is Mahlo, see Kechris-Kleinberg-Moschovakis-Woodin [A], we have that if either of the equivalent conditions (i), (ii) above hold then actually κ is also Mahlo.

As a corollary of 5.1 we see that $\kappa^{\mathbb{R}}$ is the least ${}^{b}\Pi^1_2$-indescribable Souslin cardinal.

<u>Conjecture.</u> The assumption cofinality$(\kappa) > \omega$ is not needed in 5.1.

<u>Problem.</u> Is $\kappa^{\mathbb{R}}$ the least Mahlo ${}^{b}\Pi^1_2$-indescribable cardinal?

(B) <u>The Prewellordering and scale properties for $\underset{\sim}{S}(\kappa)$.</u> By Kechris-Solovay-Steel [A], for each Souslin cardinal κ, either $\underset{\sim}{S}(\kappa)$ or $\check{\underset{\sim}{S}}(\kappa)$ has the prewellordering property. From 3.3 we see that, when $\kappa \leq \kappa^{\mathbb{R}}$, what distinguishes the first case from the second is whether cofinality$(\kappa) > \omega$ or not.

<u>Question.</u> Does this hold also for arbitrary κ?

Similarly about the scale property.

One fact that we have noticed is that if $\underset{\sim}{S}(\kappa)$ is closed under $\forall^{\mathbb{R}}$, then $\underset{\sim}{S}(\kappa)$ has the prewellordering property.

(C) <u>Properties of the Souslin cardinals.</u> <u>Question.</u> Do the properties 3.3 (iii), (iv) hold for arbitrary Souslin cardinals?

(D) About $\underset{\sim}{S}_\infty$. According to our conjecture in §2 we expect $\underset{\sim}{S}_\infty$ either to be all of power($\mathbb{R}$) or else to be $\mathbb{R}$-parametrized. In the latter case $\underset{\sim}{S}_\infty = \underset{\sim}{S}(\kappa)$, where κ is the largest Souslin cardinal, so by our remarks in (B) $\underset{\sim}{S}_\infty$ (being closed under $\forall^{\mathbb{R}}$) has also the prewellordering property i.e. is a Spector class on the structure of analysis $\mathbb{R}$. We mentioned in §1 that assuming $AD + V = L[\mathbb{R}]$ $\underset{\sim}{S}_\infty = \underset{\sim}{\Sigma}^2_1$. It is conceivable that in some reasonable theory extending AD one can prove that if $\underset{\sim}{S}_\infty$ is $\mathbb{R}$-parametrized, then $\underset{\sim}{S}_\infty \subseteq \underset{\sim}{\Sigma}^2_1$. It is already known (in AD only) that there are some forbidden values for $\underset{\sim}{S}_\infty$ when we go past $\underset{\sim}{\Sigma}^2_1$. For example, $\underset{\sim}{S}_\infty$ cannot be $\underset{\sim}{\Sigma}^2_n$ for $n > 1$ or $\underset{\sim}{\Pi}^2_n$ for $n \geq 1$, and similarly for $\underset{\sim}{\Sigma}^k_n$, $\underset{\sim}{\Pi}^k_n$ for all $k \geq 3$, $n \geq 1$. Also it cannot be $\underset{\sim}{\Delta}^k_n$ for any $k + n > 3$. These observations are based on the fact that

$$\{A : A \in \underset{\sim}{S}_\infty\} \text{ is } \underset{\sim}{\Sigma}^2_1 ,$$

(as a collection of sets of reals).

(E) Reliable cardinals. According to Becker [1979] an ordinal λ is called reliable if there is a regular scale $\{\varphi_i\}$ on a set $A \subseteq \mathbb{R}$ such that $\{\varphi_i(\alpha) : i \in \omega, \alpha \in A\} = \lambda$. Clearly every Souslin cardinal is reliable.

Conjecture. (AD). At least for cardinals $\leq \kappa^{\mathbb{R}}$, the notions of being Souslin and reliable coincide.

If one goes back to our analysis of the hyperprojective hierarchy in §3 it is easy to check that the following conjecture (which is motivated by some conjectures in Kechris [1978]) implies the preceding one.

Conjecture. (AD). If κ is a Souslin cardinal and $\underset{\sim}{S}(\kappa)$ has the scale property, then any strictly increasing wellordered sequence of sets in $\underset{\sim}{S}(\kappa)$ has length $< \kappa^+$.

This is not even known for $\kappa = \aleph_1$ i.e. for $\underset{\sim}{S}(\kappa) = \underset{\sim}{\Sigma}^1_2$. It is known by a different argument that there are no reliable cardinals between $\aleph_1$ and $\aleph_\omega$. But it is not known if there are any reliable cardinals between $\underset{\sim}{\delta}^1_3 = \aleph_{\omega+1}$ and the predecessor of $\underset{\sim}{\delta}^1_5$. (The proof that there are no reliable cardinals between $\aleph_1$ and $\aleph_\omega$ is based on the following more general fact.

Proposition (AD). Assume λ is reliable and there is a tree V on $\omega \times \lambda$, with $\sup\{\text{rank}(V(\alpha)) : V(\alpha) \text{ is wellfounded}\} = \lambda^+$. Then λ^+ is regular.)

References

H. Becker [1979], Some applications of ordinal games, Ph.D. Thesis, UCLA (1979).

A. S. Kechris [1978], On transfinite sequences of projective sets with an application to Σ_2^1 equivalence relations, Logic Colloquium '77, A. Macintyre, L. Pacholski, J. Paris (Eds.), North-Holland Publishing Co., (1978), 155-160.

A. S. Kechris, E. M. Kleinberg, Y. N. Moschovakis and H. Woodin [A], The Axiom of Determinacy, strong partition properties and nonsingular measures, this volume.

A. S. Kechris, R. M. Solovay and J. R. Steel [A], The Axiom of Determinacy and the prewellordering property, this volume.

Y. N. Moschovakis [1970], Determinacy and prewellorderings of the continuum, Math. Logic and Found. of Set Theory, Y. Bar-Hillel ed., North Holland (1970), 24-62.

Y. N. Moschovakis [1974], Elementary Induction on Abstract Structures, North Holland (1974).

Y. N. Moschovakis [1978], Inductive scales on inductive sets, Cabal Seminar 76-77, Lecture Notes in Mathematics, Springer-Verlag, Vol. 689 (1978), 185-192.

Y. N. Moschovakis [1979], Scales on coinductive sets, mimeographed notes, (1979).

Y. N. Moschovakis [1980], Descriptive Set Theory, North Holland (1980).

R. M. Solovay [1978], The independence of DC from AD, Cabal Seminar 76-77, Lecture Notes in Mathematics, Springer-Verlag, Vol. 689, (1978), 171-184.

J. R. Steel [A], Closure properties of pointclasses, this volume.

CLOSURE PROPERTIES OF POINTCLASSES

John R. Steel
Department of Mathematics
University of California
Los Angeles, California 90024

We work in ZF + AD + DC throughout this paper. Our aim is to show that certain closure and structural properties of a nonselfdual pointclass Γ follow from closure properties of the corresponding Δ together with the regularity of the Wadge ordinal of Δ. Let ${}^3E$ be the type 3 object embodying quantification over the reals, and $o({}^3E)$ the least ordinal not the order type of a prewellorder of the reals recursive in ${}^3E$. Our results imply that $o({}^3E)$ is the least regular limit point in the sequence of Suslin cardinals defined in Kechris [1980].

The methods and most of the results of the paper fall squarely within the province of "Wadge degrees," which might more informatively be titled "the general theory of arbitrary pointclasses." (Sections 1 to 3 of Van Wesep [1978] contain the necessary background material.) Surprisingly, AD is powerful enough to yield nontrivial theorems in this generality. Now, when working in such generality, it is natural to ask: which pointclasses (identified perhaps by means of their Wadge ordinals) have the closure and structural properties which make them amenable to the standard techniques of descriptive set theory. Our results bear on this question.

The author wishes to acknowledge the contribution of A. S. Kechris to this work. In a sense, the work was commissioned by him.

Some notation and terminology: We let $\mathbb{R}$ be ${}^\omega\omega$, the Baire space, and call its elements reals. If $\ell \geq 1$, then the product space $\omega^k \times ({}^\omega\omega)^\ell$ is homeomorphic to ${}^\omega\omega$, and we shall identify the two. For $A, B \subseteq {}^\omega\omega$ we say A is Wadge reducible to B, and write $A \leq_W B$, iff $\exists f : {}^\omega\omega \to {}^\omega\omega$ (f is continuous $\wedge$ $A = f^{-1}(B)$). The partial order $\leq_W$ is wellfounded; $|A|_W$ is the ordinal rank of A in $\leq_W$. A pointclass is a class of subsets of ${}^\omega\omega$ closed downward under $\leq_W$. The dual of a pointclass Γ, denoted $\breve{\Gamma}$, is $\{{-A} \mid A \in \Gamma\}$. Here, as later, complements are taken relative to ${}^\omega\omega$. If Γ is nonselfdual, i.e. $\Gamma \neq \breve{\Gamma}$, then we set $\Delta = \Gamma \cap \breve{\Gamma}$. If Γ is any pointclass, we set $o(\Gamma) = \sup\{|A|_W \mid A \in \Gamma\}$. If Γ is nonselfdual and $o(\Gamma)$ a limit ordinal, then of course $o(\Gamma) = o(\Delta)$.

We write $\mathrm{Sep}(\Gamma)$, $\mathrm{Red}(\Gamma)$, or $\mathrm{PWO}(\Gamma)$ to mean that Γ has, respectively, the separation, reduction, or prewellordering property. The closure of Γ under existential real quantification is given by

$$\exists^{\mathbb{R}}\Gamma = \{A \mid \exists B \in \Gamma\ \forall x(x \in A \Leftrightarrow \exists y(x,y) \in B)\} .$$

We let $\forall^{\mathbb{R}}\Gamma$ be the dual of $\exists^{\mathbb{R}}\breve{\Gamma}$. Similarly, the class of wellordered unions of length α of Γ sets is

$$\bigcup_\alpha \Gamma = \left\{ \bigcup_{\gamma<\alpha} A_\gamma \mid \forall \gamma < \alpha(A_\gamma \in \Gamma)\right\}$$

and $\bigcap_\alpha \Gamma$ is the dual of $\bigcup_\alpha \breve{\Gamma}$.

The selfdual pointclasses Δ which we consider will often satisfy $\exists^{\mathbb{R}}\Delta \subseteq \Delta$. In this case, Lemma 2.3.1 of Kechris-Solovay-Steel [1980] states

$$\begin{aligned} o(\Delta) &= \sup\{\text{rank}(\prec) \mid \prec \text{ is a wellfounded relation in } \Delta\} \\ &= \sup \text{rank}(\prec) \mid \prec \text{ is a prewellorder in } \Delta\} . \end{aligned}$$

§1. Consequences of the separation property. The key to the transfer of closure properties from Δ to Γ is the separation property. We begin with some simple results in this vein. Notice that the hypothesis $\mathrm{Sep}(\Gamma)$ of Theorem 1 serves only to distinguish Γ from $\breve{\Gamma}$, since by Steel [1980a] and Van Wesep [1978], exactly one of $\mathrm{Sep}(\Gamma)$ and $\mathrm{Sep}(\breve{\Gamma})$ holds.

Theorem 1.1. Let Γ be nonselfdual, and suppose $\mathrm{Sep}(\Gamma)$. Then

(a) $\bigcup_2 \Delta \subseteq \Delta \Rightarrow \bigcup_2 \Gamma \subseteq \Gamma$,
$\bigcup_\omega \Delta \subseteq \Delta \Rightarrow \bigcup_\omega \Gamma \subseteq \Gamma$;

(b) $\exists^{\mathbb{R}}\Delta \subseteq \Delta \Rightarrow \exists^{\mathbb{R}}\Gamma \subseteq \Gamma$;

(c) $(\exists^{\mathbb{R}}\Delta \subseteq \Delta \wedge \alpha < \mathrm{cof}(o(\Delta))) \Rightarrow (\bigcup_\alpha \Delta \subseteq \Delta \wedge \bigcup_\alpha \Gamma \subseteq \Gamma)$.

Proof. (a) Suppose $\bigcup_2 \Delta \subseteq \Delta$ and $\bigcup_2 \Gamma \not\subseteq \Gamma$. Since $\bigcup_2 \Gamma$ is a pointclass, i.e. closed downward under $\leq_w$, Wadge's lemma implies $\breve{\Gamma} \subseteq \bigcup_2 \Gamma$. Let $A \in \breve{\Gamma} - \Gamma$, and $A = B \cup C$ where $B,C \in \Gamma$. By $\mathrm{Sep}(\Gamma)$ we have $D,E \in \Delta$ so that

$$B \subseteq D \subseteq A$$

and

$$C \subseteq E \subseteq A .$$

But then $A = D \cup E$, so $A \in \Delta$, a contradiction. The proof that $\bigcup_\omega \Delta \subseteq \Delta \Rightarrow$

$\bigcup_\omega \Gamma \subseteq \Gamma$ is the same.

(b) It is enough to show that if A and B are disjoint sets in $\exists^{\mathbb{R}}\Gamma$, then A is separable from B by a set in Δ. We use the idea of Addison's proof of $\mathrm{Sep}(\underset{\sim}{\Sigma}^1_3)$ for this. Let

$$A(x) \Leftrightarrow \exists y P(x,y) ,$$

$$B(x) \Leftrightarrow \exists y Q(x,y) ,$$

where $P,Q \in \Gamma$. Define

$$P'(x,y,z) \Leftrightarrow P(x,y) ,$$

$$Q'(x,y,z) \Leftrightarrow Q(x,z) ,$$

and by $\mathrm{Sep}(\Gamma)$ let $D \in \Delta$ and $P' \subseteq D \subseteq -Q'$. Define

$$C(x) \Leftrightarrow \exists y \forall z D(x,y,z) .$$

Then $C \in \Delta$ since $\exists^{\mathbb{R}}\Delta \subseteq \Delta$. It is easy to check that $A \subseteq C \subseteq -B$.

(c) We extend the proof of (a). Suppose $\alpha < \mathrm{cof}(o(\Delta))$ and $\exists^{\mathbb{R}}\Delta \subseteq \Delta$, but $\bigcup_\alpha \Gamma \not\subseteq \Gamma$. Then $\breve{\Gamma} \subseteq \bigcup_\alpha \Gamma$ by Wadge's lemma. Let $A \in \breve{\Gamma} - \Gamma$, and $A = \bigcup_{\beta<\alpha} A_\beta$ where each $A_\beta \in \Gamma$. Since $\mathrm{Sep}(\Gamma)$ and $\alpha < \mathrm{cof}(o(\Delta))$, we can find a set $B \in \Delta$ so that

$$\beta < \alpha \Rightarrow \exists C \leq_W B(A_\beta \subseteq C \subseteq A) .$$

Let $\varphi : \mathbb{R} \xrightarrow{\text{onto}} \alpha$ be a Δ-norm of length α. By the Coding Lemma of Moschovakis [1970] there is a relation R in Δ so that $\forall x \exists y R(x,y)$ and $R(x,y) \wedge \varphi(x) = \beta \Rightarrow A_\beta \subseteq B_y \subseteq A$. (Here B_y is the set $\leq_W B$ via the strategy y.) But then

$$z \in A \Leftrightarrow \exists x \exists y (R(x,y) \wedge z \in B_y) ,$$

so $A \in \Delta$, a contradiction.

The proof that $\bigcup_\alpha \Delta \subseteq \Delta$ is the same. ⊠

Part (b) is due to Kechris, and part (a) to Kechris and the author independently. Notice that the proof of (c) gives slightly more: if $\exists^{\mathbb{R}}\Delta \subseteq \Delta$, $\alpha < \mathrm{cof}(o(\Delta))$, and $\langle A_\beta \mid \beta < \alpha \rangle$ is any sequence of sets each of which is Δ-separable from a fixed set A, then $\bigcup_{\beta<\alpha} A_\beta$ is Δ-separable from A. This fact will be important in the proof of Theorem 2.1.

Theorem 1.1 leaves us the question: given that $\mathrm{Sep}(\Gamma)$ and $\exists^{\mathbb{R}}\Delta \subseteq \Delta$, must $\bigcap_2 \Gamma \subseteq \Gamma$? The next theorem provides a class of examples showing the extent to which $\bigcap_2 \Gamma \subseteq \Gamma$ can in fact fail. The theorem results from analysis of the example of a Type II hierarchy given in Kechris [1977], p. 260.

If $A \subseteq \mathbb{R}$ and $\varphi : A \xrightarrow{\text{onto}} \lambda$, we say φ is Γ-bounded just in case whenever $B \subseteq A$ and $B \in \Gamma$ there is a $\beta < \lambda$ so that $\varphi''B \subseteq \beta$.

Theorem 1.2. Suppose $\mathrm{Sep}(\Gamma)$ and $\bigcup_2 \Delta \subseteq \Delta$. Let $A \in \Delta$ and $\varphi : A \xrightarrow{\text{onto}} \lambda$ be $\underset{\sim}{\Sigma}^1_1$ bounded, where $\lambda = \mathrm{cof}(o(\Delta))$. Then for some $B \in \Gamma$, $A \cap B \notin \Gamma$.

Proof. Let $\{\nu_\alpha | \alpha < \lambda\}$ be cofinal in $o(\Delta)$. Let $W \subseteq \mathbb{R}^2$ be a universal set in Γ. Consider the Solovay game:

$$\begin{array}{ll} \mathrm{I} & x \\ \mathrm{II} & \langle y,z \rangle \end{array}$$

Player II wins iff $x \in A \Rightarrow (W_y = -W_z \wedge |W_y|_w \geq \nu_{\varphi(x)})$.

Since φ is $\underset{\sim}{\Sigma}^1_1$ bounded, II must have a winning strategy σ. Let

$$R(x,y) \Leftrightarrow x \in A \wedge y \notin W_{\sigma(x)_1} .$$

Since $\breve{\Gamma}$ is closed under intersection by 1.1(a), $R \in \breve{\Gamma}$. But $\{|R_x|_w | x \in A\}$ is unbounded in $o(\Delta)$, so $R \notin \Delta$, and thus $R \notin \Gamma$. On the other hand,

$$R(x,y) \Leftrightarrow x \in A \wedge y \in W_{\sigma(x)_0} ,$$

so that $R = A \cap B$ for some $B \in \Gamma$. ⊠

Theorem 1.2 implies, for example, that if $\exists^{\mathbb{R}}\Delta \subseteq \Delta$, $\mathrm{Sep}(\Gamma)$, and $\mathrm{cof}(o(\Delta)) = \underset{\sim}{\delta}^1_n$, then Γ is not closed under intersections with Π^1_n sets. [Let $S \subseteq \mathbb{R}^3$ be universal Σ^1_n, and let $A = \{x | S_x \text{ is wellfounded}\}$. Then A is Π^1_n, and the map $\varphi : A \xrightarrow{\text{onto}} \underset{\sim}{\delta}^1_n$, where $\varphi(x) = \text{rank of } S_x$, is $\underset{\sim}{\Sigma}^1_1$-bounded.] On the other hand, Theorem 2.1 to follow implies that under these hypotheses on Γ and Δ, Γ is closed under intersections with $\underset{\sim}{\Sigma}^1_n$ sets.

On a grosser scale, one can modify the proof of 1.2 slightly (replacing the Solovay game by the Coding Lemma) to show that if $\mathrm{Sep}(\Gamma)$, $\exists^{\mathbb{R}}\Delta \subseteq \Delta$, and $o(\Delta)$ is singular, then Γ is not closed under intersections with Δ sets.

Theorem 1.2 implies that the hypothesis "$\alpha < \mathrm{cof}(o(\Delta))$" in 1.1(c) is necessary. For consider any Γ such that $\mathrm{Sep}(\Gamma)$, $\exists^{\mathbb{R}}\Delta \subseteq \Delta$, and $\mathrm{cof}(o(\Delta)) = \omega_1$. Let $\varphi : A \xrightarrow{\text{onto}} \omega_1$ be a Π^1_1 norm on a complete Π^1_1 set. Define R as in 1.2. Then $R = \bigcup_{\alpha<\omega_1} R_\alpha$ where

$$R_\alpha(x,y) \Leftrightarrow \varphi(x) \leq \alpha \wedge y \in W_{\sigma(x)_1} .$$

But each R_α is in Δ by Theorem 2.1.

Kechris and Martin have located the pointclass Γ such that $o(\Delta) = \omega_2$

with the aid of 1.2. Namely, let Γ be the class of $\omega - \underset{\sim}{\Pi}^1_1$ sets, that is, sets of the form

$$A = \bigcup_{n<\omega} A_{2n} - A_{2n+1}$$

where $\langle A_n | n < \omega \rangle$ is a decreasing sequence of $\underset{\sim}{\Pi}^1_1$ sets. Then Γ is nonselfdual, and both Γ and $\breve{\Gamma}$ are closed under intersections with $\underset{\sim}{\Pi}^1_1$ sets. By 1.2 we have $o(\Delta) \geq \omega_2$. By analyzing the ordinal games associated to Wadge games involving sets in Δ, Martin showed $o(\Delta) \leq \omega_2$. Thus $o(\Delta) = \omega_2$.

It is unpleasant to have a natural ordinal like ω_2 assigned to an unnatural class like $\omega - \underset{\sim}{\Pi}^1_1$. Solovay has shows that we get a more natural assignment if we replace $\leq_w$ by the somewhat coarser $\leq_\sigma$, where

$$B \leq_\sigma A \text{ iff } \exists \langle A_n | n < \omega \rangle (\forall n (A_n \leq_w A) \wedge B \leq_w \bigcup_{n<\omega} A_n) .$$

The ordinal ω_1 is assigned now to $\underset{\sim}{\Pi}^1_1$, ω_2 to $A(\underset{\sim}{\Pi}^1_1)$, ω_3 to $A(A(\underset{\sim}{\Pi}^1_1))$, etc. (Here "A" denotes Suslin's operation A.) The ordering $\leq_\sigma$ behaves much like the order $\leq_m$ of jump operators defined and studied in Steel [1980b]. For example, the wellfoundedness of $\leq_\sigma$ can be proved by a direct diagonal argument like that of Lemma 3 of that paper.

§2. Applications of the Martin-Monk method. We return to our closure questions. The limitations established by 1.2 clearly rely on the singularity of $o(\Delta)$. Suppose then that $\text{Sep}(\Gamma)$, $\exists^{\mathbb{R}}\Delta \subseteq \Delta$, and $o(\Delta)$ is regular; does it follow that $\bigcap_2 \Gamma \subseteq \Gamma$? We believe so, but at present have a proof only for the case that every set in Δ is κ-Suslin for some $\kappa < o(\Delta)$. This partial result will be enough for our characterization of $o({}^3E)$, since by Kechris [1980], every set A inductive over $\mathbb{R}$ is κ-Suslin for some $\kappa \leq |A|_w$. The key to our partial result is the following theorem.

Theorem 2.1. Let Γ be nonselfdual and suppose $\exists^{\mathbb{R}}\Delta \subseteq \Delta$. Let A be κ-Suslin, where $\kappa < \text{cof}(o(\Delta))$. Then for any $B \in \Gamma$, $A \cap B \in \Gamma$.

Proof. This follows from 1.1(a) if $\text{Sep}(\breve{\Gamma})$ holds, so assume $\text{Sep}(\Gamma)$. Let A, B, and κ be as in the hypotheses, and suppose for a contradiction that $A \cap B \notin \Gamma$. Let σ be a winning strategy for I in the Wadge game $G_w(A \cap B, B)$ (cf. Van Wesep [1978]). Thus whenever $\sigma(x) \in A$, we have

$$x \in B \Leftrightarrow \sigma(x) \notin B .$$

We shall use the fact that σ flips membership in B this way to get a contradiction like that in Martin's proof that $\leq_w$ is wellfounded.

Specifically, we define a sequence $\langle \sigma_n | n < \omega \rangle$ of winning strategies for I in $G_w(A \cap B, B)$. Let τ be the copying strategy for II, i.e., let $\forall x(\tau(x) = x)$. For any $x \in {}^{\omega}2$, define

$$\tau_n = \begin{cases} \sigma_n & \text{if } x(n) = 0 \\ \tau & \text{if } x(n) = 1 . \end{cases}$$

Consider the diagram of games

$$\begin{array}{lcccccc} \cdots\cdots\cdots \tau_2 & & & \tau_1 & & \tau_0 & \\ & x_2(0) & & x_1(0) & & x_0(0) \\ & & & x_1(1) & & x_0(1) \\ & \cdot & & \cdot & & x_0(2) \\ & \cdot & & \cdot & & \vdots \\ & \cdot & & \cdot & & \\ \cdots\cdots\cdots & x_2 & & x_1 & & x_0 \end{array}$$

The rule here is: $x_n = \tau_n(x_{n+1})$. If $x(n) = 0$ for infinitely many n, then because each σ_n is a strategy for I, there is a unique such sequence $\langle x_n | n < \omega \rangle$. We shall define the σ_n's so that for any $x \in {}^{\omega}2$, if $x(n) = 0$ for infinitely many n and $\langle x_n | n < \omega \rangle$ is derived from x in this way, then $x_n \in A$ for all n.

Suppose we have done this; then the standard Martin argument leads to a contradiction. For let $I = \{x \in {}^{\omega}2 | x(n) = 0$ for infinitely many $n\}$. Define

$$M = \{x \in I | x_0 \in B\} .$$

Since M has the Baire property we have a basic interval $[s]$ determined by some $s \in 2^{<\omega}$ on which M is either meager or comeager. Pick $i \notin \text{dom}(s)$, and let

$$T(x)(k) = \begin{cases} x(k) & \text{if } i \neq k \\ 1 - x(k) & \text{if } i = k . \end{cases}$$

Then T is a homeomorphism and $T''[s] = [s]$. If $x \in I$, then

$$T(x)_k = x_k \quad \text{for } k > i ,$$

and

$$T(x)_k \in B \text{ iff } x_k \notin B \text{ for } k \leq i \,.$$

Thus $T''(M \cap [s]) = -M \cap I \cap [s]$. Since I is comeager, this contradicts our choice of s.

We now define the σ_n's by induction on n. Let T be a tree on $\omega \times \kappa$ such that $A = p([T]) = \{x \in {}^\omega\omega \mid \exists f \in {}^\omega\kappa((x,f) \in [T])\}$. (Here $[T]$ is the set of infinite branches of T.) As we define σ_n we shall associate to any $\langle \tau_n \mid i \leq n\rangle$ such that $\tau_n = \sigma_n$ and $\forall i < n$ $(\tau_i = \sigma_i$ or $\tau_i = \tau)$, and any $i \leq n$ such that $\tau_i = \sigma_i$, a sequence of ordinals $\langle \xi_0, \ldots, \xi_n\rangle = \xi_{\vec{\tau},i}$. We arrange that for any $z \in {}^\omega\omega$, if the partial diagram

$$\begin{array}{ccccccccc} \vdots & \tau_n & \vdots & \tau_{n-1} & \vdots & \cdots\cdots & \vdots & \tau_0 & \vdots \\ z & & x_n & & x_{n-1} & \cdots\cdots & x_1 & & x_0 \end{array}$$

is filled in as before (i.e. $x_k = \tau_k(x_{k+1})$), but setting $x_{n+1} = z$, then

(i) $(x_i \restriction n+1, \langle \xi_0, \ldots, \xi_n\rangle) \in T$,

and

(ii) $z \notin B \Rightarrow \exists f(f \restriction n+1 = \langle \xi_0, \ldots, \xi_n\rangle \wedge (x_i, f) \in [T])$.

Moreover these sequences of ordinals cohere in the natural way, that is,

(iii) $\vec{\xi}_{\vec{\tau}\restriction k+1, i} = \vec{\xi}_{\vec{\tau},i} \restriction k+1$, for $i \leq k < n$.

It will be enough to define $\langle \sigma_i \mid i < \omega\rangle$ with associated $\vec{\xi}$'s satisfying (i)-(iii). For then suppose $x \in {}^\omega 2$ and $x(n) = 0$ for infinitely many n. Let $\tau_i = \sigma_i$ if $x(i) = 0$, and $\tau_i = \tau$ otherwise. Define x_n by: $x_n = \tau_n(x_{n+1})$. For $x(n) = 0$, let

$$f = \bigcup_{x(k)=0} \vec{\xi}_{\langle \tau_0 \ldots \tau_k\rangle, n} \,.$$

Then $f \in {}^\omega\kappa$ by (iii), and by (i), $(x_n \restriction k, f \restriction k) \in T$ for all k. Thus $x_n \in A$. If $x(n) \neq 0$, then $x_n = x_i$ for some i such that $x(i) = 0$. Thus $x_n \in A$ for all n, and we are done.

We now define σ_n. Suppose that σ_i is defined for $i < n$, together with associates satisfying (i)-(iii) above. We define σ_n in 2^n steps, one for each $\langle \tau_i \mid i < n\rangle$ with $\tau_i = \sigma_i$ or $\tau_i = \tau$ for all $i < n$. After step ℓ we have a Δ-inseparable pair $C_\ell \subseteq -B$ and $D_\ell \subseteq B$ with $D_\ell \in \Gamma$. We will have $C_{\ell+1} \subseteq C_\ell$ and $D_{\ell+1} \subseteq D_\ell$ for $\ell < 2^n$.

<u>Step 0</u>. For each $\vec{\xi} \in \kappa^n$ set

$$A_{\vec{\xi}} = \{x \mid \exists f(f \restriction (n+1) = \vec{\xi} \wedge (x,f) \in [T])\} \,.$$

Thus $A = \bigcup_{\vec{\xi} \in {}^n\kappa} A_{\vec{\xi}}$. Since $\sigma''(-B) \subseteq A$ and $\kappa < \mathrm{cof}(o(\Delta))$, the remark immediately after 1.1 implies that for some $\vec{\xi}$, $-B \cap \sigma^{-1}(A_{\vec{\xi}})$ is Δ-inseparable from B. Fix such a $\vec{\xi}$, and let

$$C_0 = -B \cap \sigma^{-1}(A_{\vec{\xi}})$$

and

$$D_0 = \{x \in B \mid (\sigma(x) \restriction n + 1, \vec{\xi}) \in T\} .$$

$D_0 \in \Gamma$ since Γ is closed under intersections with clopen sets, and one easily checks that D_0 is Δ-inseparable from C_0. Finally, at this step we set $\vec{\xi}_{\vec{\tau},n} = \vec{\xi}$ for all $\vec{\tau} = \langle \tau_i \mid i \leq n \rangle$ such that $\tau_i = \sigma_i$ or $\tau_i = \tau$ for $i < n$, and $\tau_n = \sigma_n$. (Or more precisely, we commit ourselves to doing so once we have defined σ_n.)

<u>Step $k + 1$.</u> We have (C_k, D_k) from the last step, and we are considering $\langle \tau_i \mid i < n \rangle$. Let $i < n$ be largest such that $\tau_i = \sigma_i$; if no such i exists set $C_{k+1} = C_k$, $D_{k+1} = D_k$, and go to step $k + 2$ without defining any new associates. For each $j < n$ such that $\tau_j = \sigma_j$, let $\vec{\xi}_j = \vec{\xi}_{\langle \tau_\ell \mid \ell \leq i \rangle, j}$. Besides defining C_{k+1} and D_{k+1}, we want to extend these associates.

For each $z \in {}^\omega \omega$ consider the diagram

$$\begin{array}{ccccccccc} \sigma & & \tau_{n-1} & & \cdots & & & \cdots & \tau_0 \\ \vdots & & \vdots & & \vdots & & \vdots & & \vdots \\ z & & z_n & & \cdots & z_{i+1} & z_i & \cdots & z_0 \end{array}$$

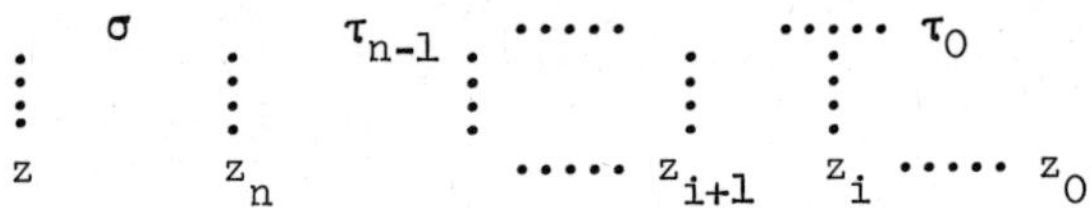

filled in as before by setting $z_{n+1} = z$.

For $j < n$ so that $\tau_j = \sigma_j$, define

$$A_{\vec{\xi}_j} = \{x \mid \exists f (f \restriction (i + 1) = \vec{\xi}_j \wedge (x,f) \in [T]\} .$$

So (ii) of our inductive hypothesis on $\langle \tau_\ell \mid \ell \leq i \rangle$ and its associates says:

$$z_{i+1} \notin B \Rightarrow z_j \in A_{\vec{\xi}_j} .$$

Now notice that if $z \in D_k$ (so $z \in B$) and $\sigma(z) \in A$, then $z_n = z_{i+1} \notin B$, and thus $z_j \in A_{\vec{\xi}_j}$ for all $j \leq i$. Define

$$X = \{z \mid \sigma(z) \in A \wedge \exists j \leq i \, (z_j \notin A_{\vec{\xi}_j})\} .$$

Then $X \in \Delta$, and $D_k \cap X = \emptyset$. Since (C_k, D_k) is Δ-inseparable, $(C_k - X, D_k)$ must be Δ-inseparable.

Notice that for $z \in C_k - X$ we have $z \notin B$, so $\sigma(z) \in A$, and thus $z_j \in A_{\vec{\xi}_j}$ for all $j \leq i$ with $\tau_j = \sigma_j$. This enables us to use the argument of step 0 to successively thin down $C_k - X$ and D_k, once for each $j \leq i$ so that $\tau_j = \sigma_j$, retaining at each step an inseparable pair (C'_j, D'_j) with $D_j \in \Gamma$. At the step for j we also define the associate $\vec{\xi}_{\langle \tau_i | i \leq n \rangle, j}$ extending $\vec{\xi}_{\langle \tau_\ell | \ell \leq i \rangle, j}$ so that

(i)' $z \in D'_j \Rightarrow (z_j \restriction (n+1), \vec{\xi}_{\langle \tau_i | i \leq n \rangle, j}) \in T$

(ii)' $z \in C'_j \Rightarrow \exists f(f \restriction (n+1) = \vec{\xi}_{\langle \tau_i | i \leq n \rangle, j} \wedge (z_j, f) \in [T])$.

We define (C_{k+1}, D_{k+1}) to be the last pair in this process, and go to step $k+2$.

Now let $(C, D) = (C_{2^n}, D_{2^n})$ be the pair we have on completion of the last step. Consider the game in which I plays y, II plays z, and II wins iff

$$y \notin B \Rightarrow z \in C$$

and

$$y \in B \Rightarrow z \in D .$$

Then I has no winning strategy in this game. For if s is such a strategy, then $C \subseteq s^{-1}(B)$ and $s^{-1}(B) \cap D = \emptyset$. Since $s^{-1}(B)$ and D are disjoint Γ sets, they can be separated by a Δ set. But such a Δ set separates C and D, a contradiction.

Fix a winning strategy s for II, and let $\sigma_n = \sigma \circ s$. Given any $\langle \tau_j | j \leq n \rangle$ with $\tau_n = \sigma_n$, and given any $j < n$ with $\tau_j = \sigma_j$, the induction hypotheses (i) and (ii) for $\vec{\tau}$ and $\vec{\xi}_{\vec{\tau}, j}$ follow at once from (i)' and (ii)' for C'_j and D'_j at the step at which $\langle \tau_i | i < n \rangle$ was considered, and the fact that

$$s''(-B) \subseteq C \subseteq C'_j$$

and

$$s''(B) \subseteq D \subseteq D'_j .$$

The construction of $\langle \sigma_n | n < \omega \rangle$, and hence the proof of the theorem, is complete. ⊠

For Γ such that $\mathrm{Sep}(\Gamma)$ and $\exists^{\mathbb{R}} \Delta \subseteq \Delta$, it should be possible to specify exactly, as a function of $\mathrm{cof}(o(\Delta))$, those pointclasses Γ' so that Γ is closed under intersections with Γ' sets. Theorems 1.2 and 2.1 do this when $\mathrm{cof}(o(\Delta)) = \underset{\sim}{\delta}^1_n$; such a Γ is closed under intersections with $\underset{\sim}{\Sigma}^1_n$ sets, but not under intersections with $\underset{\sim}{\Pi}^1_n$ sets.

By combining 1.2 and 2.1 we obtain the following curious fact: let A be κ-Suslin and let $\varphi : A \xrightarrow{\text{onto}} \lambda$ be $\underset{\sim}{\Sigma}^1_1$-bounded. Then $\mathrm{cof}(\lambda) \leq \kappa$.

[Proof. Let Δ be such that $\exists^{\mathbb{R}}\Delta \subseteq \Delta$ and $\mathrm{cof}(o(\Delta)) = \mathrm{cof}(\lambda)$. Let $\Delta = \Gamma \cap \breve{\Gamma}$, where $\mathrm{Sep}(\Gamma)$ holds. By 1.2, Γ is not closed under intersections with A. By 2.1 then, $\mathrm{cof}(o(\Delta)) \leq \kappa$.] This fact is easy to prove for natural λ, e.g. $\lambda = \underset{\sim}{\delta}^1_n$, but we see no proof for arbitrary λ which does not use 1.2 and 2.1.

The basic method in the proof of 2.1 is due to D. Martin and L. Monk; as we mentioned, Martin used it to show $\leq_w$ is wellfounded. At present this method and variants on Wadge's lemma seem to be the only tools in pure Wadge theory.

We need one further preliminary closure result. Again, the Martin-Monk method is the key.

<u>Theorem 2.2</u>. Suppose $-\mathrm{Sep}(\Gamma)$ and $\bigcup_2 \Gamma \subseteq \Gamma$. Then $\bigcup_\omega \Gamma \subseteq \Gamma$.

<u>Proof</u>. Let (A_0, A_1) be a Δ-inseparable pair of Γ sets. Suppose that $\bigcup_\omega \Gamma \nsubseteq \Gamma$, so that $\breve{\Gamma} \subseteq \bigcup_\omega \Gamma$ by Wadge's lemma. Since $\bigcup_2 \Gamma \subseteq \Gamma$, we have $-(A_0 \cup A_1) \in \breve{\Gamma}$, and hence $-(A_0 \cup A_1) = \bigcup_n C_n$ for some sequence $\langle C_n \mid n < \omega\rangle$ of Γ sets.

Now $(A_0 \cup C_n, A_1)$ is a disjoint pair of Γ sets, and so the lemma of Steel [1980a] gives Lipschitz continuous maps f_n, $n < \omega$, such that

$$f_{2n}(A_0 \cup C_n) \subseteq A_0 \wedge f_{2n}(A_1) \subseteq A_1 ,$$

and

$$f_{2n+1}(A_0 \cup C_n) \subseteq A_1 \wedge f_{2n+1}(A_1) \subseteq A_0 .$$

We proceed to the usual contradiction. For any $x \in {}^\omega\omega$ let $\langle x_n \mid n < \omega\rangle$ be the unique sequence such that $x_n = f_n(x_{n+1})$. Suppose that $\{x \mid x_0 \in C_n\}$ is nonmeager; say comeager on the interval determined by $s \in \omega^{<\omega}$. Notice that

$$x_0 \in C_n \Rightarrow (s^\frown\langle 2n\rangle^\frown x)_0 \in A_0 \cup A_1 ,$$

so that for nonmeager many $y \supseteq s$, $y_0 \notin C_n$, a contradiction. On the other hand, suppose $\{x \mid x_0 \in A_i\}$ is nonmeager; say comeager on the interval determined by s. Notice that

$$x_0 \in A_i \Rightarrow (s^\frown\langle 1\rangle^\frown x)_0 \in A_{1-i} ,$$

so that for nonmeager many $y \supseteq s$, $y_0 \in A_{1-i}$, a contradiction. ⊠

The hypothesis $-\mathrm{Sep}(\Gamma)$ in 2.2 cannot be omitted, as witnessed by the case $\Gamma = \underset{\sim}{\Pi}^0_\alpha$. Theorem 1.2 shows that the hypothesis $\bigcup_2 \Gamma \subseteq \Gamma$ cannot be omitted, even if we assume strong closure properties of Δ. However, the proof of 2.2 can be modified to show that if $\bigcup_\omega \Delta \subseteq \Delta$ and $\langle A_n | n < \omega\rangle$ is any increasing sequence of Γ sets, then $\bigcup_n A_n \in \Gamma$.

Van Wesep [1978] shows that for any nonselfdual Γ, either -Sep(Γ) or -Sep($\breve{\Gamma}$). Thus if both $\bigcup_2 \Gamma \subseteq \Gamma$ and $\bigcup_2 \breve{\Gamma} \subseteq \breve{\Gamma}$, then either $\bigcup_\omega \Gamma \subseteq \Gamma$ or $\bigcup_\omega \breve{\Gamma} \subseteq \breve{\Gamma}$. It seems quite likely that if both $\bigcup_\omega \Gamma \subseteq \Gamma$ and $\bigcup_\omega \breve{\Gamma} \subseteq \breve{\Gamma}$, then either $A(\Gamma) \subseteq \Gamma$ or $A(\breve{\Gamma}) \subseteq \breve{\Gamma}$, where "A" denotes Suslin's operation A. More vaguely, one might guess that there is always an asymmetry between the closure properties of Γ and those of $\breve{\Gamma}$.

From 2.2 we obtain a prewellordering theorem for classes closed under one but not both of $\bigcup_\omega$ and $\bigcap_\omega$. The theorem is analogous to those of Kechris-Solovay-Steel [1980].

<u>Corollary 2.3</u>. Let Γ be nonselfdual, $\bigcup_\omega \Gamma \subseteq \Gamma$ and $\bigcap_2 \Gamma \subseteq \Gamma$, but $\bigcap_\omega \Gamma \not\subseteq \Gamma$. Then PWO($\Gamma$).

<u>Proof</u>. By 2.2, Sep($\breve{\Gamma}$) holds, as otherwise $\bigcup_\omega \breve{\Gamma} \subseteq \breve{\Gamma}$. But then, by 1.1(a), $\bigcup_\omega \Delta \not\subseteq \Delta$, as otherwise $\bigcup_\omega \breve{\Gamma} \subseteq \breve{\Gamma}$. Since $\bigcup_\omega \Gamma \subseteq \Gamma$, we have $\bigcup_\omega \Delta = \Gamma$. For $A \in \Gamma$, let $A = \bigcup_n B_n$ where each $B_n \in \Delta$, and set for $x \in A$

$$\varphi(x) = \text{least } n \text{ such that } x \in B_n .$$

Since $\bigcup_2 \Delta \subseteq \Delta$, φ is a Γ norm. ⊠

The corollary generalizes the fact that $\underset{\sim}{\Sigma}^0_\alpha$ has the prewellordering property for $\alpha < \omega_1$.

§3. <u>Bounded unions and prewellordering</u>. Let Γ be a pointclass. We say that a union $\bigcup_{\alpha<\beta} A_\alpha$ is Γ-bounded iff the associated norm

$$\varphi(x) = \mu\alpha[x \in A_\alpha]$$

is Γ-bounded.

<u>Theorem 3.1</u>. Let Δ be selfdual, $\exists^{\mathbb{R}}\Delta \subseteq \Delta$, and $\mathrm{cof}(o(\Delta)) > \omega$. Then the following are equivalent:

(a) $\Delta = \Gamma \cap \breve{\Gamma}$ for some nonselfdual Γ such that PWO(Γ);

(b) $\bigcup_{o(\Delta)} \Delta \not\subseteq \Delta$.

<u>Proof</u>. (a) ⇒ (b) is clear. Suppose then that $\bigcup_{o(\Delta)} \Delta \not\subseteq \Delta$, and let $\theta \leq o(\Delta)$ be least such that $\bigcup_\theta \Delta \not\subseteq \Delta$. Since $\mathrm{cof}(o(\Delta)) > \omega$, Theorem 3.1 of Van Wesep [1978] gives a nonselfdual Γ such that $\Gamma \cap \breve{\Gamma} = \Delta$. Assume w.l.o.g. Sep($\breve{\Gamma}$). It follows that $\breve{\Gamma} \neq \bigcup_\theta \Delta$, as otherwise PWO($\breve{\Gamma}$). Thus $\Gamma \subseteq \bigcup_\theta \Delta$. Define now

$$\Gamma^* = \left\{ \bigcup_{\alpha<\theta} A_\alpha \,\middle|\, \forall\alpha(A_\alpha \in \Delta) \wedge \bigcup_{\alpha<\theta} A_\alpha \text{ is } \underset{\sim}{\Sigma}^1_1 \text{ bounded} \right\}.$$

<u>Claim.</u> $\Gamma \subseteq \Gamma^*$.

<u>Proof.</u> Let $A \in \Gamma - \breve{\Gamma}$, and let $S \subseteq \mathbb{R}^2$ be a universal $\underset{\sim}{\Sigma}^1_1$ set. Let

$$C = \{x | S_x \subseteq A\} = \{x | \forall y(y \notin S_x \text{ or } y \in A)\} .$$

Since $\mathrm{cof}(o(\Delta)) > \omega$ and every $\underset{\sim}{\Sigma}^1_1$ set is ω-Suslin, Theorems 2.1 and 1.1(b) imply that $C \in \Gamma$. (One can show that $C \in \Gamma$ without using 2.1. For C can be defined in the form "$\forall$(open $\vee\ \Gamma$)," and it is easy to see that if Γ is non-selfdual and contains the Boolean algebra generated by the open sets, then Γ is closed under unions with open sets.) Thus $C = \bigcup_{\alpha<\theta} C_\alpha$ where each $C_\alpha \in \Delta$. Let

$$A_\alpha = \{y | \exists x(x \in C_\alpha \wedge y \in S_x)\} .$$

Then each $A_\alpha \in \Delta$, $A = \bigcup_{\alpha<\theta} A_\alpha$, and $\bigcup_{\alpha<\theta} A_\alpha$ is $\underset{\sim}{\Sigma}^1_1$-bounded by construction.

<u>Claim.</u> $\Gamma^* = \Gamma$.

<u>Proof.</u> It is easy to check, using boundedness, that $\forall^{\mathbb{R}}\Gamma^* \subseteq \Gamma^*$. By Wadge then, if $\Gamma^* \not\subseteq \Gamma$, then $\forall^{\mathbb{R}}\Gamma \subseteq \Gamma^*$. It is enough for a contradiction to show that $\Gamma^* \subseteq \exists^{\mathbb{R}}\Gamma$. In fact, we show $\bigcup_\theta \Delta \subseteq \exists^{\mathbb{R}}\Gamma$. The proof is standard, granted our first claim.

Let $\langle A_\alpha | \alpha < \theta\rangle$ be a sequence of Δ sets, and let $\varphi : C \xrightarrow{\text{onto}} \theta$ be a $\underset{\sim}{\Sigma}^1_1$-bounded norm, where $C \in \Gamma$. Such a φ exists by the first claim. Let $W \subseteq \mathbb{R}^2$ be a universal set in Γ. Consider the game: I plays x, II plays y. Player II wins iff

$$x \in C \Rightarrow \exists\beta\Big(\varphi(x) \le \beta \wedge \bigcup_{\alpha\le\beta} A_\alpha = W_y\Big) .$$

Since φ is $\underset{\sim}{\Sigma}^1_1$-bounded, I has no winning strategy. Let σ be a winning strategy for II. Then

$$x \in \bigcup_{\alpha<\theta} A_\alpha \Leftrightarrow \exists y(y \in C \wedge x \in W_{\sigma(y)}) .$$

Since $\bigcap_2 \Gamma \subseteq \Gamma$, we have $\bigcup_{\alpha<\theta} A_\alpha \in \exists^{\mathbb{R}}\Gamma$.

<u>Claim.</u> $\mathrm{PWO}(\Gamma^*)$.

<u>Proof.</u> Let $A = \bigcup_{\alpha<\theta} A_\alpha$, where each A_α is in Δ and the union is $\underset{\sim}{\Sigma}^1_1$ bounded and increasing. Define for $x \in A$

$$\varphi(x) = \mu\alpha[x \in A_\alpha] .$$

To see that φ is a Γ^* norm, notice e.g. that if S is $\underset{\sim}{\Sigma}^1_1$ and

$S \subseteq \{(x,y) | x < y\}$, then $T = \{x | \exists y((x,y) \in S)\}$ is $\underset{\sim}{\Sigma}^1_1$, and $T \subseteq A$. Thus $T \subseteq A_\alpha$ for some $\alpha < \theta$, and so

$$S \subseteq \bigcup_{\beta < \alpha} (A_\beta x - A_\beta) .$$

Similarly, $\leq_\varphi$ is in Γ^*. The three claims yield the theorem. ⊠

We use "$\underset{\sim}{\mathrm{Ind}}$" to denote the class of sets definable over $\mathbb{R}$ by positive elementary induction from parameters in $\mathbb{R}$.

<u>Theorem 3.2.</u> Let Γ be nonselfdual and $\Gamma \subseteq \underset{\sim}{\mathrm{Ind}}$. Suppose that $\exists^{\mathbb{R}}\Delta \subseteq \Delta$ and $o(\Delta)$ is regular. Then $\bigcup_\omega \Gamma \subseteq \Gamma$.

<u>Proof.</u> This follows from 1.1 if $\mathrm{Sep}(\Gamma)$ holds, so assume $\mathrm{Sep}(\check{\Gamma})$. Let

$$\Gamma^* = \left\{ \bigcup_{\alpha < o(\Delta)} A_\alpha \,\middle|\, \forall \alpha (A_\alpha \in \Delta) \wedge \bigcup_{\alpha < o(\Delta)} A_\alpha \text{ is } \Delta\text{-bounded} \right\} .$$

Clearly every set in Γ^* is a $\underset{\sim}{\Sigma}^1_1$ bounded union of Δ sets, and so the proof of 3.1 implies that $\Gamma^* \subseteq \Gamma$.

<u>Claim.</u> $\Gamma^* = \Gamma$.

<u>Proof.</u> Let $A \in \Gamma - \check{\Gamma}$, and let $A = \bigcup_{\alpha < \theta} A_\alpha$ where each $A_\alpha \in \Delta$ and θ is least such that $\bigcup_\theta \Delta \not\subseteq \Delta$. We may assume the A_α's are increasing. The Coding Lemma implies that $\langle |A_\alpha|_w | \alpha < \theta \rangle$ is cofinal in $o(\Delta)$. For $\alpha < \theta$, let

$$C_\alpha = \{(x,y) | y \in A_{\alpha+1} - A_\alpha \wedge x \text{ codes a continuous function } f_x$$
$$\text{such that } f_x^{-1}(A_\alpha) \subseteq A\} .$$

Now C_α is defined in the form "$\Delta \wedge \forall z(\Delta \Rightarrow \Gamma)$." Every set in Δ is κ-Suslin for some $\kappa < o(\Delta)$ by Corollary 3.5 of Kechris [1980]. Thus our Theorems 1.1 and 2.1 imply that $C_\alpha \in \Gamma$. The proof of 3.1 now shows that if $C = \bigcup_{\alpha < \theta} C_\alpha$, then $C \in \exists^{\mathbb{R}}\Gamma$.

Notice that $\exists^{\mathbb{R}}(\bigcup_\theta \Delta) = \bigcup_\theta \exists^{\mathbb{R}}\Delta = \bigcup_\theta \Delta$. So since $\Gamma \subseteq \bigcup_\theta \Delta$, $\exists^{\mathbb{R}}\Gamma \subseteq \bigcup_\theta \Delta$, and we may write

$$C = \bigcup_{\alpha < \theta} D_\alpha$$

where each $D_\alpha \in \Delta$, and the union in increasing. Let

$$z \in B_\alpha \Leftrightarrow \exists (x,y) \in D_\alpha \, \exists \beta \leq \alpha$$
$$(y \in A_{\beta+1} - A_\beta \wedge f_x(z) \in A_\beta) .$$

Then each B_α is in Δ by 1.1(c) and the fact that $\exists^{\mathbb{R}}\Delta \subseteq \Delta$. It is easy to check that $\bigcup_{\alpha<\theta} B_\alpha = A$. Finally, the union $\bigcup_{\alpha<\theta} B_\alpha$ is Δ-bounded, since any Δ set is of the form $f_x^{-1}(A_\beta)$ for some $\beta < \theta$ and some x. This proves the claim.

It is enough now to show $\bigcup_2 \Gamma \subseteq \Gamma$; by 2.2 we then have $\bigcup_\omega \Gamma \subseteq \Gamma$. Let $A, B \in \Gamma$ towards showing $A \cup B \in \Gamma$. Since $\mathrm{Sep}(\check{\Gamma})$, we have $\mathrm{Red}(\Gamma)$, and so we may assume $A \cap B = \emptyset$. Let $A = \bigcup_{\alpha<\theta} A_\alpha$ and $B = \bigcup_{\alpha<\theta} B_\alpha$, where the unions are Δ-bounded and increasing, and each A_α and B_α is in Δ. It is enough to show that the union $\bigcup_{\alpha<\theta} (A_\alpha \cup B_\alpha)$ is Δ-bounded. So let $C \in \Delta$ and $C \subseteq \bigcup_{\alpha<\theta} (A_\alpha \cup B_\alpha)$. Then by 1.1(a), $C \cap A \in \Gamma$. On the other hand, $C \cap A = C \cap (-B)$, and $C \cap (-B) \in \check{\Gamma}$ by 2.1 and the fact that C is κ-Suslin for some $\kappa < o(\Delta)$. Thus $C \cap A \in \Delta$, and hence $C \cap A \subseteq A_\alpha$ for some $\alpha < \theta$. Similarly, $C \cap B \subseteq B_\beta$ for some $\beta < \theta$. But then $C \subseteq A_\gamma \cup B_\gamma$, where $\gamma = \max(\alpha, \beta)$, and we are done. ⊠

The hypothesis that $\Gamma \subseteq \underset{\sim}{\mathrm{Ind}}$ in Theorem 3.2 was only used to conclude, via 2.1 and Corollary 3.5 of Kechris [1980], that Γ is closed under intersection with Δ sets. Thus the conclusion of 3.2 holds for arbitrary Γ such that $\exists^{\mathbb{R}}\Delta \subseteq \Delta$, $\bigcup_{o(\Delta)} \Delta \not\subseteq \Delta$, and $o(\Delta)$ is regular, and Γ is closed under intersections with Δ sets.

One can also show that 2.1 and Corollary 3.5 of Kechris [1980] apply for Γ a bit beyond $\underset{\sim}{\mathrm{Ind}}$. For example, one can weaken the hypothesis "$\Gamma \subseteq \underset{\sim}{\mathrm{Ind}}$" of 3.2 to "every Γ set is inductive in the complete coinductive set of reals."

We now define

$$C = \{o(\Delta) \mid \Delta \text{ is selfdual} \wedge \exists^{\mathbb{R}}\Delta \subseteq \Delta\} .$$

Clearly C is cub in θ. Theorem 3.1 of Kechris [1980] implies that for $\lambda \leq o(\underset{\sim}{\mathrm{Ind}})$ such that $\omega\lambda = \lambda$, the λth element of C is the λth Suslin cardinal. Thus our next theorem is actually the characterization of $o({}^3E)$ promised in the introduction.

<u>Theorem 3.3</u>. Let $o(\Delta)$ be the least regular limit cardinal in C. Then $\Delta = \Gamma \cap \check{\Gamma}$, where Γ is the boldface 2-envelope of ${}^3E$ (i.e. the class of sets of reals semirecursive in ${}^3E$ and a real). Thus $o(\Delta) = o({}^3E)$.

<u>Proof</u>. Clearly C is cub in $o(\underset{\sim}{\mathrm{Ind}})$, and since $o(\underset{\sim}{\mathrm{Ind}})$ is regular, we have $o(\Delta) \leq o(\underset{\sim}{\mathrm{Ind}})$ and $\Delta \subseteq \underset{\sim}{\mathrm{Ind}}$. (Actually, Kechris has shown that $o(\underset{\sim}{\mathrm{Ind}})$ is Mahlo, so $o(\Delta) < o(\underset{\sim}{\mathrm{Ind}})$.) By 3.2 and its proof, we have $\Delta = \Gamma \cap \check{\Gamma}$, where Γ is the class of Δ-bounded unions of Δ sets of length $o(\Delta)$. Thus $\forall^{\mathbb{R}}\Gamma \subseteq \Gamma$, $\bigcup_\omega \Gamma \subseteq \Gamma$, and by 3.1, $\mathrm{PWO}(\Gamma)$. In order to show that Γ contains the 2-envelope of ${}^3E$ it suffices to show that Δ is "uniformly closed

under $\exists^R$," in the following sense. Let $Q(x,y)$ and $R(x,y)$ be disjoint relations in Γ. Define

$$S(x) \Leftrightarrow \forall y(R(x,y) \vee Q(x,y)) \wedge \exists y Q(x,y)$$

and

$$T(x) \Leftrightarrow \forall y(R(x,y) \vee Q(x,y)) \wedge \forall y R(x,y) .$$

Then we must show that S and T are in Γ.

To see this, let

$$Q = \bigcup_{\alpha < o(\Delta)} Q_\alpha \quad \text{and} \quad R = \bigcup_{\alpha < o(\Delta)} R_\alpha$$

be representations of Q and R as increasing Δ-bounded unions of Δ sets. Define

$$S_\alpha(x) \Leftrightarrow \forall y(R_\alpha(x,y) \vee Q_\alpha(x,y)) \wedge \exists y Q_\alpha(x,y)$$

and

$$T_\alpha(x) \Leftrightarrow \forall y(R_\alpha(x,y) \vee Q_\alpha(x,y)) \wedge \forall y R_\alpha(x,y) .$$

Clearly $S_\alpha \subseteq S$ and $T_\alpha \subseteq T$. We show simultaneously that $S \subseteq \bigcup_\alpha S_\alpha$ and that the union is Δ-bounded. For let $D \subseteq S$ and $D \in \Delta$. Let $A = (D \times {}^\omega\omega) \cap Q$, and let $B = (D \cap {}^\omega\omega) \cap R$. Then A and B are disjoint Γ sets, and complementary on $D \times {}^\omega\omega$. Thus A and B are in Δ. Let α be such that $A \subseteq Q_\alpha$ and $B \subseteq R_\alpha$. Then $D \subseteq S_\alpha$, and we are done. An identical argument shows that $T \subseteq \bigcup_\alpha T_\alpha$ and that this union is Δ-bounded. Thus S and T are in Γ, as desired.

It is well known that the class of sets of reals is recursive in ${}^3E$ and a real has the closure properties we assumed of Δ, that is, it is closed under $\exists^R$ and its ordinal is regular. Since Δ was minimal with these properties, Δ is contained in this class. Thus Γ is contained in the 2-envelope of ${}^3E$. ⊠

Theorem 3.3 implies that $o({}^3E)$ is not Mahlo. It gives some evidence for the natural conjecture, due perhaps to Moschovakis, that $o({}^3E)$ is the least regular limit cardinal. Proof of this conjecture awaits further progress in computing upper bounds for the $\underset{\sim}{\delta}_n^1$'s.

We shall close with some remarks on projective-like hierarchies which extend and simplify some of the proofs of Kechris-Solovay-Steel [1980].

For us, a projective-like hierarchy is a sequence $\langle \Gamma_i \mid i < \omega \rangle$ of nonselfdual pointclasses such that

(i) $\forall^R \Gamma_i \subseteq \Gamma_i$ or $\exists^R \Gamma_i \subseteq \Gamma_i$, but not both, for all $i < \omega$,

and

(ii) $\forall^R \Gamma_0 \subseteq \Gamma_0$,

and

(iii) For all nonselfdual $\Gamma' \subseteq \Gamma_0 \cap \breve{\Gamma}_0$, $\Gamma_0 \neq \forall^R \Gamma'$.

Our definition is slightly more liberal then that of Kechris-Solovay-Steel, mainly because we do not require $\bigcap_\omega \Gamma_i \subseteq \Gamma_i$ and $\bigcup_\omega \Gamma_i \subseteq \Gamma_i$ for all i. (Condition (iii) is slightly more liberal than theirs, too.) It is easy to see that if $\langle \Gamma_i | i < \omega \rangle$ is a projective-like hierarchy, then $\bigcap_\omega \Gamma_i \subseteq \Gamma_i$ and $\bigcup_\omega \Gamma_i \subseteq \Gamma_i$ for all $i \geq 1$. Thus our hierarchies differ from those of Kechris-Solovay-Steel only in that they sometimes have an extra class Γ_0 tacked on at the beginning. Consideration of this class seems to simplify some proofs.

Let Γ be a nonselfdual pointclass closed under one but not both of $\forall^R$ and $\exists^R$. Let $\alpha = \sup\{\beta \in C | \beta < o(\Gamma)\}$. Then $\alpha \in C$, so $\alpha = o(\Delta)$ for some Δ. It is easy to see that either Γ or $\check{\Gamma}$ is in the least projective-like hierarchy $\langle \Gamma_i | i < \omega \rangle$ such that $\Delta \subseteq \Gamma_0$. We now show that for each $i < \omega$, either $PWO(\Gamma_i)$ or $PWO(\check{\Gamma}_i)$, thereby reproving one of the main results of Kechris-Solovay-Steel [1980]. We prove this by considering cases corresponding to the types I-IV of projective-like hierarchies defined in that paper. We assume throughout that $\bigcup_{o(\Delta)} \Delta \not\subseteq \Delta$.

Case 1. $cof(\alpha) = \omega$. (Type I).

In this case, $\Gamma_0 = \bigcap_\omega \Delta$. It is easy to see, and in fact implied by 2.3, that $PWO(\check{\Gamma}_0)$. The first periodicity theorem propagates prewellordering up to hierarchy $\langle \Gamma_i | i < \omega \rangle$.

Case 2. $cof(\alpha) > \omega$.

In this case, let $\Delta = \Gamma \cap \check{\Gamma}$, where we may assume by 1.1 that $\forall^R \Gamma \subseteq \Gamma$, and by 3.1 that $PWO(\Gamma)$.

Subcase A. $\exists^R \Gamma \not\subseteq \Gamma$.

In this case $\Gamma_0 = \Gamma$, $PWO(\Gamma_0)$, and the first periodicity propagates prewellordering again. The case occurs with hierarchies of types II and III; type II in the case $\bigcup_2 \Gamma \not\subseteq \Gamma$, and type III in the case $\bigcup_2 \Gamma \subseteq \Gamma$ (and hence $\bigcup_\omega \Gamma \subseteq \Gamma$).

Subcase B. $\exists^R \Gamma \subseteq \Gamma$. (Type IV).

In this case, $\Gamma_0 = \{A \cap B | A \in \Gamma \wedge B \in \Gamma\}$. The usual difference-hierarchy proof shows that $PWO(\Gamma_0)$. Again, first periodicity propagates prewellordering.

References

A. S. Kechris [1977], Classifying projective-like hierarchies, Bulletin of Greek Math. Soc., 18 (1977), 254-275.

A. S. Kechris [1980], Souslin cardinals, κ-Souslin sets, and the scale property in the hyperprojective hierarchy, this volume.

A. S. Kechris, R. M. Solovay and J. R. Steel [1980], The axiom of determinacy and the prewellordering property, this volume.

Y. N. Moschovakis [1970], Determinacy and prewellorderings of the continuum, Math. Logic and Foundations of Set Theory (Proc. Internat. Colloq., Jerusalem 1968), North Holland, Amsterdam, (1970), 24-62.

J. R. Steel [1980a], Determinateness and the separation property, to appear in J. Symb. Logic.

J. R. Steel [1980b], A classification of jump operators, to appear in J. Symb. Logic.

R. Van Wesep [1978], Wadge degrees and descriptive set theory, Cabal Seminar 76-77, Springer Lecture Notes in Mathematics, Vol. 689 (A. S. Kechris and Y. N. Moschovakis eds.), (1978), 151-171.

A NOTE ON WADGE DEGREES

Alexander S. Kechris[(1)]
Department of Mathematics
California Institute of Technology
Pasadena, California 91125

§1. It has been shown by Martin and independently Steel (see [1], Theorem 4.2), that the wellordering of Wadge degrees of $\underset{\sim}{\Delta}^1_{2n}$ sets of reals has length $\underset{\sim}{\delta}^1_{2n+1}$. Although this is a result about projective sets, their proofs require full AD as they proceed by showing that if η_{2n} is the length of the wellordering of Wadge degrees of $\underset{\sim}{\Delta}^1_{2n}$ sets, then cofinality$(\underset{\sim}{\delta}^1_{2n+1}) \leq \eta_{2n}$, which by the regularity of $\underset{\sim}{\delta}^1_{2n+1}$ (a consequence of AD) implies that $\underset{\sim}{\delta}^1_{2n+1} \leq \eta_{2n}$. As it is easy to see that $\eta_{2n} \leq \underset{\sim}{\delta}^1_{2n+1}$, by a direct computation, we have the desired equality. Of course, the use of full AD here can be replaced, by trivial absoluteness considerations, by Determinacy($L[\omega^\omega]$), i.e. the hypothesis that all sets of reals in $L[\omega^\omega]$ are determined. Motivated by the fact that one only needs Determinacy($\underset{\sim}{\Delta}^1_{2n}$) to establish the fact that the Wadge degrees of $\underset{\sim}{\Delta}^1_{2n}$ sets are wellordered (see [1], Theorem 2.2), Martin has asked if one can compute that also $\eta_{2n} = \underset{\sim}{\delta}^1_{2n+1}$, using again only Determinacy($\underset{\sim}{\Delta}^1_{2n}$). We provide such a proof below. It is based on a method of "inverting the game quantifier" which may be also useful elsewhere.

§2. Let us take $n = 1$ for notational simplicity. From now on we assume Determinacy($\underset{\sim}{\Delta}^1_2$).

2.1. For each $A \subseteq \omega^\omega$ let $\Gamma(A)$ be a pointclass with the following properties:

(i) $A, \omega^\omega - A \in \Gamma(A)$,

(ii) $B, \omega^\omega - B \in \Gamma(A) \Rightarrow \Gamma(B) \subseteq \Gamma(A)$,

(iii) $A \in \underset{\sim}{\Delta}^1_m \Rightarrow \Gamma(A) \subseteq \underset{\sim}{\Delta}^1_m$, $\forall m \geq 2$,

(iv) $\Gamma(A)$ is ω^ω-parametrized and closed under continuous substitutions,

(v) There is a map $A \mapsto C_A$, sending each A to C_A, an ω^ω-universal set in $\Gamma(A)$ and for each $m \geq 2$ there is a total recursive function f_m such that if $\varepsilon \in \omega^\omega$ is a $\underset{\sim}{\Delta}^1_m$-code of A, then $f(\varepsilon)$

---

(1) Research partially supported by NSF Grant MCS - 17254 A01. The author is an A.P. Sloan Foundation Fellow.

is a $\underset{\sim}{\Delta}^1_m$-code of C_A.

For example, we can take $\Gamma(A) = {}_2\underset{\sim}{\mathrm{ENV}}({}^2E, A)$ = the pointclass of all pointsets semirecursive in ${}^2E$, A and a real.

2.2. Next let us recall that if $\Game$ is the game quantifier, then $\Game\underset{\sim}{\Delta}^1_2 = \underset{\sim}{\Delta}^1_3$. (Here is a quick proof due to Addison:

> Let P, Q be disjoint $\underset{\sim}{\Sigma}^1_3$ sets. Say $P(x) \Leftrightarrow \exists\alpha P'(x,\alpha)$, $Q(x) \Leftrightarrow \exists\beta Q'(x,\beta)$, where $P', Q' \in \underset{\sim}{\Pi}^1_2$. Let $P''(x,\alpha,\beta) \Leftrightarrow P'(x,\alpha)$, $Q''(x,\alpha,\beta) \Leftrightarrow Q'(x,\beta)$ and let $S(x,\alpha,\beta)$ in $\underset{\sim}{\Delta}^1_2$ separate P'', Q''. Then it is easy to check that
> $\exists\alpha(0)\,\forall\beta(0)\,\exists\alpha(1)\,\forall\beta(1)\,\ldots\, S(x,\alpha,\beta)$ separates P, Q).

Now let $W \subseteq \omega^\omega$ be Π^1_3 and universal for $\underset{\sim}{\Pi}^1_3$ and σ a Π^1_3-norm on W. For $x \in W$, put

$$H_x = \{\langle y,z\rangle : \sigma(y) \leq \sigma(z) < \sigma(x)\}.$$

Then let f be a total recursive function such that if $x \in W$ then f(x) is a $\underset{\sim}{\Delta}^1_2$-code of a set, say $\underset{\sim}{\Delta}_x$, such that $\Game\underset{\sim}{\Delta}_x = H_x$. (The existence of such an f is clear from the proof in the preceding paragraph.) Put finally for $x,y \in W$:

$$x \lesssim y \Leftrightarrow \Gamma(\underset{\sim}{\Delta}_x) \subseteq \Gamma(\underset{\sim}{\Delta}_y).$$

<u>Lemma</u> 1. $\lesssim$ is a prewellordering.

<u>Proof</u>. $\lesssim$ is obviously reflexive and transitive.

$\lesssim$ is connected: Let $x,y \in W$. Then by Wadge, $\underset{\sim}{\Delta}_x \leq_W \underset{\sim}{\Delta}_y$ or $\underset{\sim}{\Delta}_y \leq_W \omega^\omega - \underset{\sim}{\Delta}_x$, where $X \leq_W Y$ iff X is reducible to Y via a continuous function. Say the first case occurs. Then $\underset{\sim}{\Delta}_x \in \Gamma(\underset{\sim}{\Delta}_y)$. But also $\omega^\omega - \underset{\sim}{\Delta}_x \leq_W \omega^\omega - \underset{\sim}{\Delta}_y$, therefore, $\omega^\omega - \underset{\sim}{\Delta}_x \in \Gamma(\underset{\sim}{\Delta}_y)$, thus $\Gamma(\underset{\sim}{\Delta}_x) \subseteq \Gamma(\underset{\sim}{\Delta}_y)$ i.e. $x \lesssim y$.

$\lesssim$ is wellfounded: Given $\emptyset \subsetneq A \subseteq W$ let $x \in A$ be such that $\underset{\sim}{\Delta}_x$ has least Wadge ordinal. Then for any $y \in A$, $\underset{\sim}{\Delta}_x \leq_W \underset{\sim}{\Delta}_y$ or $\underset{\sim}{\Delta}_x \leq_W \omega^\omega - \underset{\sim}{\Delta}_y$, therefore, as above $x \lesssim y$. ⊣

Let $\varphi : W \twoheadrightarrow \lambda$ be the norm associated with $\lesssim$.

<u>Lemma</u> 2. φ is a Π^1_3-norm.

<u>Proof</u>. Fix $y \in W$. We want to express $x \in W \wedge x \lesssim y$ in a Δ^1_3 way uniformly in y. When both x, y are in W the condition $x \lesssim y$ is

equivalent to

$$\underset{\sim}{\Delta}_x \lesssim_W C_{\underset{\sim}{\Delta}_y} \wedge \omega^\omega - \underset{\sim}{\Delta}_x \lesssim_W C_{\underset{\sim}{\Delta}_y},$$

which is clearly Δ^1_3 uniformly in x, y. So it is enough to find a total recursive function g such that

(i) $y \in W \Rightarrow g(y) \in W$,

(ii) $x,y \in W \wedge x \underset{\sim}{\leq} y \Rightarrow \sigma(x) \leq \sigma(g(y))$.

Because then for $y \in W$:

$$x \in W \wedge x \underset{\sim}{\leq} y \Leftrightarrow \sigma(x) \leq \sigma(g(y)) \wedge x \underset{\sim}{\leq} y,$$

which by our preceding remarks is Δ^1_3 uniformly in y.

In order to construct g we use the following

Sublemma. There is a total recursive function h such that

(i) $y \in W \Rightarrow h(y) \in W$

(ii) $x,y \in W \wedge H_x \lesssim_W H_y \Rightarrow \sigma(x) \leq \sigma(h(y))$.

Proof. Let h be a total recursive function such that if $y \in W$, then $h(y) \in W$ and

$$\begin{aligned} z \in C_{H_y} &\Leftrightarrow \langle a,z\rangle \in W \\ &\Leftrightarrow \sigma(\langle a,z\rangle) \leq \sigma(h(y)), \end{aligned}$$

for some $a \in \omega^\omega$. Then if $x,y \in W \wedge H_x \lesssim_W H_y$, but $\sigma(x) > \sigma(h(y))$, towards a contradiction, we have $z \in C_{H_y} \Leftrightarrow \sigma(\langle a,z\rangle) \leq \sigma(h(y)) < \sigma(x)$, for some $a \in \omega^\omega$, therefore, $C_{H_y} \lesssim_W H_x \lesssim_W H_y$, a contradiction.

To complete the proof of Lemma 2, we construct now g as follows: Let f^1 be total recursive such that if $y \in W$ then $f^1(y)$ is a $\underset{\sim}{\Delta}^1_2$-code of $C_{\underset{\sim}{\Delta}_y}$ and let f^2, f^3 be total recursive such that if $y \in W$ then $f^3(y) \in W$ and

$$\exists\alpha(\langle z,\langle t,\alpha\rangle\rangle \in C_{\underset{\sim}{\Delta}_y}) \Leftrightarrow \sigma(\langle f^2(y),\langle z,t\rangle\rangle) < \sigma(f^3(y)).$$

Let $g = h \circ f_3$. Assume now $x,y \in W$ and $x \underset{\sim}{\leq} y$. Then $\Gamma(\underset{\sim}{\Delta}_x) \subseteq \Gamma(\underset{\sim}{\Delta}_y)$, so

$$\langle t,\alpha\rangle \in \underset{\sim}{\Delta}_x \Leftrightarrow \langle z_0,\langle t,\alpha\rangle\rangle \in C_{\underset{\sim}{\Delta}_y}, \quad \text{for some } z_0,$$

thus

$$t \in H_x \Leftrightarrow t \in \mathfrak{D}\underset{\sim}{\Delta}_x \Leftrightarrow \mathfrak{D}\alpha(\langle t,\alpha\rangle \in \underset{\sim}{\Delta}_x)$$
$$\Leftrightarrow \mathfrak{D}\alpha(\langle z_0, \langle t,\alpha\rangle\rangle \in C_{\underset{\sim}{\Delta}_y})$$
$$\Leftrightarrow \sigma(\langle f^2(y), \langle z_0, t\rangle\rangle) < \sigma(f^3(y)).$$

So

$$H_x \lesssim_W H_{f^3(y)}, \quad \text{thus} \quad \sigma(x) \leq h(f^3(y)) = g(y). \qquad \dashv$$

Using Lemma 2 we complete the proof of the result as follows: By Lemma 2 we have $\lambda = \underset{\sim}{\delta}^1_3$. Since by direct computation we can easily see that $\eta_2 \leq \underset{\sim}{\delta}^1_3$ it is enough to show $\lambda \leq \eta_2$. For that define for $\xi < \lambda$:

$$f(\xi) = \text{Wadge ordinal of } C_{\underset{\sim}{\Delta}_x},$$
$$\text{where } \varphi(x) = \xi.$$

Since

$$\varphi(x) = \varphi(y) \Rightarrow \Gamma(\underset{\sim}{\Delta}_x) = \Gamma(\underset{\sim}{\Delta}_y)$$
$$\Rightarrow C_{\underset{\sim}{\Delta}_x} \lesssim_W C_{\underset{\sim}{\Delta}_y} \wedge C_{\underset{\sim}{\Delta}_y} \lesssim_W C_{\underset{\sim}{\Delta}_x},$$

this is well-defined. Also $f : \lambda \to \eta_2$, so it is enough to show that f is order preserving. Indeed, let $\xi < \zeta < \lambda$ and $\varphi(x) = \xi$, $\varphi(y) = \zeta$. Then $x \underset{\sim}{\leq} y$ and $y \not\leq x$, so $\Gamma(\underset{\sim}{\Delta}_x) \subsetneq \Gamma(\underset{\sim}{\Delta}_y)$. Consequently, $C_{\underset{\sim}{\Delta}_x} \lesssim_W C_{\underset{\sim}{\Delta}_y}$ but $C_{\underset{\sim}{\Delta}_y} \not\lesssim_W C_{\underset{\sim}{\Delta}_x}$, therefore, by Wadge $C_{\underset{\sim}{\Delta}_x} \lesssim_W \omega^\omega - C_{\underset{\sim}{\Delta}_y}$ i.e. the Wadge ordinal of $C_{\underset{\sim}{\Delta}_x}$, which is $f(\xi)$, is smaller than the Wadge ordinal of $C_{\underset{\sim}{\Delta}_y}$, which is $f(\zeta)$. Q.E.D.

References

[1] R. Van Wesep, Wadge degrees and descriptive set theory, Cabal Seminar 76-77, Proceedings, Caltech-UCLA Logic Seminar 1976-77, Ed. by A.S. Kechris and Y.N. Moschovakis, Lecture Notes in Mathematics, 689, 151-170, Springer-Verlag.

ORDINAL GAMES AND PLAYFUL MODELS

Yiannis N. Moschovakis(1),(2)
Department of Mathematics
University of California
Los Angeles, California 90024

Consider the game G in which the first player I picks a countable ordinal ξ (and makes no other moves) and then player II picks bit by bit an infinite sequence of integers $\alpha = (n_0, n_1, n_2, \ldots)$, as in the diagram

$$\begin{array}{lll} \text{I} & \xi & \\ \text{II} & n_0, n_1, n_2, \ldots & \alpha; \end{array}$$

at the end of the game, II wins if the sequence α codes a wellordering of ω of rank ξ.

It is clear that I cannot win this game. At the same time, any winning strategy for II would produce an uncountable, wellorderable set of irrationals; we know then that we cannot prove in ZFC that II has a definable winning strategy, and with full determinacy (on ω) we can prove outright that this game (on $\aleph_1$) is not determined.

Despite this well-known and trivial example, one can show (using AD) that many ordinal games are in fact determined. Our main purpose here is to establish a result in this direction and to apply it to the study of the inner models H_Γ which we associated with certain pointclasses Γ in section 8G of Moschovakis [1980]. If we postpone for a while the (natural) technical definitions that we need, we can state our main result in the following form.

---

(1) During the preparation of this paper, the author was partially supported by NSF Grant # MCS78-02989.

(2) I want to express my sincerest appreciation to my student Howard Becker whose advice was instrumental in improving the final version of this paper.

<u>Theorem.</u> Assume AD, let Γ be a pointclass which resembles Π^1_1, let H_Γ be the associated playful model and suppose that

$$\lambda < \underset{\sim}{\delta} = \text{the ordinal associated with } \Gamma;$$

then

$$\text{Power}(\lambda) \cap H_\Gamma = \text{the largest wellorderable collection of subsets of } \lambda \text{ which is } \exists^{\mathcal{N}}\Gamma(\lambda) \text{ in the codes.}$$

When $\Gamma = \Pi^1_1$, then $H_\Gamma = L$, and for $\lambda = \omega$ this reduces to Solovay's well-known characterization of the constructible power of ω as the largest wellorderable, Σ^1_2 collection of subsets of ω.

In addition to this consequence for playful models, our basic determinacy theorem has many other applications and we will indicate some of them here. Becker [A], [B] has also used these techniques in several ways, and in particular to show that the continuum hypothesis holds for many ordinals in H_Γ. It is still an open question whether the generalized continuum hypothesis holds in these models, and it may be that this problem is related to many difficult questions about combinatorial properties of ordinals under determinacy hypotheses.

The games which we will use are variations of games invented by Martin for a different but related purpose. Our proofs will combine this idea of Martin with the powerful methods of Harrington-Kechris [A] and the basic technique of the proof of the Third Periodicity Theorem, 6E.1 of Moschovakis [1980].

We assume ZF + DC throughout and list all additional hypotheses. We will also use without apology or explicit mention the notation of Moschovakis [1980] which we will cite as DST. In particular, $\mathcal{N} = {}^\omega\omega$ is the set of <u>irrationals</u>, sometimes called <u>reals</u> and denoted by R in other papers.

§1. <u>Ordinal games and the Harrington-Kechris approximation</u>. To see how ordinal games come up naturally in descriptive set theory, let us consider first the possible extensions of the <u>perfect set theorem</u> to ordinals $\lambda > \omega$.

Fix then some $\lambda \geq \omega$ and suppose $\mathcal{P} \subseteq \text{Power}(\lambda)$ is a collection of subsets of λ. An <u>imbedding of the full binary tree into</u> $\mathcal{P}$ is a function

$$(t_0,\dots,t_{n-1}) \mapsto \xi(t_0,\dots,t_{n-1})$$

which assigns ordinals below λ to all finite binary sequences such that the following holds: for each infinite binary sequence α, there exists some $X \in \mathcal{P}$ such that for all n,

$$\xi(\alpha(0),\alpha(1),\ldots,\alpha(n-1)) \in X \Leftrightarrow \alpha(n) = 1.$$

In effect we are *tagging* the nodes of the full binary tree with ordinals below λ, such that for each infinite branch there is some member of $\mathcal{P}$ which realizes the countably many membership decisions determined by the branch.

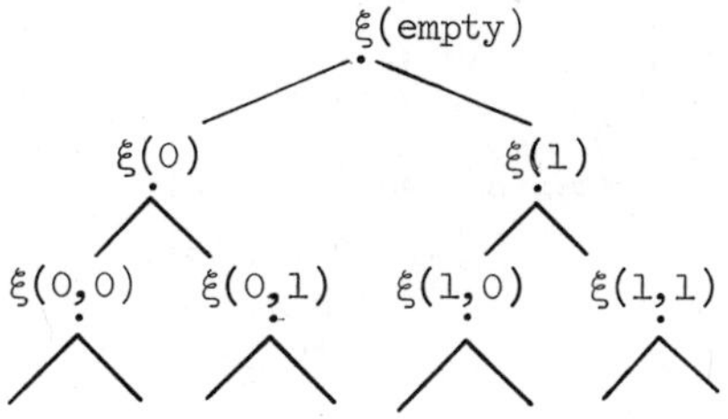

For example, if

$$\alpha(0) = 1,\ \alpha(1) = 0,\ \alpha(2) = 0,\ldots,$$

then there must be some $X \in \mathcal{P}$ such that

$$\xi(\text{empty}) \in X,\ \xi(1) \notin X,\ \xi(1,0) \notin X,\ldots .$$

Without bothering to define precisely *perfect subsets* of $\text{Power}(\lambda)$, we will often say that "$\mathcal{P}$ has a perfect subset" if $\mathcal{P}$ admits an imbedding of the full binary tree. In the opposite case, when there is no such imbedding, we will say that $\mathcal{P}$ *is thin.*

Notice that if $\mathcal{P}$ is wellorderable and admits an imbedding $u \mapsto \xi(u)$ of the full binary tree, then the mapping

$$\alpha \mapsto \text{least member } X \text{ of } \mathcal{P} \text{ such that for all}$$
$$n,\ \xi(\alpha(0),\ldots,\alpha(n-1)) \in X \Leftrightarrow \alpha(n) = 1$$

imbeds the Cantor set ${}^{\omega}2$ injectively into $\mathcal{P}$, so that ${}^{\omega}2$ is also wellorderable; thus with AD, *sets which have perfect subsets are not wellorderable.*

The converse should also be true with AD, so that every thin collection of subsets of λ should be wellorderable and in fact should have cardinality no more than that of λ. To try to prove this, consider the following game on λ:

I	ξ_0		ξ_1		...
II		t_0		t_1	

Here I plays ordinals below λ, II plays $t_i = 1$ or 0 and at the end

I wins $\Leftrightarrow$ there exists some $X \in \mathcal{P}$ such that for all i, $\xi_i \in X \Leftrightarrow t_i = 1$.

If I has a winning strategy σ in this game, let

$$\xi(t_0,\ldots,t_{n-1}) = \sigma(t_0,\ldots,t_{n-1})$$

and check easily that this function is an imbedding of the full binary tree into $\mathcal{P}$.

Thus, if the game is determined and $\mathcal{P}$ is thin, then II must have a winning strategy τ. Now fix $X \in \mathcal{P}$ and call a sequence

$$\xi_0, t_0, \ldots, \xi_n, t_n$$

good (relative to X and τ) if it is an initial part of a run of the game where II plays according to τ and

$$\xi_i \in X \Leftrightarrow t_i = 1 \qquad (i \leq n).$$

By definition, the empty sequence is good, and it cannot be that every good sequence has a proper good extension - or else we would get a complete run of the game where II has played to win but nevertheless X verifies that I has won. Thus there is a maximal good sequence $\xi_0, t_0, \ldots, \xi_n, t_n$ and then clearly for all ξ,

$$\xi \in X \Leftrightarrow \tau(\xi_0,\ldots,\xi_n,\xi) = 0,$$

so that the sequence $\xi_0,\ldots,\xi_n$ completely determines X. Thus the map

$X \mapsto$ the least (hexicographically) maximal good sequence relative to X and τ

maps $\mathcal{P}$ injectively into the tuples from λ, and $\mathcal{P}$ is wellorderable with cardinality at most that of λ.

This is a familiar and appealing argument, but of course, we cannot prove this ordinal game to be determined except in the trivial case where we already know that $\mathcal{P}$ admits an imbedding of the full binary tree.

One way to get around such non-determined games is to appeal to a fundamental theorem of Harrington and Kechris [A] which we now explain.

Suppose

$$\pi : A \twoheadrightarrow \kappa \qquad (A \subseteq \mathfrak{N})$$

is a regular norm which we can use to code ordinals below some fixed κ and

$\lambda < \kappa$; to simplify notation we will write

$$\pi(\alpha) = |\alpha| = \text{the ordinal coded by } \alpha.$$

With P and π we associate the <u>Harrington-Kechris</u> game P^π on $\mathfrak{N}$, where I and II successivley choose irrationals $\alpha_0, \alpha_1, \alpha_2, \ldots$ which must code ordinals below λ, i.e. each player must choose α_n so that

$$\alpha_n \in A \ \& \ |\alpha_n| < \lambda;$$

if they both obey this rule, then at the end

$$\text{I wins} \Leftrightarrow (|\alpha_0|, |\alpha_1|, |\alpha_2|, \ldots) \in P.$$

The main result of Harrington-Kechris [A], and the basis of all we will do in this paper is the following theorem, where

$$\begin{aligned}\kappa^{\mathbb{R}} &= \text{the closure ordinal of the continuum} \\ &= \text{supremum } \{|\underset{\sim}{\leq}| : \underset{\sim}{\leq} \text{ is a hyperprojective prewellordering of } \mathfrak{N}\}.\end{aligned}$$

<u>The Harrington-Kechris Theorem</u>. Assume AD, let $\lambda \leq \kappa^{\mathbb{R}}$, let $\pi : A \twoheadrightarrow \kappa$ be an inductive norm onto some ordinal $\kappa \geq \lambda$ and suppose $P \subseteq {}^{\omega}\lambda$; then the associated Harrington-Kechris game P^π on $\mathfrak{N}$ is determined. ⊣

Notice that in this result there is no definability condition on P -which is why we must include full AD among the hypotheses.

When we are confronted with a non-determined ordinal game P, we may think of the associated Harrington-Kechris game P^π (relative to some coding) as an <u>intensional approximation</u> to P, which is determined and may be useful. In the case of the perfect set theorm, it does give some information.

<u>Theorem</u> 1.1. Assume that the inner model $L(\mathbb{R})$ satisfies AD, fix an infinite $\lambda \leq \kappa^{\mathbb{R}}$ and suppose $\mathcal{P} \subseteq \text{Power}(\lambda)$ is a thin collection of subsets of λ which lies in $L(\mathbb{R})$. If the axiom of choice holds in the world, then (in the world) we can find an injection

$$f : \mathcal{P} \rightarrowtail \lambda.$$

<u>Proof</u>. Let P be the game described above and let P^π be the associated Harrington-Kechris game relative to some inductive norm π. Now P^π is determined in $L(\mathbb{R})$ and easily (as above) I cannot win it or else $\mathcal{P}$ would have a perfect subset. Thus II wins P^π via some τ^π and it is clear that τ^π is

winning for P^π in the world. Using choice, fix a function

$$\xi \mapsto \alpha^\xi$$

which assigns to each $\xi < \lambda$ some code α^ξ of ξ $(|\alpha^\xi| = \xi)$ and define

$$\tau(\xi_0,\ldots,\xi_n) = \tau^\pi(\alpha^{\xi_0},\ldots,\alpha^{\xi_n}).$$

Now check easily that τ wins P for II in the world and can be used as above to inject $\mathcal{P}$ into λ. $\dashv$

This is an unsatisfactory result, not so much for its dependence on the axiom of choice (which, after all is true), but because the resulting well-ordering of the thin set $\mathcal{P}$ lives in the world and not in $L(\mathbb{R})$. It is an open problem whether AD implies straight out that every thin collection of subsets of $\lambda \geq \omega$ is wellorderable with cardinality no bigger than that of λ. In §3 we will establish a partial but interesting result in this direction.

We will end this section with another application of ordinal games and the Harrington-Kechris approximation to the question of measurability of cardinals in the model $L(\mathbb{R})$. The next result was proved jointly with Harrington and Kechris.

Theorem 1.2. Assume that the inner model $L(\mathbb{R})$ satisfies AD and that the world (V) satisfies the axiom of choice. If $\lambda \leq \kappa^{\mathbb{R}}$ is a regular cardinal in V, then

$$L(\mathbb{R}) \models \lambda \text{ is measurable.}$$

Proof. Suppose $A \subseteq \lambda$ is any subset of λ which lies in $L(\mathbb{R})$ and consider the following ordinal game P:

$$\begin{array}{lllll} \text{I} & \xi_0 & & \xi_2 & \ldots \\ \text{II} & & \xi_1 & & \xi_3 \end{array}$$

I and II successively choose ordinals

$$\xi_0 < \xi_1 < \xi_2 < \cdots < \lambda$$

and at the end

$$\text{I wins} \Leftrightarrow \sup\{\xi_i : i \in \omega\} \in A.$$

This game of course need not be determined, but its Harrington-Kechris approximation P^π is determined, since it is easily defined by a payoff set which lies in $L(\mathbb{R})$.

Lemma. For each set $A \subseteq \lambda$ which lies in $L(\mathbb{R})$ and the associated game P,

$$\begin{aligned} \text{I wins } P^{\pi} \text{ in } L(\mathbb{R}) &\Leftrightarrow \text{I wins } P \text{ in } V \\ &\Leftrightarrow \text{there is a closed, unbounded set } C \subseteq \lambda \text{ (in } V\text{) such that } \\ &\quad \{\zeta \in C : \text{cofinality}(\zeta) = \omega\} \subseteq A. \end{aligned}$$

Proof of the lemma. If I wins P^{π} in $L(\mathbb{R})$, then I easily wins P in V using a choice function $\xi \mapsto \alpha^{\xi}$ as in the proof of Theorem 1.1.

If σ is a winning strategy for I (in V), put

$$\begin{aligned} C &= \{\zeta < \lambda : \zeta \text{ is closed under } \sigma\} \\ &= \{\zeta < \lambda : \text{if } \xi_1, \xi_3, \ldots, \xi_{2n-1} < \zeta, \\ &\qquad \text{then } \sigma(\xi_1, \xi_3, \ldots, \xi_{2n-1}) < \zeta\}. \end{aligned}$$

Clearly C is closed, and since λ is regular, C is easily unbounded. If $\zeta \in C$ and ζ has cofinality ω, consider a run of the game P where I plays to win by σ and II plays $\xi_1, \xi_3, \ldots$ below ζ, so that $\sup\{\xi_n : n \in \omega\} = \zeta$; then I wins and $\zeta \in A$.

Finally, suppose there is (in V) a closed, unbounded set $C \subseteq \lambda$ such that its points of cofinality ω lie in A, but I does not win P^{π}. Then II wins P^{π}, so (as above) II wins P in V, and (as above, again), there is a closed unbounded $C' \subseteq \lambda$, such that if $\zeta \in C'$ and cofinality$(\zeta) = \omega$, then $\zeta \notin A$. But $C \cap C'$ contains some point of cofinality ω, which must then belong to both A and $\lambda - A$.

We now define in $L(\mathbb{R})$

$$A \in \mathcal{U} \Leftrightarrow \text{I wins the game } P^{\pi} \text{ associated with } A \text{ as above.}$$

The set $\mathcal{U}$ is obviously an ultrafilter, by its definition and the determinacy of P^{π}. If $\{A_{\xi}\}_{\xi<\mu<\lambda}$ is a sequence of sets in $\mathcal{U}$, then (in V) we can find closed, unbounded sets $C_{\xi} \subseteq A_{\xi}$, such that $[\xi \in C_{\xi}$ & cofinality$(\zeta) = \omega]$ $\Rightarrow \zeta \in A_{\xi}$. Now $C = \bigcap_{\xi<\mu} C_{\xi}$ is easily closed and unbounded below λ, with all its points of coninality ω in $\bigcap_{\xi<\mu} A_{\xi}$, so by the lemma, $\bigcap_{\xi<\mu} A_{\xi} \in \mathcal{U}$ and $\mathcal{U}$ is λ-complete. ⊣

This result too is unsatisfactory, partly because the only ordinal which is known to satisfy the hypothesis (under Determinacy$(L(\mathbb{R}))$) is $\aleph_1$ - and the measurability of $\aleph_1$ is well-known. We wrote up the simple proof in considerable detail, to indicate all the places where the axiom of choice (in V)

is used. In §3 we will prove a similar result under AD, with no use of choice.

§2. The determinacy of scaled ordinal games. Our plan for proving improved versions of Theorems 1.1 and 1.2 is to replace the Harrington-Kechris approximation to an ordinal game by a finer approximation which will be a *determined ordinal game*. In this section we will establish the determinacy of the ordinal games which are involved.

Fix again an infinite ordinal λ. We will be looking at spaces of the form

$$\mathfrak{X} = X_1 \times \cdots \times X_k,$$

where each X_i is λ or ${}^{\omega}\lambda$, topologized as products with λ taken discrete. A *pointset on* λ is any subset $P \subseteq \mathfrak{X}$ of such a product space, i.e. any relation with arguments in λ and in ${}^{\omega}\lambda$. Since $\omega \subseteq \lambda$ and $\mathfrak{N} = {}^{\omega}\omega \subseteq {}^{\omega}\lambda$, we can obviously view every ordinary pointset (on ω) as a pointset on λ - which holds of a sequence of arguments only when they all lie in the appropriate subspaces. In the same way, it makes sense to apply *quantification over* ω or $\mathfrak{N}$ to pointsets on λ, e.g. if $Q \subseteq \mathfrak{X} \times {}^{\omega}\lambda$, we can set

$$\begin{aligned} P(x) &\Leftrightarrow (\exists\alpha)Q(x,\alpha) \\ &\Leftrightarrow (\exists f \in {}^{\omega}\lambda)[f \in \mathfrak{N} \ \& \ Q(x,f)]. \end{aligned}$$

As in the classical situation of descriptive set theory where $\lambda = \omega$, a *semiscale* on a set $P \subseteq \mathfrak{X}$ is a sequence $\overline{\varphi} = \{\varphi_i\}$ of norms

$$\varphi_i : P \to \text{Ordinals}$$

such that if $x_0, x_1, x_2, \ldots$ is any sequence of points in P, if $\lim_{n\to\infty} x_n = x$ and if for each i, the sequence of ordinals

$$\varphi_i(x_0), \varphi_i(x_1), \varphi_i(x_2), \ldots$$

is ultimately constant, then $x \in P$; if in addition we can conclude that for each i,

$$\varphi_i(x) \leq \lim_{n\to\infty} \varphi_i(x_n),$$

then $\overline{\varphi}$ is a *scale* on P.

It will be convenient to call these semiscales and scales *uniform* since they are defined directly on pointsets on λ and not on their *code sets* which we will define below.

If $\overline{\varphi}$ is a uniform semiscale on P, we put

$$x \leq^*_{\varphi_n} y \Leftrightarrow P(x) \;\&\; [\neg P(y) \vee \varphi_n(x) \leq \varphi_n(y)],$$
$$x <^*_{\varphi_n} y \Leftrightarrow P(x) \;\&\; [\neg P(y) \vee \varphi_n(x) < \varphi_n(y)].$$

A semiscale $\overline{\varphi}$ on P is <u>very good</u> if

$$\varphi_n(x) \leq \varphi_n(y) \;\&\; i < n \Rightarrow \varphi_i(x) \leq \varphi_i(y)$$

and whenever $x_0, x_1, \ldots$ are in P and for each n the sequence

$$\varphi_n(x_0), \varphi_n(x_1), \varphi_n(x_2), \ldots$$

is ultimately constant, then in fact there is some $x \in P$ such that $\lim_{i\to\infty} x_i = x$. As with pointsets on ω, it is easy to check that if P admits a semiscale, then P also admits a very good semiscale, in fact it admits a very good scale.

<u>Theorem</u> 2.1. Fix an infinite ordinal λ, let Γ be the collection of all pointsets on λ which admit semiscales and assume AD. Then Γ contains all open and closed pointsets on λ, it contains all inductive pointsets (on ω) and it is closed under continuous substitutions, &, $\vee$, countable conjunctions and disjunctions, $\exists^{\mathbb{R}}$, $\forall^{\mathbb{R}}$ and $(\exists f \in {}^{\omega}\lambda)$.

Similarly with "scales" in place of "semiscales."

<u>Proof</u>. For example, if $P \subseteq {}^{\omega}\lambda$ is open, put

$$\varphi_n(f) = \langle f(0), \ldots, f(m)\rangle, \quad \text{for the least } m \text{ such that } g \upharpoonright m = f \upharpoonright m \Rightarrow g \in P,$$

and if P is closed put

$$\varphi_n(f) = \langle f(0), \ldots, f(n-1)\rangle,$$

where $(\xi_1, \ldots, \xi_n) \mapsto \langle \xi_1, \ldots, \xi_n\rangle$ is some one-to-one mapping of tuples of ordinals into the ordinals. For the closure properties we use the arguments of DST and Moschovakis [1978] for the case $\lambda = \omega$. It is important to notice that in the proof of the Second Periodicity Theorem in DST, when we construct a scale for

$$P(x) \Leftrightarrow (\forall\alpha) Q(x, \alpha)$$

from a given scale for Q, the variable x in Q is treated as a parameter and remains fixed throughout the argument - so it may as well include ordinals or sequences from λ. ⊣

Notice that by this theorem, all Borel subsets of ${}^{\omega}\lambda$ are scaled as are all relations of the form

$$P(f,\alpha) \Leftrightarrow C(f,\alpha) \;\&\; Q(\alpha),$$

where $C \subseteq {}^{\omega}\lambda \times \mathcal{N}$ is Borel and $Q \subseteq \mathcal{N}$ is inductive. It is mostly sets like these which we will use in the applications.

The next theorem is the main result of this paper.

<u>Theorem</u> 2.2. Assume AD, let $\lambda \leq \kappa^{\mathbb{R}}$, let $\pi : A \twoheadrightarrow \kappa$ be an inductive norm onto some ordinal $\kappa \geq \lambda$ and suppose $P \subseteq {}^{\omega}\lambda$.

(1) If P admits a uniform semiscale and I wins the associated Harrington-Kechris game P, then I wins P.

(2) If ${}^{\omega}\lambda - P$ admits a uniform semiscale and II wins the associated Harrington-Kechris game P, then II wins P.

In particular, if both P and ${}^{\omega}\lambda - P$ admit uniform semiscales, then P is determined.

<u>Proof</u>. Fix $P \subseteq {}^{\omega}\lambda$ as in the hypothesis, and suppose I wins P^{π} and $\overline{\varphi}$ is a very good uniform semiscale on P. We will apply the basic idea in the proof of the Third Periodicity Theorem 6E.1 of DST to get a best winning (multiple-valued) strategy for I in P^{π} (relative to the semiscale that $\overline{\varphi}$ induces on P^{π}) and then use that best strategy to get a winning strategy for I in P.

Suppose then that $\alpha_0,\ldots,\alpha_n$ and $\beta_0,\ldots,\beta_n$ are winning positions for I in P^{π}, with n even. With any two such sequences we associate the game $H_n(\alpha_0,\ldots,\alpha_n;\beta_0,\ldots,\beta_n)$ (on $\mathcal{N}$) in which players F (first) and S (second) move as follows).

$$\begin{array}{ll|lllll}
\text{I} & \alpha_0,\ldots,\alpha_n & F\alpha_{n+1} & & & S\alpha_{n+2} \to F\alpha_{n+3} & \\
 & & \searrow & & \nearrow & \searrow & \\
\text{II} & \beta_0,\ldots,\beta_n & & S\beta_{n+1} \to F\beta_{n+2} & & & \ldots
\end{array}$$

At the end of the run,

$$S \text{ wins} \Leftrightarrow (|\alpha_0|,\ldots,|\alpha_n|,|\alpha_{n+1}|,\ldots) \leq^{*}_{\varphi_n} (|\beta_0|,\ldots,|\beta_n|,|\beta_{n+1}|,\ldots).$$

The clear motivation is that F and S play simultaneously two runs of the game P^{π}, where on the top board F makes the move for II and S makes the moves for I while on the bottom board S makes the moves for II and F makes the moves for I. We know that S can win on the top board and F can win on the bottom board; we call S a winner in the joint game if he can win on the top board <u>with no bigger ordinal in the norm</u> φ_n than F can win on the bottom.

The important observation is that this game $H_n(\alpha_0,\ldots,\alpha_n;\beta_0,\ldots,\beta_n)$ is determined - simply because it is the Harrington-Kechris game Q^π associated with

$$\begin{aligned} Q(\xi_0,\xi_1,\ldots) \Leftrightarrow\ & (|\alpha_0|,\ldots,|\alpha_n|,\xi_0,\xi_3,\xi_4,\xi_7,\xi_8,\ldots) \in P \\ & \& \ \{(|\beta_0|,\ldots,|\beta_n|,\xi_1,\xi_2,\xi_5,\xi_6,\ldots) \notin P \\ & \vee \varphi_n(|\alpha_0|,\ldots,|\alpha_n|,\xi_0,\xi_3,\xi_4,\xi_7,\xi_8,\ldots) \\ & \leq \varphi_n(|\beta_0|,\ldots,|\beta_n|,\xi_1,\xi_2,\xi_5,\xi_6,\ldots)\} . \end{aligned}$$

This is where we use (crucially) the hypothesis that $\overline{\varphi}$ is a uniform semiscale on P.

As in the proof of the Third Periodicity Theorem, we can easily check that the relation

$$(\alpha_0,\ldots,\alpha_n) \leq_n (\beta_0,\ldots,\beta_n) \Leftrightarrow \text{S wins } H_n(\alpha_0,\ldots,\alpha_n;\beta_0,\ldots,\beta_n)$$

is a prewellordering on the winning positions for I (of length $n+1$) in the game P^π. (This argument uses only the determinacy of these games and depended choices.)

We call $\alpha_0,\ldots,\alpha_n$ <u>minimal</u> if it is a winning position for I in the game P^π (with n even) and if for every other position $\beta_0,\ldots,\beta_n$,

$$|\alpha_0| = |\beta_0|,\ldots,|\alpha_{n-1}| = |\beta_{n-1}| \Rightarrow (\alpha_0,\ldots,\alpha_{n-1},\alpha_n) \leq_n (\beta_0,\ldots,\beta_{n-1},\beta_n).$$

We call $\xi_0,\ldots,\xi_n$ <u>minimal</u> if there exists a minimal sequence $\alpha_0,\ldots,\alpha_n$ such that

$$\xi_0 = |\alpha_0|,\xi_1 = |\alpha_1|,\ldots,\xi_n = |\alpha_n|.$$

<u>Lemma</u>. If $\xi_0,\xi_1,\xi_2,\ldots$ is a sequence of ordinals below λ and for each even n, $\xi_0,\ldots,\xi_n$ is minimal, then $(\xi_0,\xi_1,\xi_2,\ldots) \in P$.

<u>Proof of the lemma</u>, is quite similar to the proof of the corresponding lemma in the Third Periodicity Theorem.

For each even n choose $\alpha_0^n,\alpha_1^n,\ldots,\alpha_n^n$ which is minimal and such that

$$\xi_0 = |\alpha_0^n|,\ldots,\xi_n = |\alpha_n^n| .$$

and construct the following diagram of games (on $\mathfrak{h}$) in the obvious way.

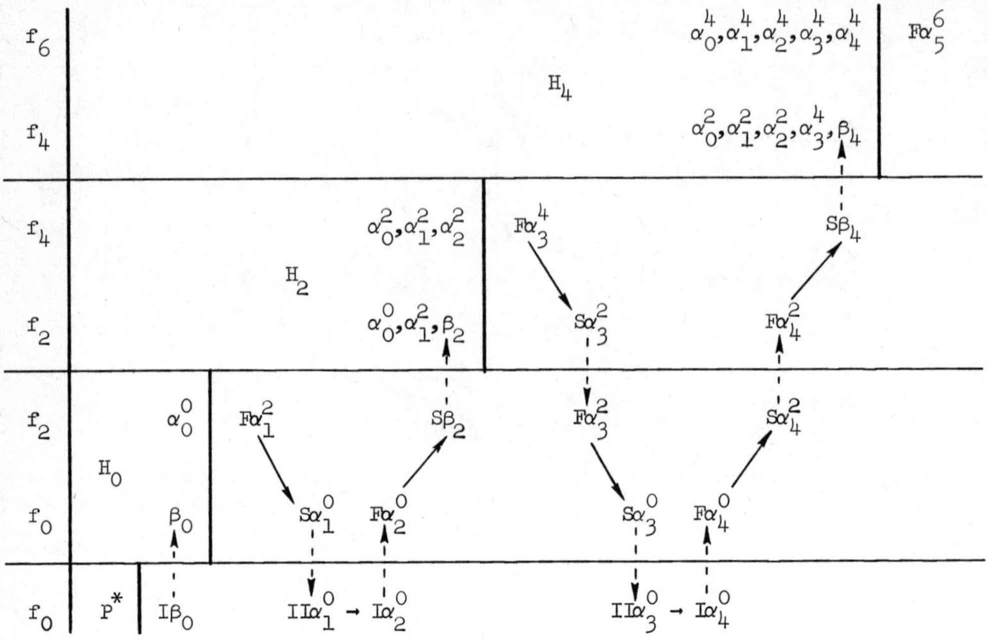

Following the conventions of DST for putting down diagrams of this kind, we use dotted arrows to indicate copied moves and solid arrows to indicate responses by winning strategies. On the bottom board, I plays to win P^{π}. On the boards above, S plays to win each of the games $H_0, H_2, H_4, \ldots$ and we need the axiom of depended choices to choose winning strategies for the appropriate sequences as they are generated in the diagram.

From this diagram we obtain sequences of irrationals $f_0, f_2, f_4, \ldots$ and all of them are in P^{π}, since I wins the bottom game and S wins all the others. If we let $\bar{f}_0, \bar{f}_2, \bar{f}_4, \ldots$ be the corresponding sequences of ordinals,

$$\bar{f}_n(i) = \pi(f_n(i)),$$

then obviously

$$\lim_{n\to\infty} \bar{f}_n = (\xi_0, \xi_1, \xi_2, \ldots)$$

and

$$\varphi_0(\bar{f}_2) \leq \varphi_0(\bar{f}_0),$$

$$\varphi_2(\bar{f}_4) \leq \varphi_2(\bar{f}_2);$$

This implies as usual that all the norms $\varphi_i(\bar{f}_n)$ are ultimately constant, since $\bar{\varphi}$ is a very good semiscale, so that $(\xi_0, \xi_1, \xi_7, \dots) \in P$.

Using this lemma, I can win P by playing successively so that $(\xi_0), (\xi_0, \xi_1, \xi_2), (\xi_0, \xi_1, \xi_2, \xi_3, \xi_4), \dots$ are all minimal. ⊣

§3. Martin games. Our determined approximations to ordinal games on $\lambda \geq \omega$ are particularly useful when λ can be coded by a scale in the following very strong sense.

Suppose $\bar{\varphi}$ is a regular scale on some set $W \subseteq \mathcal{N}$, with values into some ordinal κ. We will call $\bar{\varphi}$ a scale - coding of (the ordinals below) λ, if for every $\xi < \lambda$ there is some $\alpha \in W$ such that

$$(1) \quad \varphi_0(\alpha) = \xi,$$

$$(2) \quad \text{for all } n, \ \varphi_n(\alpha) < \lambda.$$

An ordinal λ is reliable if it admits a scale - coding and Γ - reliable if it admits a scale - coding $\bar{\varphi}$ which is a Γ - scale. (The term "reliable" is due to Becker.)

Notice that if $\bar{\varphi}$ is a scale - coding of λ on W and we put

$$\alpha \in W^* \Leftrightarrow \alpha \in W \ \& \ (\forall n)[\varphi_n(\alpha) < \lambda],$$

then the restriction $\bar{\varphi}^*$ of $\bar{\varphi}$ to W^* is also a scale - coding of λ which has the additional property

$$(3) \quad \text{for all } \alpha \in W^* \text{ and } n, \ \varphi^*_n(\alpha) < \lambda.$$

We are using the wider definition because it may be for some Γ that $\bar{\varphi}$ is a Γ - scale while $\bar{\varphi}^*$ is not. When questions of definability are not important it is convenient to assume that scale - codings also satisfy (3).

Some examples of reliable ordinals (under AD) are $\aleph_1$, $\underset{\sim}{\delta}^1_{2n+1}$ and $\aleph_\omega$. Kechris has verified that the successor cardinals

$$\aleph_2, \aleph_3, \aleph_4, \dots$$

are not reliable, but the general problem of classifying the reliable ordinals appears quite hopeless at this time.

If $\bar{\varphi}$ is a scale - coding for λ and $\xi < \lambda$, then an honest code for ξ (relative to $\bar{\varphi}$) is any sequence

$$(a_0, \xi_0, a_1, \xi_1, \dots) = (\alpha, f)$$

such that

$$\alpha \in W,\ \varphi_0(\alpha) = \xi = \xi_0,\ \varphi_n(\alpha) = \xi_1,\ \varphi_2(\alpha) = \xi_2, \ldots .$$

In some sense, an honest code for ξ gives us some more information about ξ than just its irrational part α which codes ξ via the first norm of the scale. If

$$T^{\overline{\varphi}} = \{(\alpha(0), \varphi_0(\alpha), \alpha(1), \varphi_1(\alpha), \ldots, \alpha(n), \varphi_n(\alpha)) : \alpha \in W\}$$

is the tree associated with the scale $\overline{\varphi}$, then an honest code for ξ is nothing but an infinite branch of that tree

$$\alpha(0), \xi, \alpha(1), \xi_1, \alpha(2), \xi_2, \ldots$$

which is the left-most branch in the tree with irrational part α.

Suppose now that $\lambda < \kappa^{\mathbb{R}}$ is a hyperprojective ordinal, $\overline{\varphi}$ is a scale - coding of λ and P is a game on λ: to make the results more directly applicable we will assume that the game P allows auxiliary moves in ω, i.e. $P \subseteq {}^{\omega}(\lambda \times \omega)$ and the game looks like this:

$$\begin{array}{lcccccc} \mathrm{I} & \xi_0, c_0 & & \xi_2, c_2 & & \ldots \\ \mathrm{II} & & \xi_1, c_1 & & \xi_3, c_3 & \end{array}$$

The <u>Martin game</u> associated with P and the scale coding $\overline{\varphi}$ will be an expansion of P, in which the players put down <u>alleged honest codes</u> for the ordinals $\xi_0, \xi_1, \ldots,$ by making additional <u>side</u> moves.

We will give the precise definition in a more general context, where instead of dealing with some $P \subseteq {}^{\omega}(\lambda \times \omega)$, we are given instead a set

$$Q \subseteq {}^{\omega}(\mathcal{N} \times \omega);$$

most commonly there will be some $P \subseteq {}^{\omega}(\lambda \times \omega)$ around and we will have

$$\begin{aligned} Q(\alpha_0, c_0, \alpha_1, c_1, \ldots) &\Leftrightarrow P^{\varphi_0}(\alpha_0, c_0, \alpha_1, c_1, \ldots) \\ &\Leftrightarrow \alpha_0, \alpha_1, \ldots \in W \\ &\qquad \& \ P(\varphi_0(\alpha_0), c_0, \varphi_0(\alpha_1), c_1, \varphi_0(\alpha_2), c_2, \ldots), \end{aligned}$$

but there are important cases where it is necessary to deal with <u>intensional ordinal games</u>, where the payoff $Q(\alpha_0, c_0, \alpha_1, \ldots)$ depends not only on the <u>ordinals</u> $\xi_0, \xi_1, \ldots$ played but also on some <u>codes</u> $\alpha_0, \alpha_1, \ldots$ for these ordinals.

Suppose then that $\lambda < \kappa^{\mathbb{R}}$ and a scale - coding $\overline{\varphi}$ of λ are given and let $Q \subseteq {}^{\omega}(n \times \omega)$. The <u>Martin game</u> $Q^{\overline{\varphi}}$ associated with Q and $\overline{\varphi}$ is played as follows.

$$\begin{array}{llll}
\text{I} & \begin{array}{l}\xi_0, c_0 \\ \alpha_0(0), f_0(0)\end{array} & & \begin{array}{l}\xi_2, c_2 \\ \alpha_0(1), f_0(1), \alpha_1(0), f_1(0), \alpha_2(0), f_2(0)\end{array} \\
\text{II} & & \begin{array}{l}\xi_1, c_1 \\ \beta_0(0), g_0(0), \beta_1(0), g_1(0)\end{array} & \qquad \cdots
\end{array}$$

Here I plays ordinals below λ and integers

$$\xi_0, c_0, \xi_2, c_2, \xi_4, c_4, \ldots,$$

irrationals (bit by bit)

$$\alpha_0, \alpha_1, \alpha_2, \alpha_3, \ldots$$

and sequences of ordinals below λ (bit by bit)

$$f_0, f_1, f_2, \ldots;$$

similarly, II plays $\xi_1, c_1, \xi_3, c_3, \ldots,$ irrationals $\beta_0, \beta_1, \beta_2, \ldots$ (bit by bit) and sequences of ordinals below λ, $g_0, g_1, g_2, \ldots$ (bit by bit). The order of moves is indicated by the diagram which is quite messy; in the future we will indicate it symbolically as follows.

$$\begin{array}{llll}
\text{I} & \xi_0, c_0, \begin{pmatrix}\alpha_0 \\ f_0\end{pmatrix}, & & \xi_2, c_2, \begin{pmatrix}\alpha_1 \\ f_1\end{pmatrix}, \begin{pmatrix}\alpha_2 \\ f_2\end{pmatrix}, \\
\text{II} & & \xi_1, c_1, \begin{pmatrix}\beta_0 \\ g_0\end{pmatrix}, \begin{pmatrix}\beta_1 \\ g_1\end{pmatrix}, & \qquad \cdots
\end{array}$$

The rules of the game are as follows.

(1) We must have

$$f_0(0) = g_0(0) = \xi_0,$$

$$f_1(0) = g_1(0) = \xi_1,$$

$$\cdots$$

or else the first player to violate this loses.

(2) For each n, k we must have

$$(\alpha_n(0),f_n(0),\alpha_n(1),f_n(1),\ldots,\alpha_n(k),f_n(k)) \in T^{\overline{\varphi}},$$

$$(\beta_n(0),g_n(0),\beta_n(1),g_n(1),\ldots,\beta_n(k),g_n(k)) \in T^{\overline{\varphi}},$$

or else the first player to violate this loses.

These two conditions obviously imply that in each run of the game where they are obeyed, we have at the end (by the scale property) for each n

$$\alpha_n \in W \ \& \ \varphi_0(\alpha_n) \leq \xi_n,$$

$$\beta_n \in W \ \& \ \varphi_0(\beta_n) \leq \xi_n.$$

In effect, each player has played on <u>alleged</u> <u>honest</u> <u>code</u> α_n, f_n or β_n, g_n, for each ordinal ξ_n played in the main part of the game.

(3) If (1) and (2) hold and there is some n such that

$$i < n \Rightarrow \varphi_0(\alpha_i) = \varphi_0(\beta_i),$$

$$\varphi_0(\alpha_n) \neq \varphi_0(\beta_n),$$

then I wins if $\varphi_0(\alpha_n) > \varphi_0(\beta_n)$ and II wins if $\varphi_0(\alpha_n) < \varphi_0(\beta_n)$.

(4) If (1) and (2) hold and for every n we have

$$\varphi_0(\alpha_n) = \varphi_0(\beta_n),$$

then

$$\text{I wins} \Leftrightarrow Q(\alpha_0,c_0,\alpha_1,c_1,\ldots).$$

The rules of the game insure that each player <u>verifies</u> bit by bit by the <u>witnesses</u> $f_n(i),g_n(i)$ that he is playing members of W which code ordinals below the ξ_n in question; then because of rule (3), each player can <u>keep his opponent honest</u> at any n, simply by making sure that his own play is an honest code for ξ_n. Thus in the case where Q comes from an ordinal game P ($Q = P^{\varphi_0}$), then the game $Q^{\overline{\varphi}}$ can often replace P, even though the payoff for $Q^{\overline{\varphi}}$ is expressed in terms of the codes $\alpha_0,\alpha_1,\ldots$ etc. for the ordinals played.

In the case

$$Q = P^{\varphi_0}$$

comes from an ordinal game, we will also call

$$Q^{\overline{\varphi}} = P^{\overline{\varphi}}$$

the Martin game associated with P and $\overline{\varphi}$.

It is obviously important here that the coding $\overline{\varphi}$ is a scale and not just a semiscale.

Of course the main result is that these Martin games are often determined.

Theorem 3.1. Assume AD, let $\lambda < \kappa^{\mathbb{R}}$, let $\overline{\varphi}$ be an inductive scale - coding of λ and suppose that $Q \subseteq {}^{\omega}(\mathcal{N} \times \omega)$ is hyperprojective in the codes, i.e. such that both

$$Q_1^{\#}(\alpha,\gamma) \Leftrightarrow (\forall n)[(\alpha)_n \in W \ \& \ \varphi_0((\alpha)_n) < \lambda]$$
$$\& \ Q((\alpha)_0, \gamma(0)_1, (\alpha), \gamma(1), \ldots)$$

$$Q_2^{\#}(\alpha,\gamma) \Leftrightarrow (\forall n)[(\alpha)_n \in W \ \& \ \varphi_0((\alpha)_n) < \lambda]$$
$$\& \ \neg\, Q((\alpha)_0, \gamma(0), (\alpha)_1, \gamma(1), \ldots)$$

are inductive. Then the associated Martin game $Q^{\overline{\varphi}}$ is determined. (For projective λ and $Q_1^{\#}, Q_2^{\#}$, this result is due to Martin.)

Proof is quite direct from 2.1 and 2.2 once the definitions are mastered, because both $Q^{\overline{\varphi}}$ and its complement admit uniform scales. ⊣

Let us now go back and apply this Martin approximation to the perfect set game.

Fix as in §1 an inductive

$$\pi : A \twoheadrightarrow \kappa^{\mathbb{R}} \qquad (A \subseteq \mathcal{N}).$$

For each $\lambda < \kappa^{\mathbb{R}}$ and each $X \subseteq \lambda$, put

$$\text{Code}(X) = \{\alpha \in A \,.\, |\alpha| = \pi(\alpha) \in X\}$$

and recall that by the coding lemma of DST, Code(S) is a hyperprojective set (granting AD). If $G \subseteq \mathcal{N} \times \mathcal{N}$ is a universal inductive set, then there exists some $\varepsilon \in \mathcal{N}$ such that

$$\text{Code}(x) = \{\alpha : G((\varepsilon)_0, \alpha)\}$$
$$= \{\alpha : \neg\, G((\varepsilon)_1, \alpha)\};$$

we call any such ε a hyperbrojective code for X. Finally, we call a collection $\mathcal{P} \subseteq \mathrm{Power}(\lambda)$, <u>hyperprojective in the codes</u>, if $\mathcal{P}^{\#}$ is hyperprojective, where

$$\mathcal{P}^{\#}(\varepsilon) \Leftrightarrow \varepsilon \text{ is a hyperprojective code of some } X \in \mathcal{P}.$$

It is not hard to show that this notion is independent of the particular π and G used to define it - for example by using 8G.20 of DST.

<u>Theorem</u> 3.2. Assume AD, let $\lambda < \kappa^{\mathbb{R}}$ and let $\mathcal{P} \subseteq \mathrm{Power}(\lambda)$ be a thin collection of subsets of λ which is hyperprojective in the codes. Then $\mathcal{P}$ is wellorderable and has cardinality no more than λ^{*}, where

$$\lambda^{*} = \text{the least reliable ordinal} \geq \lambda.$$

<u>Proof</u>. Let $\mathcal{P}^{\#}$ be the code set of $\mathcal{P}$ as above and consider the game Q where the players move as follows.

$$\begin{array}{lllll} \mathrm{I} & \xi_0, \varepsilon(0) & & \xi_1, \varepsilon(1), & \cdots \\ \mathrm{II} & & t_0, & & t_1 \end{array}$$

Here each $\xi_i < \lambda$, each t_i is 1 or 0 and at the end of the game,

$$\begin{aligned} \mathrm{I} \text{ wins} \Leftrightarrow\ & \varepsilon \in P^{\#} \\ & \ \&\text{ if } X \text{ is the subset of } \lambda \text{ coded by} \\ & \ \varepsilon, \text{ then } (\forall i)[\xi_i \in X \Leftrightarrow t_i = 1]. \end{aligned}$$

This game need not be determined, but of course we can look at the associated Martin game $Q^{\overline{\Psi}}$, where the players attempt also to product honest codes $\alpha_n, f_n, \beta_n, g_n$ for the ordinals ξ_n, relative to a scale - coding of λ^{*}, the least reliable ordinal above λ. An easy prewellordering argument shows that $\lambda^{*} < \kappa^{\mathbb{R}}$ and then the Coding Lemma implies that the scale - coding on λ^{*} is hyperprojective; thus Theorem 3.1 implies easily that the Martin game $Q^{\overline{\Psi}}$ is determined.

We can now give a minor modification of the perfect set argument to prove that I cannot win $Q^{\overline{\Psi}}$.

Suppose I wins by σ and construct a tree of runs of $Q^{\overline{\Psi}}$ as follows. Given an initial segment of a run in which the last ordinal played by I in the main part of the game is ξ_n (and I of course has also played a finite set of appropriate witnesses), have II choose an honest code β_n, g_n for ξ_n and resolve that from then on he will give witnesses for ξ_n using β_n, g_n;

now get two extensions of this initial segment by pretending that II plays $t_n = 0$ and $t_n = 1$ and having I respond by his winning strategy.

The axiom of depended choices is obviously used here to construct this tree of runs of $Q^{\overline{\psi}}$. At the end, each infinite branch from this tree represents a run of $Q^{\overline{\psi}}$ in which II has played honest codes for all the ordinals and I has won - so I too has played honest codes for the ordinals. But also, along each branch, I has played some ε which codes a set $X \subseteq \lambda$, $X \in \mathcal{P}$ and if the codes played by I are $\alpha_0, \alpha_1, \alpha_2, \ldots,$ then

$$(\forall i)[\,|\alpha_i| \in X \Leftrightarrow t_i = 1];$$

since the α_i then are honest codes, we have

$$(\forall i)[\xi_i \in X \Leftrightarrow t_i = 1].$$

It is now obvious that the reduct of the tree of runs to the main ordinals of the game ξ_n gives an imbedding of the full binary tree into $\mathcal{P}$, which contradicts the hypothesis that $\mathcal{P}$ is thin.

Once we know that I cannot win, we may infer that II wins $Q^{\overline{\psi}}$ by some strategy τ, and we can modify slightly the perfect set argument to get an injection of $\mathcal{P}$ into the finite sequences from λ^*. The idea is to fix some $X \in \mathcal{P}$ and some ε which codes X and call a sequence

$$\xi_0, \begin{pmatrix} \alpha_0 \\ f_0 \end{pmatrix}, t_0, \xi_1, \begin{pmatrix} \alpha_1 \\ f_1 \end{pmatrix}, t_1, \ldots, \xi_n, \begin{pmatrix} \alpha_n \\ f_n \end{pmatrix}, t_n$$

good for X and ε relative to τ if the following hold.

(1) Each $\xi_i < \lambda$ and each α_i, f_i is an honest code for ξ_i.

(2) $\xi_i \in X \Leftrightarrow t_i = 1$.

(3) In the initial part of the Martin game $Q^{\overline{\psi}}$ where I plays the ξ_i, ε, and witnesses following α_i, f_i, and II plays τ, player II gives the integers t_i.

It is now clear that there is a maximal good sequence; if u is the finite sequence of ordinals and integers which is the initial part of the Martin game corresponding to a maximal good sequence as in (3) above, then for each $\xi < \lambda$, easily

$$(*) \qquad \xi \in X \Leftrightarrow \tau(u, \xi) = 0.$$

Thus u determines X completely and the map

$$X \mapsto \text{the least (lexicographically) } u \text{ from } \lambda^* \text{ which determines } X \text{ by } (*)$$

gives an injection of $\mathcal{P}$ into the finite sequences from λ^*. ⊣

Becker [B] has improved this theorem considerably by showing that the hypothesis "$\mathcal{P}$ is hyperprojective in the codes" is not needed.

As a second application of Martin games, we give a considerable improvement of Theorem 1.2.

Theorem 3.3. Assume AD, let $\lambda < \kappa^{\mathbb{R}}$ be reliable and regular and put

$$A \in \mathcal{U} \Leftrightarrow \text{there exists a closed, unbounded set } C \subseteq \lambda, \text{ such that } \{\zeta \in C : \text{cofinality}(\zeta) = \omega\} \subseteq A.$$

Then $\mathcal{U}$ is a λ-complete ultrafilter, so in particular λ is measurable.

Proof. Fix a scale - coding $\overline{\varphi}$ for λ on a set $W \subseteq \mathcal{N}$ and for $\alpha \in W$ such that $(\forall n)[\varphi_n(\alpha) < \lambda]$, put

$$f(\alpha) = \text{supremum}\{\varphi_n(\alpha) : n \in \omega\}.$$

By the regularity of λ, each $f(\alpha)$ is $< \lambda$, and hence for each $\xi < \lambda$ we can set

$$c(\xi) = \text{infimum}\{f(\alpha) : \alpha \in W \ \& \ \varphi_0(\alpha) = \xi \ \& \ (\forall n)[\varphi_n(\alpha) < \lambda]\}.$$

Fix now some $A \subseteq \lambda$ and consider the ordinal game P in the proof of Theorem 1.2, where I and II successively choose the terms of an increasing sequence

$$\xi_0 < \xi_1 < \cdots$$

and at the end,

$$\text{I wins} \Leftrightarrow \sup\{\xi_i : i \in \omega\} \in A.$$

By the Coding Lemma, the set

$$\text{Code}(A) = \{\alpha \in W : \varphi_0(\alpha) \in A\}$$

is hyperprojective, so the associated Martin game $P^{\overline{\varphi}}$ is determined. Assume I wins via σ and put

$$C = \{\zeta < \lambda : \zeta \text{ is closed under } \sigma \text{ and } c\};$$

this simply means that for $\zeta \in C$,

$$\xi < \zeta \Rightarrow c(\xi) < \zeta$$

and in any run of $P^{\overline{\varphi}}$ where II plays below ζ, I's responses by σ are also below ζ.

The regularity of λ implies immediately that C is closed and unbounded below λ.

Fix now $\zeta \in C$ of cofinality ω and consider a run of the Martin game $P^{\overline{\varphi}}$ where I plays to win by σ and II plays as follows.

(1) In choosing the "main moves" $\xi_1, \xi_3, \ldots$, II makes sure that at the end

$$\sup\{\xi_i : i \in \omega\} = \zeta;$$

II can do this, since cofinality$(\zeta) = \omega$.

(2) For his subsidiary moves in $P^{\overline{\varphi}}$, where II plays codes for the main ordinals played by either I or II, have II choose honest codes below ζ; II can do this, since ζ is closed under c.

In this run of $P^{\overline{\varphi}}$ all moves are below ζ, since ζ is closed under σ, and all codes played are honest, since I wins and II has played honest codes. Thus the payoff insures that $\zeta \in A$.

This and the corresponding argument in the case II wins $P^{\overline{\varphi}}$, show that for each $A \subseteq \lambda$ either A or $\lambda - A$ contains all the ω-points in some closed, unbounded set C, so that $\mathcal{U}$ is an ultrafilter. To complete the proof, we must verify that $\mathcal{U}$ is λ-complete, without using the axiom of choice.

Suppose then that $\{A_\xi : \xi < \mu\}$ is a family of sets in $\mathcal{U}$ indexed by some $\mu < \lambda$ and consider the following ordinal game Q.

I	ξ		ξ_1		$\cdots$
II		ξ_0		ξ_2	

Here I plays any $\xi < \mu$ and then II and I successively choose the terms of an increasing sequence

$$\xi_0 < \xi_1 < \xi_2 < \cdots .$$

At the end,

$$\text{II wins} \Leftrightarrow \sup\{\xi_i : i \in \omega\} \in A_\xi.$$

By the Coding Lemma again, the set

$$\mathrm{Code}(\{A_\xi : \xi < \mu\}) = \{(\alpha,\beta) : \alpha,\beta \in W \ \&\ \varphi_0(\alpha) < \mu \ \&\ \varphi_0(\beta) \in A_{\varphi_0(\alpha)}\}$$

is hyperprojective, so the corresponding Martin game $Q^{\overline{\varphi}}$ is determined.

It is clear that I cannot win $Q^{\overline{\varphi}}$ because as soon as the first move ξ is put down, II need only play an honest code for ξ and then use a winning strategy for I in the game associated with A_ξ, which exists since $A_\xi \in \mathcal{U}$. Thus II wins, say by τ. It is now easy to check as above that if

$$C = \{\zeta : \zeta \text{ is closed under } \tau \text{ and } c\},$$

then

$$\zeta \in C \ \&\ \mathrm{cofinality}(\zeta) = \omega \Rightarrow \zeta \in \textstyle\bigcap_{\xi<\mu} A_\xi,$$

so that $\bigcap_{\xi<\mu} A_\xi \in \mathcal{U}$ and $\mathcal{U}$ is λ-complete. ⊣

This theorem too is not very satisfactory, since the only ordinals that we know (with AD) to satisfy the hypotheses are like the $\underset{\sim}{\delta}^1_{2n+1}$, and they have been known to be measurable since the early days of the theory. It should be the case under AD, that <u>every regular λ below $\kappa^{\mathbb{R}}$ is measurable</u>, but it is not clear at this time how to go about proving this.

We put down one move simple "choice-type" result which follows directly from Theorem 3.1 and illustrates the applicability of these methods.

<u>Theorem</u> 3.4. Assume AD, let $\lambda < \kappa^{\mathbb{R}}$ be reliable and suppose $R \subseteq \lambda \times {}^{\omega}\kappa$ is a relation on ordinals below λ and infinite sequences from κ. Then

$$(\forall \xi < \lambda)(\exists f \in {}^{\omega}\kappa)R(\xi,f) \Rightarrow (\exists A \subseteq \kappa)\{\mathrm{card}(A) = \mathrm{card}(\lambda)$$
$$\&\ (\forall \xi < \lambda)(\exists f \in {}^{\omega}A)R(\xi,f)\}.$$ ⊣

§4. <u>Definable winning strategies</u>. Once we know that a certain ordinal game is determined, it is natural to ask if the winner has a definable winning strategy. In this section we will prove some useful results of this type.

First we must set up notation and recall some basic facts from section 8G of DST.

In §3 we defined the code set $\mathrm{Code}(X)$ of some $X \subseteq \lambda$, relative to some norm

$$\pi : A \twoheadrightarrow \kappa$$

onto an ordinal $\kappa \geq \lambda$,

$$\text{Code}(X) = \{\alpha \in A : |\alpha| = \pi(\alpha) \in X\}.$$

In the same way we can define code sets for more complicated subsets of λ^n, ${}^{\omega}\lambda$, etc,. For example, if $X \subseteq {}^{\omega}\lambda$, then

$$\text{Code}(X) = \{\alpha : \text{for all } n, (\alpha)_n \in A \text{ and } X(|(\alpha)_0|, |(\alpha)_1|, \ldots)\},$$

if $X \subseteq \bigcup \lambda^n$, then

$$\text{Code}(X) = \{(n,\alpha) : \text{for } i < n, (\alpha)_i \in A \text{ and } (|(\alpha)_0|, \ldots, |(\alpha)_{n-1}|) \in X\},$$

etc. If $\text{Code}(X)$ is in a pointclass Γ, we will say that X _is in_ Γ _in the codes, relative to_ π.

A uniform semiscale $\overline{\varphi}$ on some pointset on λ, P _is in_ Γ _in the codes relative to_ π, if the associated pointsets on λ, $\leq^*$ and $<^*$ are in Γ in the codes relative to π.

We will need to deal with the relativization $\Gamma(\overline{\nu})$ of a pointclass Γ to a sequence of ordinals $\overline{\nu} = \nu_1, \ldots, \nu_n$ relative to a norm $\pi : A \twoheadrightarrow \kappa$, where $\overline{\nu} = \nu_1, \ldots, \nu_n < \kappa$: $P \subseteq \mathcal{X}$ is in $\Gamma(\overline{\nu})$ if for some $Q \subseteq \mathcal{N}^n \times \mathcal{X}$ and all $\overline{\beta} = \beta_1, \ldots, \beta_n$ such that

$$\pi(\beta_1) = \nu_1, \ldots, \pi(\beta_n) = \nu_n,$$

we have

$$P(x) \Leftrightarrow Q(\overline{\beta}, x).$$

Thus we will often say that $X \subseteq {}^{\omega}\lambda$ (for example) is in $\Gamma(\overline{\nu})$ in the codes, relative to some π.

Recall from the exercises in section 8G of DST, that a pointclass Γ _resembles_ Π^1_1 if Γ is a Spector pointclass closed under $\forall^{\mathcal{N}}$, with the scale property, and such that whenever $P \subseteq \mathcal{X} \times \mathcal{Y}$ is in some relativized ambiguous class $\Delta(\alpha)$, then the relation

$$Q(x) \Leftrightarrow \{y : P(x,y)\} \text{ is meager}$$

is also in $\Delta(\alpha)$. The most important examples (with AD) are Π^1_{2n+1} $(n = 0,1,2,\ldots)$, the collection IND of all (absolutely) inductive relations and (for the recursion theorists) the pointclass K of all relations which are semirecursive in ${}^3E$ in the sense of Kleene.

The collection IND is the largest pointclass which is currently known to resemble Π^1_1, from any hypotheses.

One of the basic facts that we will use is the Harrington-Kechris <u>Ordinal Quantification Theorem</u>, 8G.20 of DST. This was not given a fancy name in DST, but it seems wise to separate it here from the Harrington-Kechris theorem which is the basis of this paper. Both results are now written up in Harrington-Kechris [A].

One of the consequences of the Ordinal Quantification Theorem is that if Γ <u>resembles</u> Π^1_1, $\lambda < \underset{\sim}{\delta}$, $\pi : A \twoheadrightarrow \underset{\sim}{\delta}$ <u>is a</u> Γ-<u>norm and</u> $P \subseteq \bigcup_n \lambda^n$ (or $P \subseteq {}^{\omega}\lambda$) <u>is in</u> $\exists^{\mathfrak{n}} \Gamma(\bar{\nu})$ <u>in the codes relative to</u> π, <u>for some</u> $\bar{\nu} < \underset{\sim}{\delta}$, <u>then</u> P <u>is also in</u> $\exists^{\mathfrak{n}} \Gamma(\bar{\nu})$ <u>in the codes relative to any other</u> Γ-<u>form</u> $\pi' : A' \twoheadrightarrow \underset{\sim}{\delta}$, (granting AD of course). This basic invariance of the notion "in $\exists^{\mathfrak{n}} \Gamma(\bar{\nu})$ in the codes" will be used explicitly on many occasions, and is also implicitly assumed in some of our choices of notions.

For the main result in this section we will need a more refined version of the Harrington-Kechris Theorem than that we put down in §1.

<u>The Harrington-Kechris Theorem (strong version)</u>. Assume AD, suppose Γ is a pointclass which resembles Π^1_1 and let $\lambda < \underset{\sim}{\delta}$ be the order-type of a prewellordering of $\mathfrak{n}$ in $\underset{\sim}{\Delta}$, let $P \subseteq {}^{\omega}\lambda$ and let

$$\pi : A \twoheadrightarrow \lambda$$

be any $\underset{\sim}{\Delta}$-norm. Then the associated Harrington-Kechris game P on $\mathfrak{n}$ is determined, and whechever player wins it has a winning strategy (with graph) in $\underset{\sim}{\Delta}$. ⊣

In the applications, π is often the restriction of some Γ-norm

$$\pi' : A' \to \underset{\sim}{\delta}$$

to the set

$$A = \{\alpha \in A' : \pi(\alpha) < \lambda\}.$$

It is the definability (or uniformization) conclusion in this theorem which becomes important at this stage of the game. For example, taking $\Gamma = \Pi^1_3$, if $\lambda < \underset{\sim}{\delta}^1_3$ (e.g. if $\lambda = \aleph_\omega$) and $P \subseteq {}^{\omega}\lambda$, and if I (say) wins the Harrington-Kechris game P^π relative to some Γ-norm π, then I has a winning strategy σ such that its "graph"

$$G_\sigma(n,\alpha,\beta) \Leftrightarrow \sigma((\alpha)_1,(\alpha)_3,\ldots,(\alpha)_{2n-1}) = \beta$$

is in $\underset{\sim}{\Delta}^1_3$. It is important (and suprising) that <u>there is no definability assumption on</u> P <u>in this result</u>.

After these preliminary remarks we can formulate and prove the sharp version of Theorem 2.2 which we need. The proof below is due to Becker and is simpler than my original proof of this theorem which involved reworking the "minimal moves" argument at the level of ordinal games.

Theorem 4.1. Assume AD, let Γ be a pointclass which resembles Π^1_1, let $\lambda < \underset{\sim}{\delta}$ and suppose $P \subseteq {}^{\omega}\lambda$ and $\pi : A \twoheadrightarrow \underset{\sim}{\delta}$ is a regular Γ-norm.

(1) If for some $\bar{\nu} < \underset{\sim}{\delta}$, P is in $\Gamma(\bar{\nu})$ in the codes and admits a uniform semiscale $\bar{\varphi}$ which is in $\Gamma(\bar{\nu})$ in the codes, and if I wins the Harrington-Kechris game P^{π}, then I has a winning strategy (for the ordinal game P) which is in $\exists^{\eta}\,\Gamma(\bar{\nu},\lambda)$ in the codes.

(2) Similarly, if ${}^{\omega}\lambda - P$ is in $\Gamma(\bar{\nu})$ in the codes and admits a uniform semiscale which is in $\Gamma(\bar{\nu})$ in the codes and if II wins the Harrington-Kechris game P^{π}, then II has a winning strategy for P which is in $\exists^{\eta}\,\Gamma(\bar{\nu},\lambda)$ in the codes.

Proof. We will establish (1), the argument for (2) being similar.

Assume the hypothesis on P and recall the games $H_n(\alpha_0,\ldots,\alpha_n;\beta_0,\ldots,\beta_n)$ in the proof of Theorem 2.2. Put

$$\begin{aligned} R(n,\alpha,\beta) \Leftrightarrow\ & (\alpha)_0,\ldots,(\alpha)_n,(\beta)_0,\ldots,(\beta)_n \in A \\ & \& \ |(\alpha)_0| < \lambda \ \& \cdots \& \ |(\alpha)_n| < \lambda \\ & \& \ |(\beta)_0| < \lambda \ \& \cdots \& \ |(\beta)_n| < \lambda \\ & \& \ [\text{both sequences } (\alpha)_0,\ldots,(\alpha)_n \text{ and} \\ & \quad (\beta)_0,\ldots,(\beta)_n \text{ are winning for I in the game } P^{\pi}] \\ & \& \ [\text{player } S \text{ wins the game } H_n((\alpha)_0,\ldots,(\alpha)_n; \\ & \quad (\beta)_0,\ldots,(\beta)_n)]. \end{aligned}$$

Lemma. The relation R is in $\exists^{\eta}\,\Gamma(\bar{\nu},\lambda)$.

Proof of the lemma. Let us compute first one part of the conjunction which defines R,

$$\begin{aligned} Q(n,\alpha) \Leftrightarrow\ & (\alpha)_0,\ldots,(\alpha)_n \in A \ \& \text{ the sequence } (\alpha)_0,\ldots,(\alpha)_n \\ & \quad \text{is winning for I in } P^{\pi} \text{ (with } n \text{ even)} \\ \Leftrightarrow\ & n \text{ is even } \& \ (\alpha)_0,\ldots,(\alpha)_n \in A \ \& \ |(\alpha)_0| < \lambda \ \& \cdots \& \ |(\alpha)_n| < \lambda \\ & \& \ (\exists\sigma)\{\sigma \text{ is a winning strategy for I in the subgame} \\ & \quad P^{\pi}((\alpha)_0,\ldots,(\alpha)_n) \text{ of } P^{\pi}\}. \end{aligned}$$

Now the strong version of the Harrington-Kechris Theorem implies that if a winning σ exists in this equivalence, then in fact a winning σ with "graph" in $\underset{\sim}{\Delta}$ exists. Choose $G \subseteq \mathcal{N} \times (\omega \times \mathcal{N} \times \mathcal{N})$ to be universal in Γ let χ be a Γ-norm on G and as always, code the $\underset{\sim}{\Delta}$-subsets of $\omega \times \mathcal{N} \times \mathcal{N}$ by setting

$$Z = \mathcal{N} \times \mathcal{N} \times \omega \times \mathcal{N} \times \mathcal{N},$$

$$I = \{(\varepsilon,\alpha^*,n^*,\beta^*,\gamma^*) : G(\alpha^*,n^*,\beta^*,\gamma^*)\}$$

and for

$$z = (\varepsilon,\alpha^*,n^*,\beta^*,\gamma^*) \in I$$

putting

$$A_z = \{(n,\beta,\gamma) : G(\varepsilon,n,\beta,\gamma) \;\&\; \chi(\varepsilon,n,\beta,\gamma) \leq \chi(\alpha^*,n^*,\beta^*,\gamma^*)\}.$$

It is clear that I is in Γ, each A_z is in $\underset{\sim}{\Delta}$ (and uniformly for $z \in I$) and by the Boundedness Theorem, every $\underset{\sim}{\Delta}$-subset of $\omega \times \mathcal{N} \times \mathcal{N}$ is A_z for some $z \in I$.

Continuing the computation then, we have

$$\begin{aligned} Q(n,\alpha) \Leftrightarrow\; & n \text{ is even } \& \; (\alpha)_0,\dots,(\alpha)_n \in A \;\&\; |(\alpha)_0| < \lambda \;\&\; \cdots \;\&\; |(\alpha)_n| < \lambda \\ & \& \; (\exists z)\{z \in I \;\&\; (\forall k)(\forall \beta)(\exists !\gamma)A_z(2k,\beta,\gamma) \\ & \qquad \& \; (\forall \beta)(\forall k)[(\beta)_{2k} \in A \;\&\; |(\beta)_{2k}| < \lambda] \\ & \qquad \& \; (\forall k)A_z(2k,\langle(\beta)_0,\dots,(\beta)_{2k}\rangle,(\beta)_{2k+1})] \\ & \Rightarrow P(|(\alpha)_0|,\dots,|(\alpha)_n|,|(\beta)_0|,|(\beta)_1|,\dots); \end{aligned}$$

this simply says that $Q(n,\alpha)$ holds exactly when there is a winning $\underset{\sim}{\Delta}$-strategy for I in P^π, in the position $(\alpha)_0,\dots,(\alpha)_n$.

Now the key to the result is that in the presence of the hypotheses and the side conditions, clearly

$$\begin{aligned} (\forall \beta)(\exists !\gamma)A_z(2k,\beta,\gamma) \Leftrightarrow\; & (\forall \beta)[(\exists \gamma \in \Delta(z,\beta))A_z(2k,\beta,\gamma) \\ & \& \; (\forall \gamma')[A_z(2k,\beta,\gamma') \Rightarrow \gamma = \gamma']] \end{aligned}$$

This is because if for some β there is exactly one γ satisfying the $\Delta(z,\beta)$-relation $A_z(2k,\beta,\gamma)$, then this γ clearly is in $\Delta(z,\beta)$. Once the quantification on γ becomes restricted in this way, then by the theorem on restricted quantification 4D.3 of DST, this second clause in the equivalence for $Q(n,\alpha)$ is in Γ and then the whole expression (easily) is in $\exists^{\mathcal{N}}\Gamma(\bar{\nu})$.

The remaining computation which is required to prove the lemma is quite similar and we will omit it; the point is that like P^{π}, the game H_n is also in $\Gamma(\bar{\nu})$, and it is the Harrington-Kechris game associated with an ordinal game, so that the strong version of the Harrington-Kechris Theorem applies.

To prove the theorem from this lemma, notice first that if $\alpha_0,\ldots,\alpha_n,\beta_0,\ldots,\beta_n,\alpha'_0,\ldots,\alpha'_n,\beta'_0,\ldots,\beta'_n$ are all in A and

$$|\alpha_0| = |\alpha'_0|,\ldots,|\alpha_n| = |\alpha'_n|,|\beta_0| = |\beta'_0|,\ldots,|\beta_n| = |\beta'_n|,$$

then

$$S \text{ wins } H_n(\alpha_0,\ldots,\alpha_n;\beta_0,\ldots,\beta_n) \Leftrightarrow S \text{ wins } H_n(\alpha'_0,\ldots,\alpha'_n;\beta'_0,\ldots,\beta'_n);$$

this is simply because the payoff for H_n depends only on the ordinals $|\alpha_0|,\ldots,|\alpha_n|,|\beta_0|,\ldots,|\beta_n|$. Thus

$$\begin{aligned} R^*(n,\alpha) &\Leftrightarrow (\alpha)_0,\ldots,(\alpha)_n \text{ is minimal} \\ &\Leftrightarrow (\forall\xi < \lambda)(\exists\beta)\{|\beta| = \xi \;\&\; R(n,\alpha,\langle(\alpha)_0,\ldots,(\alpha)_{n-1},\beta\rangle)\} \end{aligned}$$

and by the ordinal quantification theorem the relation R^* is easily in $\exists^n\,\Gamma(\bar{\nu},\lambda)$. Finally, by Theorem 2.2 the set

$$\begin{aligned} \Sigma = \{(\xi_1,\xi_3,\ldots,\xi_{2n-1},\xi_{2n}) : (\exists\alpha)[R^*(2n,\alpha) \\ \& \; |(\alpha)_1| = \xi_1 \;\&\; \cdots \;\&\; |(\alpha)_{2n-1}| = \xi_{2n-1} \\ \& \; |(\alpha)_{2n}| = \xi_{2n}]\} \end{aligned}$$

is the graph of a winning strategy for I in P and it is obviously in $\exists^n\,\Gamma(\bar{\nu},\lambda)$ in the codes. ⊣

As a corollary of this result, we can establish several improved versions of the determinacy of Martin games, Theorem 3.1. The next lemma, due to Becker, simplifies considerably these and many other applications of Theorem 4.1.

<u>Lemma</u> 4.2. (Becker). Assume AD, let Γ be a pointclass which resembles Π^1_1, let $\lambda < \underset{\sim}{\delta}$ and suppose $\bar{\varphi}$ is a Γ-scale on some W which is a scale-coding of λ; then the tree

$$T = T^{\bar{\varphi},\lambda} = T^{\bar{\varphi}} \text{ restricted to } \lambda$$

is in $\Delta(\lambda,\bar{\nu})$ in the codes (relative to φ_0) for suitable ordinals $\bar{\nu} = \nu_1,\nu_2 < \underset{\sim}{\delta}$.

<u>Proof</u>. By definition,

$$(c_0,\xi_0,\dots,c_{n-1},\xi_{n-1}) \in T \Leftrightarrow \xi_0,\dots,\xi_{n-1} < \lambda \ \&\ (\exists\alpha)\{\alpha \in W$$
$$\&\ \alpha(0) = c_0 \ \&\ \cdots\ \&\ \alpha(n-1) = c_{n-1}$$
$$\&\ \varphi_0(\alpha) = \xi_0 \ \&\ \cdots\ \&\ \varphi_{n-1}(\alpha) = \xi_{n-1}\}$$

and relative to the Γ-norm φ_0,

$$(n,u,\beta) \in \mathrm{Code}(T) \Leftrightarrow \mathrm{Seq}(u)\ \&\ \ell h(u) = n$$
$$\&\ (\beta)_0,\dots,(\beta)_{n-1} \in W$$
$$\&\ \varphi_0((\beta)_0) < \lambda \ \&\ \cdots\ \&\ \varphi_0((\beta)_{n-1}) < \lambda$$
$$\&\ (\exists\alpha)\{\alpha \in W$$
$$\&\ \alpha(0) = (u)_0 \ \&\ \cdots\ \&\ \alpha(n-1) = (u)_{n-1}$$
$$\&\ \varphi_0(\alpha) = \varphi_0((\beta)_0)\ \&\ \varphi_1(\alpha) = \varphi_0((\beta)_1)$$
$$\&\ \cdots\ \&\ \varphi_{n-1}(\alpha) = \varphi_0((\beta)_{n-1})\};$$

this is the set whose complexity we must compute.

Now by the Ordinal Quantification Theorem 8G.2 of DST, Code(T) is clearly in $\exists^{\mathcal{N}}\Gamma(\lambda)$, so that for some Q in Γ and every δ such that

$$\varphi_0(\delta) = \lambda,$$

we have

$$(n,u,\beta) \in \mathrm{Code}(T) \Leftrightarrow (\exists\alpha)Q(\delta,n,u,\beta,\alpha).$$

Let

$$\pi : Q \to \underset{\sim}{\delta}$$

be a Γ-norm and for each $(c_0,\xi_0,\dots,c_{n-1},\xi_{n-1}) \in T$, let

$$f(c_0,\xi_0,\dots,c_{n-1},\xi_{n-1}) = \mathrm{infimum}\{\pi(\delta',n,u,\beta',\alpha) : \varphi_0(\delta') = \lambda$$
$$\&\ u = \langle c_0,\dots,c_{n-1}\rangle$$
$$\&\ (\beta')_0,\dots,(\beta')_{n-1} \in W$$
$$\&\ \varphi_0((\beta')_0) = \xi_0 \ \&\ \cdots\ \&\ \varphi_0((\beta')_{n-1}) = \xi_{n-1}$$
$$\&\ Q(\delta',n,u,\beta',\alpha)\}.$$

Finally put

$$\nu_1 = \text{supremum}\{f(c_0,\xi_0,\ldots,c_{n-1},\xi_{n-1}) : (c_0,\xi_0,\ldots,c_{n-1},\xi_{n-1}) \in T\}$$

and notice that by the regularity of $\underset{\sim}{\delta}$,

$$\nu_1 < \underset{\sim}{\delta}.$$

Now clearly

$$\begin{aligned}(n,u,\beta) \in \text{Code}(T) \Leftrightarrow (\exists\alpha)(\exists\delta')(\exists\beta')\{&\varphi_0(\delta') = \lambda \\ &\& \ \pi(\delta',n,u,\beta',\alpha) < \nu_1 \\ &\& \ \varphi_0((\beta)_0) = \varphi_0((\beta')_0) \\ &\& \cdots \& \ \varphi_0((\beta)_{n-1}) = \varphi_0((\beta')_{n-1})\}\end{aligned}$$

and hence $\text{Code}(T)$ is in $\exists^{\mathfrak{n}} \Delta(\lambda,\nu_1) \subseteq \neg\Gamma(\lambda,\nu_1)$ in the codes, since Γ is closed under $\forall^{\mathfrak{n}}$ (and here $\neg\Gamma$ is closed under $\exists^{\mathfrak{n}}$).

From this it follows that $\neg\,\text{Code}(T)$ is in $\Gamma(\lambda,\nu_1)$ and by repeating the argument, we find some ν_2 so that $\neg\,\text{Code}(T)$ is in $\neg\Gamma(\lambda,\nu_1,\nu_2)$ in the codes and hence $\text{Code}(T)$ is in $\Delta(\lambda,\nu_1,\nu_2)$ in the codes. ⊣

Let us first put down one strong version of the determinacy of Martin games.

<u>Theorem</u> 4.3. Assume AD, let Γ be a pointclass which resembles Π^1_1, let $\lambda < \underset{\sim}{\delta}$ and suppose $\overline{\varphi}$ is a Γ-scale on some $W \subseteq \mathfrak{n}$ which is a scale-coding of λ. Suppose $Q \subseteq {}^{\omega}(\mathfrak{n} \times \omega)$ is in Δ in the codes, i.e. both Q_1 and Q_2 are in Γ, where

$$Q_1(\alpha,\gamma) \Leftrightarrow (\forall n)[(\alpha)_n \in W] \ \& \ Q((\alpha)_0,\gamma(0),\ldots,(\alpha)_n,\gamma(n),\ldots),$$

$$Q_2(\alpha,\gamma) \Leftrightarrow (\forall n)[(\alpha)_n \in W] \ \& \neg\, Q((\alpha)_0,\gamma(0),\ldots,(\alpha)_n,\gamma(n),\ldots).$$

Then the Martin game $Q^{\overline{\varphi}}$ on λ associated with Q and $\overline{\varphi}$ is determined and whichever player wins it has a winning strategy in $\exists^{\mathfrak{n}}\,\Gamma(\lambda,\overline{\nu})$ in the codes for some $\overline{\nu} = \nu_1,\ldots,\nu_k < \underset{\sim}{\delta}$.

<u>Proof</u>. The lemma implies that the payoff of either player is $Q^{\overline{\varphi}}$ admits a uniform scale which is in $\exists^{\mathfrak{n}}\,\Gamma(\lambda,\overline{\nu}')$ in the codes, so the game is determined by Theorem 2.1 and the winner can win in the appropriate pointclass by Theorem 4.1. ⊣

This result is not as immediately applicable as one might hope, because in many cases (as we will see) one cannot use simple Martin games. Sometimes we must consider <u>Martin-like</u> games associated with a given $Q \subseteq {}^{\omega}(\mathfrak{n} \times \omega)$, λ

and $\overline{\varphi}$, where the players make additional moves which purport to be honest codes of their main moves, as in $Q^{\overline{\varphi}}$, but where the payoff is not defined as simply as in $Q^{\overline{\varphi}}$ - e.g. one may interchange the order in which clauses (1), (2), (3) and (4) of §3 are applied. It is then not hard to use Lemma 4.2 as a step in applying the Basic Theorem 4.1. Becker [A], [B] has used very effectively Martin-like games of this type and we will illustrate their use in the next section.

§5. The power function in the models H_Γ. The inner model H_Γ of ZFC associated with each pointclass Γ that resembles Π^1_1 was defined in section 8G of DST and its basic properties were developed in 8G.22-8G.32 of DST. One of the most important facts about H_Γ, is that if $\lambda < \underset{\sim}{\delta}$ and $\Sigma \subseteq \bigcup_n \lambda^n$, then

$$\Sigma \in H_\Gamma \Leftrightarrow \Sigma \text{ is in } \exists^{\mathcal{N}} \Gamma(\overline{\nu}) \text{ in the codes, for}$$
$$\text{some } \overline{\nu} = \nu_1,\ldots,\nu_n < \underset{\sim}{\delta}.$$

This of course ties up with the theorems we proved in §4.

If $\pi : \to \underset{\sim}{\delta}$ is any Γ-norm and $\lambda < \underset{\sim}{\delta}$, then the Coding Lemma implies again (granting AD) that for each set $X \subseteq \lambda$,

$$\text{Code}(X) = \{\alpha \in A : \pi(\alpha) \in X\}$$

is in $\underset{\sim}{\Delta}$; a code for X in any ε such that

$$\text{Code}(X) = \{\alpha : G((\varepsilon_0,\alpha)\}$$
$$= \{\alpha : \neg G((\varepsilon)_1,\alpha)\},$$

where G is some fixed Γ-universal set. A set $\mathcal{P} \subseteq \text{Power}(\lambda)$ is in some Γ^* in the codes (relative to π and G), if the relation

$$\mathcal{P}^\#(\varepsilon) \Leftrightarrow \varepsilon \text{ codes some } X \in \mathcal{P}$$

is in Γ^*. The Ordinal Quantification Theorem implies directly that if Γ resembles Π^1_1 and $\Gamma^* = \exists^{\mathcal{N}} \Gamma(\overline{\nu})$, then this notion is independent of the choice of π and G.

Theorem 5.1. Assume AD, let Γ be a pointclass which resembles Π^1_1 and suppose $\lambda < \underset{\sim}{\delta}$; then

$$\text{Power}(\lambda) \cap H_\Gamma = \text{the largest wellorderable collection of subsets of } \lambda \text{ which is in } \exists^{\mathcal{N}} \Gamma(\lambda) \text{ in the codes}$$

= the largest thin collection of subsets of λ which is in $\exists^{\mathbb{R}}\Gamma(\bar{\nu})$ in the codes, for any $\bar{\nu} = \nu_1,\dots,\nu_n < \underset{\sim}{\delta}$.

Proof. That $\text{Power}(\lambda) \cap H_\Gamma$ is in fact in $\exists^{\mathbb{R}}\Gamma(\lambda)$ in the codes (relative to any Γ-norm onto $\underset{\sim}{\delta}$) and of course wellorderable can be proved easily, as in 8G.29 of DST.

To prove the converse, suppose $\mathcal{P} \subseteq \text{Power}(\lambda)$ is thin and in $\exists^{\mathbb{R}}\Gamma(\nu)$ in the codes, where we have allowed (without loss of generality) for only one ordinal parameter. (We will fix a Γ-scale $\bar{\varphi}$ on W and take all codings relative to the first Γ-norm φ_0.) By definition then, there is some $R(\beta,\varepsilon,\alpha)$ in Γ such that whenever

$$\beta \in W \;\&\; \varphi_0(\beta) = \nu,$$

then

$$\varepsilon \text{ codes some } X \in \mathcal{P} \Leftrightarrow (\exists\alpha)R(\beta,\varepsilon,\alpha).$$

Now fix a Γ-norm

$$\chi : R \to \underset{\sim}{\delta},$$

fix any ordinal μ and let λ^* be any ordinal greater than λ, ν and μ for which $\bar{\varphi}$ is a scale-coding; ordinals like this exist easily, because $\underset{\sim}{\delta}$ is regular.

We now play an ordinal game P which is a minor variation of the perfect set game we played in the proof of Theorem 3.2. The moves are as follows:

I	$\lambda', \nu', \mu', \xi_0, \varepsilon(0), \alpha(0)$	$\xi_1, \varepsilon(1), \alpha(1)$	$\dots$
II	t_0	t_1	

In the first three moves, player I must put down the parameters $\lambda' = \lambda$, $\nu' = \nu$ and $\mu' = \mu$ or else he loses - the purpose of these seemingly silly moves will become clear in a moment. After this, I plays ordinals $\xi_0, \xi_1, \dots$ below λ and (bit by bit) irrationals ε and α and II plays $t_i = 1$ or $t_i = 0$. At the end of the game,

$$\begin{aligned}
\text{I wins} \Leftrightarrow\ & \lambda' = \lambda \;\&\; \nu' = \nu \;\&\; \mu' = \mu \\
& \& \; R(\nu', \varepsilon, \alpha) \\
& \& \; \chi(\nu', \varepsilon, \alpha) \leq \mu' \\
& \& \; (\forall i)[\xi_i \in \text{the set coded by } \varepsilon \Leftrightarrow t_i = 1].
\end{aligned}$$

The definition of this game P is not precise, because although we know what $R(\beta,\varepsilon,\alpha)$ means, we do not know how to interpret $R(\nu,\varepsilon,\alpha)$ or $\chi(\nu,\varepsilon,\alpha)$ for that matter. Of course what we have in mind is the intentional ordinal game whose payoff will depend not only on the ordinal moves indicated above, but also on additional side moves which give alleged honest codes for these ordinals.

To make this precise, consider the Martin-like game where the players make these additional moves, as in §3; in the notation of that section, we indicate these moves as follows.

$$\begin{array}{ll} \text{I} & \lambda', \begin{pmatrix} \alpha_{\lambda'} \\ f_{\lambda'} \end{pmatrix}, \nu', \begin{pmatrix} \alpha_{\nu'} \\ f_{\nu'} \end{pmatrix}, \mu', \begin{pmatrix} \alpha_{\mu'} \\ f_{\mu'} \end{pmatrix}, \xi_0, \begin{pmatrix} \alpha_0 \\ f_0 \end{pmatrix}, \varepsilon(0), \alpha(0), \qquad \ldots \\ \text{II} & \qquad\qquad\qquad\qquad\qquad\qquad\qquad\qquad\qquad\qquad t_0 \end{array}$$

The payoff is determined as follows:

(0) I must play $\lambda' = \lambda$, $\nu' = \nu$ and $\mu' = \mu$, or else he loses.

(1) Exactly like (1) in §3, we must have

$$f_{\lambda'}(0) = \lambda',\ f_{\nu'}(0) = \nu',\ f_{\mu'}(0) = \mu',\ f_0(0) = \xi_0, \ldots .$$

$$g_{\lambda'}(0) = \lambda',\ g_{\nu'}(0) = \nu',\ g_{\mu'}(0) = \mu',\ g_0(0) = \xi_0, \ldots .$$

(2) Exactly like (2) in §3, for $f_{\lambda'},\ldots,g_{\lambda'},\ldots,f_0,g_0,\ldots$.

(3) Exactly like (3) in §3, for $\alpha_{\lambda'},\beta_{\lambda'},\ldots,\alpha_0,\beta_0,\alpha_1,\beta_1,\ldots$.

(4) If no one wins by (0)-(3), then

$$\begin{aligned} \text{I wins} \Leftrightarrow\ & R(\alpha_{\nu'},\varepsilon,\alpha) \\ & \& \ \chi(\alpha_{\nu'},\varepsilon,\alpha) \leq \varphi_0(\alpha_{\mu'}) \\ & \& \ (\forall i)[\alpha_i \in \text{the set coded by } \varepsilon \\ & \quad \Leftrightarrow t_i = 1]. \end{aligned}$$

If we call the payoff of this game P, it is clear from Lemma 4.2 that both P and its complement are in $\Gamma(\overline{\nu'})$ for some $\overline{\nu'}$ (including λ, ν, μ) and they admit uniform scales which are in $\Gamma(\overline{\nu'})$ in the codes (relative to φ_0). Thus by Theorem 4.1, whichever player wins the associated Harrington-Kechris game can win P with a strategy in $\exists^{\mathbb{R}}\Gamma(\overline{\nu'})$ in the codes.

From this point on the proof is very similar to that of Theorem 3.2 and we will only outline it briefly.

First check that player I simply cannot win; because against his winning strategy (which must start with λ, ν, and μ) II can play honest codes for λ, ν, μ and all the later ordinals and then play both $t_i = 1$ and $t_i = 0$ to construct an imbedding of the full binary tree into $\wp$ contradicting the hypotheses.

Now II wins with a strategy τ which is in $\exists^{\mathbb{R}}\Gamma(\bar{\nu}')$ in the codes for some $\bar{\nu}'$, so that $\tau \in H_\Gamma$. For any μ, put

$$X \in \mathcal{P}_\mu \Leftrightarrow X \in \mathcal{P} \ \&\ \text{infimum}\{\chi(\beta,\varepsilon,\alpha) : R(\beta,\varepsilon,\alpha) \ \&\ \varphi_0(\beta) = \nu\} < \mu$$

and repeat the usual perfect set argument to show the obvious: if $X \in \mathcal{P}_\mu$, then there is some initial part u of the play in the game (associated with μ) such that for all $\xi < \lambda$.

$$\xi \in X \Leftrightarrow \tau(u,\xi) = 0.$$

Thus $\mathcal{P}_\mu \subseteq H_\Gamma$ for each $\mu < \underset{\sim}{\delta}$, and here

$$\mathcal{P} = \bigcup_{\mu < \underset{\sim}{\delta}} \mathcal{P}_\mu \subseteq H_\Gamma,$$

which is what we set out to prove. ⊣

References

H. S. Becker [A], AD and the supercompactness of $\aleph_1$, to appear.

H. S. Becker [B], Thin collections of sets of projective ordinals and analogs of L, to appear.

L. A. Harrington and A. S. Kechris [A], On the determinacy of games on ordinals, to appear.

A. S. Kechris [A], Homogeneous trees and projective scales, this volume.

Y. N. Moschovakis [1978], Inductive scales on inductive sets, Cabal Seminar 76-77, Springer Lecture Notes in Mathematics, Vol. 689, 1978, 185-192.

Y. N. Moschovakis [1980, DST], Descriptive Set Theory, Studies in Logic, Vol. 100, North Holland Publishing Co., Amsterdam, 1980.

MEASURABLE CARDINALS IN PLAYFUL MODELS

Howard S. Becker and Yiannis N. Moschovakis
Department of Mathematics
University of California
Los Angeles, California 90024

The axiom of determinacy (AD) implies that there are transitive models of ZFC in which measurable cardinals exist. Indeed this follows from the much weaker assumption that all games in the difference hierarchy on $\underset{\sim}{\Pi}^1_1$ are determined; the assumption that all $\underset{\sim}{\Delta}^1_2$ games are determined gives models with measures that concentrate on measurables. Martin [A] contains proofs of these and of many similar results. These models were originally constructed for the sole purpose of producing examples of models containing measurable cardinals, and thus proving relative consistency results. They appear to be rather unnatural models, of little interest for any other purpose.

In [1980], Moschovakis introduced a family of inner models of ZFC, one model, known as H_Γ, for each pointclass Γ that resembles Π^1_1. These models are quite natural from the perspective of descriptive set theory, for on questions of descriptive set theory H_Γ is an analog of L for the pointclass Γ. They satisfy AC but also a certain amount of definable determinacy; that is, they are "playful models." They have many pleasant properties, some of which we will describe below.

The models H_Γ which were originally introduced as "analogs of L" turned out to have one property which is not at all L-like - they have a lot of measurable cardinals in them. In this paper we will investigate measurable cardinals in H_Γ. Our purpose in doing so is not to prove any new or better theorems on the consistency strength of determinacy and large cardinal axioms. Our purpose is to better understand the models H_Γ, which we believe to be important and interesting. We work in ZF + DC, and sometimes we assume AD as an additional hypothesis.

§1. The models H_Γ. A pointclass Γ resembles Π^1_1 if

(i) Γ is a Spector pointclass with the scale property and closed under $\forall^{\mathcal{N}}$.

(ii) For each $\alpha \in \mathcal{N}$, if $P \subset \mathcal{X} \times \mathcal{Y}$ is in $\Delta(\alpha)$ and

$$Q(x) \Leftrightarrow \{y : P(x,y)\} \text{ is not meager}$$

then Q is also in $\Delta(\alpha)$.

Assuming AD, the pointclasses resembling Π^1_1 include Π^1_{2n+1} for all n, the class IND of inductive sets, the class of sets semirecursive in ^3E, $(\Sigma^2_1)^{L[\mathbb{R}]}$, the above classes relativized to any irrational, and many others.

In section 8G of Moschovakis [1980], an inner model H_Γ of ZFC associated with each pointclass Γ that resembles Π^1_1 was defined. For $\Gamma = \Pi^1_{2n+1}$ we use the simpler notation $H_{2n+1} = H_{\Pi^1_{2n+1}}$. The basic properties of the H_Γ's were developed in 8G.22-8G.32 of Moschovakis [1980]. Additional results on these models appear in Moschovakis [A], in this volume and in Becker [A].

We fix some notation, following Moschovakis [1980]:

$$\Gamma^* = \exists^{\mathbb{R}}\Gamma$$

$$\Delta = \Gamma \cap \breve{\Gamma}$$

$$\Delta^* = \Gamma^* \cap \breve{\Gamma^*}$$

$$\underset{\sim}{\delta} = \sup\{\xi : \text{there is a } \underset{\sim}{\Delta} \text{ prewellordering of } \mathbb{R} \text{ of length } \xi\}.$$

The model H_Γ is an analog of L, particularly with respect to subsets of $\underset{\sim}{\delta}$. The relationship between H_Γ and the pointclasses Γ and Γ^* is the same as the relationship between L and the pointclasses Π^1_1 and Σ^1_2. H_1 is L.

If $\xi < \underset{\sim}{\delta}$, a set of irrationals is $\Gamma(\xi)$ if it is $\Gamma(\alpha)$, uniformly in all codes α for ξ, and similarly for $\Delta(\xi)$, $\Gamma^*(\xi)$, etc. One of the important L-like properties of H_Γ is:

1.1 (AD) Let $A \subset \underset{\sim}{\delta}$. If A is $\Gamma^*(\xi_0,\ldots,\xi_k)$-in-the-codes for some $\xi_0,\ldots,\xi_k < \underset{\sim}{\delta}$ then $A \in H_\Gamma$.

In Moschovakis [1980], H_Γ is defined to be L[P], where P is a subset of $\underset{\sim}{\delta}$, and thus it has a canonical constructibility ordering, which we denote by $\leq_\Gamma$ (or by $\leq_{2n+1}$ if $\Gamma = \Pi^1_{2n+1}$). For ξ an ordinal, $H_\Gamma \restriction \xi$ denotes the set of all elements of H_Γ whose rank in the wellordering $\leq_\Gamma$ is less than ξ. Another L-like property of H_Γ is that for $\xi < \underset{\sim}{\delta}$, virtually anything one would ever want to know about $H_\Gamma \restriction \xi$ is $\Delta^*(\xi)$. For example, if $\xi, \eta < \underset{\sim}{\delta}$ then:

1.2 (AD) Membership in $\mathrm{Pow}(\eta) \cap (H_\Gamma \restriction \xi)$ is $\Delta^*(\xi,\eta)$.

1.3 (AD) Truth in $H_\Gamma \restriction \xi$ is $\Delta^*(\xi)$.

1.4 (AD) The ordering $\leq_\Gamma \restriction (\mathrm{Pow}(\eta) \cap (H_\Gamma \restriction \xi))$ is $\Delta^*(\xi,\eta)$.

We will not formally define H_Γ here or prove any of the above properties.

We refer the reader to Moschovakis [1980] for details.

The purpose of this paper is to investigate just one aspect of these models, namely the measurable cardinals in them. We have two results:

(1) There are a lot of measurable cardinals in H_Γ, some with measures of high order.

(2) The least measurable cardinal in H_{2n+1} is $\aleph_1^V$.

The first is due to Moschovakis and is proved in sections 2 and 3 of this paper. The second, which is due to Becker, is proved in section 4. In section 5 we list some open problems.

§2. The λ-closed unbounded measure. If λ is a regular cardinal and κ an ordinal such that $\operatorname{cof} \kappa > \lambda$, then let $F(\lambda,\kappa)$ be the λ-closed unbounded filter on κ. That is,

$$F(\lambda,\kappa) = \{A \subset \kappa : \exists B \subset A \text{ such that } \begin{array}{l} (1) \text{ if } \operatorname{cof} \xi = \lambda \ \& \ \xi = \sup(B \cap \xi) \text{ then } \xi \in B \\ (2) \ B \text{ is unbounded in } \kappa\} . \end{array}$$

It is well known that if $F(\lambda,\kappa)$ is a normal ultrafilter then κ must be a regular cardinal; in a universe without the axiom of choice, κ may be a successor cardinal. Without choice it is also not necessarily the case that $F(\lambda,\kappa)$ is normal, i.e. that it satisfies Fodor's Theorem. But assuming AD, there are many examples of λ, κ for which $F(\lambda,\kappa)$ is known to be a normal ultrafilter. It is an open question whether $F(\lambda,\kappa)$ is always a normal ultrafilter, for any regular cardinals λ, κ such that $\lambda < \kappa < \Theta$.

2.1 Theorem (AD). Let Γ be a pointclass which resembles Π^1_1, let λ be a regular cardinal, and let κ be an ordinal such that $\operatorname{cof} \kappa > \lambda$ and $\kappa < \underset{\sim}{\delta}$. Then the set $(F(\lambda,\kappa) \cap H_\Gamma)$ is in H_Γ. And in fact the set

$$\{(\zeta,A) : \operatorname{cof} \zeta > \lambda \ \& \ \zeta < \underset{\sim}{\delta} \ \& \ A \in (F(\lambda,\zeta) \cap H_\Gamma)\}$$

is in H_Γ.

Proof. The ordinal $\underset{\sim}{\delta}$ associated with the class Γ is a regular cardinal (Moschovakis [1980], 7D.8), and $H_\Gamma = L[P]$ where P is a subset of $\underset{\sim}{\delta}$. So a standard constructibility argument will show that $(\lneq_\Gamma \restriction \operatorname{Pow}(\kappa))$ has order type at most $\underset{\sim}{\delta}$, where $\lneq_\Gamma$ is the constructibility ordering of H_Γ, which is of course in H_Γ. So to prove the first assertion of the theorem it is enough to show that E is in H_Γ, where E is the following set of ordinals:

$$\{\xi < \underset{\sim}{\delta} : \text{The } \xi\text{th subset of } \kappa, \text{ with respect to } \lneq_\Gamma, \text{ is in } F(\lambda,\kappa)\} .$$

By 1.1, to prove E is in H_Γ it will suffice to show that E is a $\Gamma^*(\lambda,\kappa,\nu)$ set of ordinals, for some $\nu < \underset{\sim}{\delta}$. This is proved by the techniques in 8G of Moschovakis [1980] (techniques also used in Moschovakis [A]). We will give an outline of the proof here, but will omit the detailed quantifier counting. Two key facts used in this pointclass computation are the Moschovakis Coding Lemma, that every bounded subset of $\underset{\sim}{\delta}$ is $\underset{\sim}{\Delta}$-in-the-codes, and the Harrington-Kechris Ordinal Quantification Theorem [A], that $\Gamma^*(\eta)$ is closed under quantification of the form $(\forall\, \xi < \eta)$. Both of these are proved in Moschovakis [1980] and used in pointclass computations in Moschovakis [1980], 8.G and in Moschovakis [A]. The reader who understands those pointclass computations will have no trouble following our proof that E is $\Gamma^*(\lambda,\kappa,\nu)$, and will also have no trouble understanding why we choose to leave the details out of this paper.

First of all, the set S of ordinals less than κ which have cofinality λ is easily $\Gamma^*(\lambda,\kappa)$-in-the-codes. Now Γ has the prewellordering property and $\kappa < \underset{\sim}{\delta}$, $\underset{\sim}{\delta}$ regular, so the codes for ordinals in S must be bounded in the prewellordering below some $\nu < \underset{\sim}{\delta}$. So S is $\Delta^*(\lambda,\kappa,\nu)$ for any such ν, that is, for sufficiently large ν. Since S is $\Delta^*(\lambda,\kappa,\nu)$, the set

$$\{\alpha \in \mathcal{N} : \alpha \text{ is a } \underset{\sim}{\Delta}\text{-code for a set } A \subset \kappa \;\&\; A \in F(\lambda,\kappa)\}$$

is a $\Gamma^*(\lambda,\kappa,\nu)$ set. From this and 1.2-1.4 (proved in Moschovakis [1980]) it follows that E is $\Gamma^*(\lambda,\kappa,\nu)$.

Next let

$$\hat{E} = \{(\zeta,\xi) \in \underset{\sim}{\delta} \times \underset{\sim}{\delta} : \text{cof } \zeta > \lambda \;\&\; \text{the } \xi\text{th subset of } \zeta, \text{ with respect to } \lneq_\Gamma, \text{ is in } F(\lambda,\zeta)\} .$$

The second assertion of the theorem is proved in a similar manner, by showing that there is a Γ^* set $D \subset \underset{\sim}{\delta} \times \underset{\sim}{\delta} \times \underset{\sim}{\delta}$ such that for all $\zeta,\xi < \underset{\sim}{\delta}$

$$(\zeta,\xi) \in \hat{E} \Leftrightarrow \text{for sufficiently large } \nu < \underset{\sim}{\delta},\ (\zeta,\xi,\nu) \in D .$$

Then D is in H_Γ by 1.1, so $\hat{E}$ is also in. $\dashv$

2.2 <u>Corollary</u> (AD): Let Γ be a pointclass which resembles Π^1_1, let λ be a regular cardinal, and let κ be an ordinal such that $\lambda < \kappa < \underset{\sim}{\delta}$. If κ is a regular cardinal and $F(\lambda,\kappa)$ is a normal ultrafilter in V, then

$$H_\Gamma \models \text{"}\kappa \text{ is a measurable cardinal"} .$$

Assuming AD, the filters $F(\omega,\aleph_1)$, $F(\omega,\aleph_2)$, and $F(\omega_1,\aleph_2)$ are all normal ultrafilters (Kechris [1978]). By 2.1, if $\Pi^1_3 \subset \Gamma$ then the restriction of these filters to H_Γ is in H_Γ, and so

$$H_\Gamma \models \text{"}\aleph^V_1 \text{ and } \aleph^V_2 \text{ are measurable cardinals"} .$$

(This result appears without proof in Moschovakis [1980], page 574.) Of course since choice is true in H_Γ, H_Γ does not know that the ultrafilter is the ω-closed (or ω_1-closed) unbounded filter; a set $A \subset \aleph^V_1$ of measure 1 in H_Γ does contain an ω-closed unbounded set B, but B need not be in H_Γ.

The next measurable cardinal is $\aleph_{\omega+1}$ $(= \underset{\sim}{\delta}^1_3)$ which has three normal measures, $F(\omega,\underset{\sim}{\delta}^1_3)$, $F(\omega_1,\underset{\sim}{\delta}^1_3)$, and $F(\omega_2,\underset{\sim}{\delta}^1_3)$ (Kechris [1978]). When $\Gamma = \Pi^1_3$, $\underset{\sim}{\delta} = \underset{\sim}{\delta}^1_3$, so by Scott's Theorem $\underset{\sim}{\delta}^1_3$ is not a measurable cardinal in H_3. But if $\Pi^1_5 \subset \Gamma$, then the restrictions of all three filters are in H_Γ and so

$$H_\Gamma \models \text{"}\aleph^V_{\omega+1} \text{ is a measurable cardinal"} .$$

Higher up in the projective hierarchy most set theoretic questions are still unanswered. We do not even know exactly what the measurable cardinals less than $\underset{\sim}{\delta}^1_5$ are. It is known however that for all n, odd or even, $\underset{\sim}{\delta}^1_n$ is a regular cardinal and $F(\omega,\underset{\sim}{\delta}^1_n)$ is a normal ultrafilter. Thus $\underset{\sim}{\delta}^1_3$ and $\underset{\sim}{\delta}^1_4$ are measurable in $H_5, H_7, \ldots$, $\underset{\sim}{\delta}^1_5$ and $\underset{\sim}{\delta}^1_6$ are measurable in $H_7, H_9, \ldots$, and so on. The cardinality of the $\underset{\sim}{\delta}^1_n$'s is also an open question.

Theorem 2.1 applies to pointclasses far beyond the projective hierarchy and thus gives us models with many measures and many measurable cardinals. For example, consider $\kappa^{\mathbb{R}}$, the ordinal associated with the pointclass IND. The ordinal $\kappa^{\mathbb{R}}$ is a weakly Mahlo cardinal and for every regular $\lambda < \kappa^{\mathbb{R}}$, $F(\lambda,\kappa^{\mathbb{R}})$ is a normal ultrafilter (Kechris, Kleinberg, Moschovakis [A]). Moreover the set

$$\{\kappa < \kappa^{\mathbb{R}} : \kappa \text{ is a regular cardinal } \& \ F(\omega,\kappa) \text{ is a normal ultrafilter}\}$$

is unbounded in $\kappa^{\mathbb{R}}$. So if Γ is large, i.e. $(\text{IND} \cup \breve{\text{IND}}) \subset \Gamma$, then in H_Γ $\kappa^{\mathbb{R}}$ has at least $\kappa^{\mathbb{R}}$ normal measures and there are $\kappa^{\mathbb{R}}$ measurable cardinals less than $\kappa^{\mathbb{R}}$.

§3. Measures of high order. The models have measurable cardinals beyond the ones we have just described. There are many ordinals which are not cardinals of V but which are measurable cardinals in H_Γ. For example, there exists an ω_1-closed unbounded set of κ's below $\aleph_2$ such that κ is measurable in H_3, the measure on κ being the restriction of $F(\omega,\kappa)$. Thus in H_3, $\aleph^V_2$ is a measurable cardinal of order at least 2 - this notion will be defined below.

In this section we will prove a general theorem (3.3) about models in which the restrictions of filters $F(\lambda,\kappa)$ to the model are in the model, assuming the $F(\lambda,\kappa)$ are normal ultrafilters in V. We show that these measures in the model stack up to produce measures of high order. This result is due to Kunen (unpublished, but similar to the results published in Kunen [1971]). The theorem itself has nothing to do with the H_Γ's or with determinacy. But assuming AD, the theorem can be applied to the H_Γ's to prove the existence of higher order measures in them, such as the measure $F(\omega_1,\aleph_2)$ on $\aleph_2$, which concentrates on measurable cardinals.

3.1 <u>Definition</u>. Let μ be a normal measure (ultrafilter) on a cardinal κ. We call μ a <u>measure of order</u> ξ if for μ-a.e. $\lambda < \kappa$, for all $\zeta < \xi$, there is a measure of order ζ on λ. The <u>order</u> of a measurable cardinal κ is

$$\sup\{\xi + 1 : \text{there is a measure of order } \xi \text{ on } \kappa\} .$$

3.2 <u>Lemma</u>. Let M be a transitive model of ZFC, let ξ and ν be regular cardinals, and let κ be an ordinal in M such that

(1) $\nu > \xi$

(2) $\operatorname{cof} \kappa > \nu$

(3) The set $\{(\zeta,A) : \zeta < \kappa \ \&\ \operatorname{cof} \zeta > \xi \ \&\ A \in (F(\xi,\zeta) \cap M)\}$ is in M.

Suppose that either $F(\xi,\kappa)$ and $F(\nu,\kappa)$ are both normal ultrafilters in V, or else $(F(\xi,\kappa) \cap M)$ and $(F(\nu,\kappa) \cap M)$ are both in M and are both normal ultrafilters in M. Then for $F(\nu,\kappa)$ - a.e. η,

$$M \models \text{"}(F(\xi,\eta) \cap M) \text{ is a normal ultrafilter on } \eta\text{"} .$$

<u>Proof</u>. Since $F(\nu,\kappa)$ - a.e. η has cofinality ν, by (1) and (3) for $F(\nu,\kappa)$ - a.e. η, $(F(\xi,\eta) \cap M)$ is in M, and clearly M thinks that it is a filter on η. What remains to be shown is that for $F(\nu,\kappa)$ - a.e. η, this filter on η is an ultrafilter in M and is normal in M.

Suppose that for $F(\nu,\kappa)$ - a.e. η it is not an ultrafilter in M. Using (3) and the axiom of choice in the model, for each such η there is an $A_\eta \subset \eta$ such that $A_\eta \in M$ and neither A_η nor $\eta \backslash A_\eta$ is in $F(\xi,\eta)$. Let

$$A = \{\zeta < \kappa : \text{for } F(\nu,\kappa) \text{ - a.e. } \eta,\ \zeta \in A_\eta\} .$$

By the normality of $F(\nu,\kappa)$, for $F(\nu,\kappa)$ - a.e. η, $A_\eta = A \cap \eta$. Since $F(\xi,\kappa)$ is an ultrafilter, either A or $\kappa \backslash A$ contains a ξ-closed unbounded set B; without loss of generality, suppose $B \subset A$. Let

$$C = \{\eta < \kappa : \text{cof } \eta > \xi \ \& \ (B \cap \eta) \text{ is } \xi\text{-closed unbounded}\} .$$

By (1) and (2), C is ν-closed unbounded. For $F(\nu,\kappa)$ - a.e. η, $B \cap \eta \subset A_\eta$ and $\eta \in C$, hence $A_\eta \in F(\xi,\eta)$, a contradiction.

Next suppose that for $F(\nu,\kappa)$ - a.e. η it is not normal in M. Using choice in the model, for each such η choose a pressing down function $g_\eta : \eta \to \eta$ that is not constant $F(\xi,\eta)$ - a.e. For each $\zeta < \kappa$, let $h_\zeta : (\kappa \backslash \zeta) \to \zeta$ be the function $h_\zeta(\eta) = g_\eta(\zeta)$. The function h_ζ is pressing down, so is constant $F(\nu,\kappa)$ - a.e. Let $G : \kappa \to \kappa$ be the function $G(\zeta) = h_\zeta(\eta) = g_\eta(\zeta)$ for $F(\nu,\kappa)$ - a.e. η. Again using normality, for $F(\nu,\kappa)$ - a.e. η, $g_\eta = G \upharpoonright \eta$. But G is pressing down and $F(\xi,\kappa)$ is a normal ultrafilter, so G is constant on a ξ-closed unbounded set. Hence $F(\nu,\kappa)$ - a.e. g_η is also constant on a ξ-closed unbounded set, a contradiction. $\dashv$

3.3 <u>Theorem.</u> Let M be a transitive model of ZFC, let κ be an ordinal in M, and let $\{\lambda_\xi : \xi \leq \nu\}$ be a sequence of regular cardinals such that

(1) $\xi < \xi' \leq \nu \Rightarrow \lambda_{\xi'} > \lambda_\xi$

(2) $\text{cof } \kappa > \lambda_\nu$

(3) For all $\xi < \nu$, the set

$$E_\xi = \{(\zeta, A) : \zeta < \kappa \ \& \ \text{cof } \zeta > \lambda_\xi \ \& \ A \in (F(\lambda_\xi, \zeta) \cap M)\}$$

is in M; and in fact the map $\xi \mapsto E_\xi$ is in M.

Suppose that either for all $\xi \leq \nu$, $F(\lambda_\xi,\kappa)$ is a normal ultrafilter in V, or else for all $\xi \leq \nu$, $(F(\lambda_\xi,\kappa) \cap M)$ is in M and is a normal ultrafilter in M.

(a) Then for $F(\lambda_\nu,\kappa)$ - a.e. η, for all $\xi < \nu$,

$$M \models \text{"}(F(\lambda_\xi,\eta) \cap M) \text{ is a normal ultrafilter of order } \xi\text{"} .$$

(b) Suppose, in addition, that $(F(\lambda_\nu,\kappa) \cap M)$ is in M. Then

$$M \models \text{"}(F(\lambda_\nu,\kappa) \cap M) \text{ is a normal ultrafilter of order } \nu\text{"} .$$

<u>Proof.</u> The proof is by induction on ν. For any given ν, (b) follows directly from (a) and Definition 3.1. So we assume (b) is true for all $\nu' < \nu$ and prove that (a) holds for ν.

By Lemma 3.2, for any fixed $\xi < \nu$, for $F(\lambda_\nu,\kappa)$ - a.e. $\eta < \kappa$,

$$M \models \text{"}(F(\lambda_\xi,\eta) \cap M) \text{ is a normal ultrafilter on } \eta\text{"} .$$

By the normality of $F(\lambda_\nu,\kappa)$, there is an $A \subset \kappa$ such that $A \in F(\lambda_\nu,\kappa)$, and for all $\eta \in A$, for all $\xi < \nu$,

$$M \models \text{"}(F(\lambda_\xi, \eta) \cap M) \text{ is a normal ultrafilter on } \eta\text{"} .$$

That is, for any $\kappa' \in A$ and for any $\nu' < \nu$, κ' and ν' satisfy the hypothesis of part (b) of this theorem. So by the induction hypothesis

$$M \models \text{"}(F(\lambda_{\nu'}, \kappa') \cap M) \text{ is a normal ultrafilter of order } \nu'\text{"} ,$$

which proves the theorem. ⊣

We now assume AD and apply this theorem to the models H_Γ. In doing so we are of course also using Theorem 2.1. Let $\Pi^1_3 \subset \Gamma$. Letting $\lambda_0 = \omega$, $\lambda_1 = \omega_1$, and $\kappa = \aleph_2$, we see that $\aleph_2^V$ is a measurable cardinal of order at least 2 in H_Γ. Now let $\lambda_0 = \omega$, $\lambda_1 = \omega_1$, $\lambda_2 = \omega_2$, and $\kappa = \underset{\sim}{\delta}^1_3$. Although $\underset{\sim}{\delta}^1_3$ itself is not measurable in H_3, by part (a) of 3.3, we know that in H_3 there is a set of measurables of order 2 which is unbounded in $\underset{\sim}{\delta}^1_3$. For Γ larger than Π^1_3, $\underset{\sim}{\delta}^1_3$ is measurable of order at least 3 in H_Γ - the ω_2-closed unbounded filter gives the order 2 measure, while the ω_1-closed and ω-closed unbounded filters give measures of order 1 and 0, respectively. If we let Γ be a pointclass larger than IND, let $\kappa = \kappa^{\mathbb{R}}$ and let $\{\lambda_\xi : \xi < \kappa^{\mathbb{R}}\}$ be the sequence of regular cardinals below $\kappa^{\mathbb{R}}$, then 2.1 and 3.3 show that, in H_Γ, κ has measures of all orders less than κ. So κ is a measurable cardinal of order at least κ.

§4. The smallest measurable cardinal. In this section we prove that no ordinal which is countable in V is a measurable cardinal in H_{2n+1}. Our proof does not go through for all the H_Γ's, for arbitrary Γ resembling Π^1_1; however the theorem may be true in this generality.

Since $H_{2n+1} = L[P]$ where P is a subset of $\underset{\sim}{\delta}^1_{2n+1}$, by Scott's Theorem no ordinal greater than or equal to $\underset{\sim}{\delta}^1_{2n+1}$ can be measurable in H_{2n+1}. Thus

4.1 $\quad H_{2n+1} \models \text{"}\forall\kappa(\kappa \text{ a measurable cardinal} \rightarrow \aleph_1^V \leq \kappa < (\underset{\sim}{\delta}^1_{2n+1})^V)\text{"}$.

Since $\underset{\sim}{\delta}^1_1 = \aleph_1$ and $H_1 = L$, 4.1 gives an "explanation" of why there are no measurable cardinals in L.

For which cardinals (of V) κ does there exist an ordinal λ such that $\text{card}(\lambda) = \kappa$ and $H_{2n+1} \models$ "λ is a measurable cardinal"? While 4.1 restricts the class of such κ's it does not completely answer the question. Notice that the measurable cardinals of H_{2n+1} whose existence we proved in sections 2 and 3 all have cardinality (in V) $\aleph_1 = \underset{\sim}{\delta}^1_1$, $\aleph_2 = \underset{\sim}{\delta}^1_2$, $\aleph_\omega = \kappa_3$, $\aleph_{\omega+1} = \underset{\sim}{\delta}^1_3$, $\aleph_{\omega+2} = \underset{\sim}{\delta}^1_4$, κ_5, $\underset{\sim}{\delta}^1_5$, $\underset{\sim}{\delta}^1_6$, κ_7, $\underset{\sim}{\delta}^1_7$, $\underset{\sim}{\delta}^1_8$, etc., where κ_{2n+1} is the predecessor of $\underset{\sim}{\delta}^1_{2n+1}$. The cardinals $\aleph_3, \aleph_4, \aleph_5, \ldots$ themselves are all singular in H_{2n+1} $(n \geq 1)$, hence not measurable in the model. But Martin has shown that for any $n \geq 1$ and $m \geq 1$, there is a λ such that $\text{card}(\lambda) = \aleph_m$ and

$$H_{2n+1} \models \text{"}\lambda \text{ is a measurable cardinal"} .$$

Martin's proof uses the Kunen measures (see Solovay [1978]) and probably cannot be extended above $\aleph_\omega$ - for these larger cardinals the situation is still unclear.

Before beginning the proof that $\aleph_1^V$ is the least measurable cardinal, let us establish some notation. Fix some $n \geq 1$ and consider the model H_{2n+1}. Let κ denote the least measurable cardinal of H_{2n+1}. By 2.2 $\kappa \leq \aleph_1^V$; a priori, it may be countable. Let U be a normal ultrafilter on κ, in H_{2n+1}, and let

$$j : H_{2n+1} \to M$$

be the elementary embedding associated with U. That is, M is the transitive collapse of the ultrapower of H_{2n+1} modulo U, and j is the canonical embedding. The ultrapower and the map j both live inside the model H_{2n+1}.

Assuming AD, by 8G.24 of Moschovakis [1980]:

4.2 For any Σ^1_{2n+2} set A, there is a tree T_A on $\omega \times \underset{\sim}{\delta}^1_{2n+1}$ such that $T_A \in H_{2n+1}$ and such that (in V)

$$\begin{aligned} A &= p[T_A] \\ &= \text{the projection onto the first coordinate of the set of branches through } T_A. \end{aligned}$$

Let $G \subset \mathfrak{n}$ be a complete Π^1_{2n+1} set. Let S be the tree in H_{2n+1} such that $G = p[S]$. Let $T = j(S)$. Clearly T is a tree in M.

4.3 Lemma (AD). In V, $G = p[T]$.

Proof. First suppose that $\alpha \in G$. Then $\alpha \in p[S]$, say $(\alpha(0),\xi_0,\alpha(1),\xi_1,\dots) \in [S]$. Therefore $(\alpha(0),j(\xi_0),\alpha(1),j(\xi_1),\dots) \in [T]$ and $\alpha \in p[T]$. Conversely, suppose that $\alpha \in p[T]$; we must show that $\alpha \in G$. Suppose, towards a contradiction, that it is not. Now $\mathfrak{n}\backslash G$ is Σ^1_{2n+1} and so by 4.2 there is a tree S' in H_{2n+1} such that $\mathfrak{n}\backslash G = p[S']$. In particular, $\alpha \in p[S']$. Let R be the tree on $\omega \times \underset{\sim}{\delta}^1_{2n+1} \times \underset{\sim}{\delta}^1_{2n+1}$ defined as follows:

$$\begin{aligned} R = \{((a_0,\xi_0,\eta_0) \dots (a_k,\xi_k,\eta_k)) : ((a_0,\xi_0),\dots,(a_k,\xi_k)) \in S' \ \& \\ ((a_0,\eta_0) \dots (a_k,\eta_k)) \in T\} . \end{aligned}$$

Then $p[R]$ is precisely the set of irrationals in both $p[S']$ and $p[T]$. So $\alpha \in p[R]$. The trees S' and T are both in H_{2n+1} - for T we are using the fact that j is (a class) in H_{2n+1}. The definition of R is clearly absolute, so R is in H_{2n+1}. Since $\alpha \in p[R]$, by the absoluteness of well-foundedness there must be some $\beta \in H_{2n+1}$ such that $\beta \in p[R]$. Hence $\beta \in p[S']$

and $\beta \in p[T]$. Since $\beta \in p[S']$, $\beta \notin G$.

Now work inside H_{2n+1}. We know that $\beta \in p[T]$ and T is the ultrapower of S mod U. Say that $(\beta(0),[f_0],\beta(1),[f_1],\ldots) \in [T]$, where $[f_i]$ is the equivalence class of f_i mod U, and $f_i : \kappa \to \underset{\sim}{\delta}^1_{2n+1}$. Since the ultrafilter U is countably complete, for a.e. $\xi < \kappa$, $(\beta(0),f_0(\xi),\beta(1),f_1(\xi),\ldots) \in [S]$. So $\beta \in p[S]$. Going back from H_{2n+1} to V, we still have that $\beta \in p[S]$, hence $\beta \in G$, a contradiction. ⊣

4.4 <u>Lemma</u> (AD). If N is any transitive model of ZFC such that $M \subset N$ then N is Σ^1_{2n+2}-correct, that is, Σ^1_{2n+2} formulas are absolute for N.

<u>Proof</u>. Any model containing a tree whose projection is a complete Π^1_{2n+1} set is a Σ^1_{2n+2}-correct model; this is essentially 8G.15 of Moschovakis [1980]. This fact plus 4.3 clearly implies 4.4. ⊣

4.5 <u>Lemma</u> (AD). Let ξ be any countable (in V) ordinal and let B be any subset of ξ. If $B \in H_{2n+1}$ then $B \in M$.

<u>Proof</u>. Suppose that B is in H_{2n+1}. Let ν be the ordinal such that B is the νth subset of ξ with respect to $\leq_{2n+1}$. Since ξ is countable and AD implies that every well orderable set of reals is countable, ν must be countable. Using 1.2-1.4 it is easy to prove that B is $\Sigma^1_{2n+2}(\xi,\nu)$-in-the-codes. That is, the following set of irrationals is $\Sigma^1_{2n+2}(\xi,\nu)$:

$$B^* = \{\gamma \in WO : |\gamma| \in B\} .$$

We now start with M as the ground model, collapse the countable ordinals ξ and ν by forcing, and let N be the generic extension of M. Generic objects exist since there are only countably many sets of conditions in M; this is again a consequence of the fact that ξ and ν are countable and every wellorderable set of reals is countable. By 4.4 N is Σ^1_{2n+2}-correct, and N contains codes α_ξ and α_ν for ξ and ν. So the $\Sigma^1_{2n+2}(\alpha_\xi,\alpha_\nu)$ definition of B^* is absolute for N. N contains codes for all ordinals less than ξ, so B is in N. Since B is in every generic extension of M, B must be in M. ⊣

If $\kappa = \aleph_1$ then 4.5 is trivially true, since (in H_{2n+1}) the measure is ξ^+-additive. But if $\kappa < \aleph_1$ and $\xi \geq (\kappa^+)^{H_{2n+1}}$, then 4.5 is very strange - in fact it is strange enough to derive a contradiction.

4.6 <u>Theorem</u> (AD). Let $n \geq 1$. The least measurable cardinal in H_{2n+1} is $\aleph_1^V$.

Proof. Let κ be the least measurable cardinal in H_{2n+1}. We already know that $\kappa \leq \aleph_1^V$.

Since

$$H_{2n+1} \models \text{"}\kappa \text{ is the least measurable cardinal"} ,$$

$$M \models \text{"}j(\kappa) \text{ is the least measurable cardinal"} .$$

And $j(\kappa) > \kappa$ so κ is not measurable in M. Since the measure is normal, $\text{Pow}(\kappa) \cap H_{2n+1} = \text{Pow}(\kappa) \cap M$. But U, the normal ultrafilter on κ (in H_{2n+1}), cannot be in M. We will show that if $\kappa < \aleph_1^V$ then U is in M.

So suppose $\kappa < \aleph_1^V$, and let $\xi = (2^\kappa)^M$. Then $\xi < \aleph_1^V$. Let $\leq_M$ be $j(\leq_{2n+1})$. Let

$$U^* = \{\lambda < \xi : \text{The } \lambda\text{th subset of } \kappa, \text{ with respect to } \leq_M, \text{ is in } U\} .$$

Since j and hence $\leq_M$ are classes in H_{2n+1}, and $U \in H_{2n+1}$, clearly $U^* \in H_{2n+1}$. So by Lemma 4.5, $U^* \in M$. This means $U \in M$. ⊣

§5. Questions. We conclude this paper with a list of open problems, which we will state in the form of conjectures. We are assuming AD as a hypothesis in all the conjectures.

1. $H_3 \models$ "There is exactly one normal measure on $\aleph_1^V$" .
2. $H_3 \models$ "$\aleph_2^V$ is the least measurable cardinal of order 2" .
3. $H_3 \models$ "$\aleph_2^V$ is the least measurable cardinal with two different normal measures" .
4. There is an ordinal λ such that $\text{card}(\lambda) = \aleph_{\omega+3}$ and $H_5 \models$ "λ is a measurable cardinal" .
5. $H_{IND} \models$ "$\aleph_1^V$ is the least measurable cardinal" .
6. $H_3 \models$ "There does not exist a cardinal κ such that κ is κ^+-compact" .

References

H. Becker [A], Thin collections of sets of projective ordinals and analogs of L, to appear.

L. A. Harrington and A. S. Kechris [A], On the determinacy of games on ordinals, to appear.

A. S. Kechris [1978], AD and projective ordinals, Cabal Seminar 76-77, Springer Lecture Notes in Mathematics #689, (1978), 91-132.

A. S. Kechris, E. M. Kleinberg and Y. N. Moschovakis [A], The axiom of determinacy, strong partition relations and non-singular measures, this volume.

K. Kunen [1971], On the GCH at measurable cardinals, in: R. O. Gandy and C. M. E. Yates, ed., Logic Colloquium '69, North-Holland (Amsterdam 1971), 107-110.

D. A. Martin [A], Borel and Projective Games, to appear.

Y. N. Moschovakis [1980], Descriptive Set Theory, Studies in Logic #100, North-Holland (Amsterdam 1980).

Y. N. Moschovakis [A], Ordinal games and playful models, this volume.

R. M. Solovay [1978], A Δ^1_3 coding of subsets of ω_ω, Cabal Seminar 76-77, Springer Lecture Notes in Mathematics #689, (1978), 133-150.

Π^1_2 MONOTONE INDUCTIVE DEFINITIONS

Donald A. Martin
Department of Mathematics
University of California, Los Angeles
Los Angeles, California 90024

§1. Introduction. In this paper we study the closure ordinal of Π^1_2 monotone inductive definitions. Most of our results hold for the dual of an arbitrary Spector pointclass, and they are presented in terms of Spector pointclasses. If the reader is interested only in Π^1_2, he may pretend that we are always talking about Π^1_2.

Let $\Phi : P(\omega) \to P(\omega)$. Φ is regarded as an inductive definition as follows. For ordinals α we define $\Phi^\alpha \subseteq \omega$. Suppose Φ^β has been defined for each $\beta < \alpha$. Set $\Phi^{<\alpha} = \cup_{\beta<\alpha} \Phi^\beta$. Now let

$$\Phi^\beta = \Phi(\Phi^{<\alpha}) \cup \Phi^{<\alpha} .$$

Since Φ^α increases with α, there is a countable α such that $\Phi^\alpha = \Phi^{<\alpha}$. The least such α is called $o(\Phi)$. We let $\Phi^{o(\Phi)} = \Phi^\infty$. If Γ is a class of relations in finite products of $P(\omega)$ and ω, $o(\Gamma) = \sup_{\Phi\in\Gamma} o(\Phi)$, where $\Phi \in \Gamma$ means that $\{\langle X,n\rangle : n \in \Phi(X)\} \in \Gamma$.

Definition. Φ is monotone if $X \subseteq Y \to \Phi(X) \subseteq \Phi(Y)$.

If Φ is monotone, Φ^∞ is the smallest fixed point of Φ, i.e., the smallest X such that $\Phi(X) \subseteq X$ (in fact, $= X$). To see this, suppose X is a fixed point of Φ. $\Phi^{<0} \subseteq X$ and, if $\Phi^{<\alpha} \subseteq X$, then $\Phi^\alpha = \Phi(\Phi^{<\alpha}) \cup \Phi^{<\alpha} \subseteq \Phi(X) \cup X \subseteq X$. (Since $\Phi(\Phi(X)) \subseteq \Phi(X)$, $\Phi(X)$ is a fixed point; so, if X is the smallest fixed point, $\Phi(X) = X$.)

Definition. If Γ is a class of relations, $o(\Gamma \text{ mon}) = \sup\{o(\Phi) : \Phi \in \Gamma$ and Φ monotone$\}$.

Γ monotone inductive definitions are well-understood for a variety of classes Γ. However, a certain kind of class, typified by Π^1_2, has resisted study. One normally thinks of inductive definitions as defining a complicated set by iterating a simple operation. But for Π^1_2 this is not at all the case.

Proposition. If Φ is Π^1_2 monotone, Φ^∞ is Π^1_2.

Proof. $n \in \Phi^\infty \leftrightarrow \forall X(\Phi(X) \subseteq X \to n \in X) \leftrightarrow \forall X(\exists m(m \in \Phi(X) - X) \vee n \in X)$.

Thus Π^1_2 monotone indcutive definitions are extremely inefficient and can - with respect to the set they define - all be replaced by Π^1_2 monotone Φ with $o(\Phi) = 1$.

Our aim in this paper is to study $o(\Gamma \text{ mon})$ for Γ a class such as Π^1_2. However, our methods almost all apply equally well to a more general situation If Γ is a class of subsets of finite products of $P(\omega)$ and ω (i.e., Γ is a <u>pointclass</u> in the sense of Moschovakis [1980], except that we do not allow perfect Polish spaces other than $P(\omega)$), $\breve{\Gamma}$, the <u>dual of</u> Γ, is the class of all sets $\breve{G}$, where $G \in \Gamma$. Here if, say, $G \subseteq (P(\omega))^m \times \omega^n$, $\breve{G}$ is the set of all $\langle X_1,\ldots,X_m,e_1,\ldots,e_n\rangle$ such that $\langle X_1,\ldots,X_m,e_1,\ldots,e_n\rangle \notin G$. Now Π^1_2 is $(\Sigma^1_2)^\smile$ and Σ^1_2 is a <u>Spector pointclass</u> as defined in Moschovakis [1980]. A Spector pointclass is a pointclass Γ which is closed under union, intersection, and number quantification, contains all Σ^0_1 sets, has the prewellordering property, is ω-parameterized, and has the <u>substitution property</u>. ω-parameterization means that, for each of our product spaces $\mathcal{Y}$, there is an $G \subseteq \omega \times \mathcal{Y}$ such that $G \in \Gamma$ and such that G enumerates the Γ subsets of $\mathcal{Y}$. G can be chosen so that the s-m-n theorem and the recursion theorem hold (see Moschovakis [1980]), and we assume in the sequel that this has been done. That Γ has the substitution property means that, if $R(X,\ldots)$, $P(n,\ldots)$ and $Q(n,\ldots)$ are Γ relations, there is a Γ relation $S(\ldots)$ such that, for any values of the other parameters such that $\forall n(P(n,\ldots) \leftrightarrow \neg Q(n,\ldots))$

$$S(\ldots) \leftrightarrow R(\{n : P(n,\ldots)\},\ldots) .$$

Thus we wish to study $o(\breve{\Gamma} \text{ mon})$, where Γ is a Spector pointclass. The reader not familiar with Spector pointclasses may read the paper by simply reading "Σ^1_2" whenever he sees "Γ" and "Π^1_2" whenever he sees "$\breve{\Gamma}$".

In §2, we show that, for Γ a Spector pointclass,

$$o(\breve{\Gamma} \text{ mon}) > o(\Gamma) .$$

$o(\breve{\Gamma}) > o(\Gamma)$ is a result of Aandera. (His result applies also to some non-Spector pointclasses.)

In §3 we show that, if Φ is $\breve{\Gamma}$ monotone and $\eta < o(\Phi)$, then there is a $\breve{\Gamma}$ monotone Θ with $o(\Theta) > \eta$ such that $\alpha < o(\Theta)$ implies that $\Theta^{<\alpha}$ is a prewellordering of length α. As a corollary to the proof of this technical result, we deduce that there is a $\breve{\Gamma}$ monotone Φ such that

$$o(\Phi) = o(\breve{\Gamma} \text{ mon}) .$$

In §4 we use the results of §2 and §3 to show that, if Φ is $\breve{\Gamma}$ monotone and $\eta < o(\Phi)$, then $o(\breve{\Gamma} \text{ mon}) > o(\Gamma(\Phi^{<\eta}))$, i.e., that the result of §2 remains true when we allow a <u>stage</u> in a $\breve{\Gamma}$ monotone induction to be used as a parameter in Γ inductions. In §5 we restrict our attention to Γ such as Σ^1_2, where the relativized version of the assertion that if Φ is $\check{\Gamma}$ monotone the $\Phi^\infty \in \check{\Gamma}$ holds. We show in this case that $o(\check{\Gamma}\text{ mon})$ is admissible. In §6 we show that the results of §5 apply to a variety of classes defined via the game quantifier.

§2. Γ non-monotone inductive definitions.

Theorem A. Let Γ be a Spector pointclass. $o(\Gamma) < o(\check{\Gamma}\text{ mon})$. Indeed, if Φ is a Γ operation, there is a $\check{\Gamma}$ monotone operation Θ such that $o(\Phi) < o(\Theta)$ and $\Theta^\infty = \omega$.

Proof. The second assertion of the theorem implies the first. By ω-parameterization for Γ, there is a universal Γ operation Φ, satisfying $o(\Phi) = o(\Gamma)$.

Let Φ be a Γ operation. Let φ be a norm on $\{\langle X,n\rangle : n \in \Phi(X)\}$ as given by the fact that Γ is a Spector class.

To describe the idea of the proof, we give a slightly eccentric definition:

Definition. Let $A \subseteq \omega$ and $R \subseteq \omega \times \omega$. $\langle A,R\rangle$ is a well-founded relation just in case R is, in the usual sense, a well-founded relation. The field of $\langle A,R\rangle$, which we denote by A', is

$$\{n : n \in A \text{ or } \exists m\langle m,n\rangle \in R \text{ or } \exists m\langle n,m\rangle \in R\} .$$

If $e \in A'$, $\langle A,R\rangle_e = \langle A_e,R_e\rangle$, where $A_e = \{n : n \in A' \;\&\; \langle n,e\rangle \in R\}$ and $R_e = \{\langle n,m\rangle : n \in A_e \;\&\; m \in A_e \;\&\; \langle n,m\rangle \in R\}$. The height of $\langle A,R\rangle$ is the unique ordinal α such that there is a surjection $f : A' \to \alpha$ satisfying $\langle n,m\rangle \in R \to f(n) < f(m)$. It is easily seen that $\text{height}(\langle A,R\rangle) = \sup_{e\in A'}\{\text{height}(\langle A,R\rangle_e) + 1\}$.

Let us adopt some reasonable way of coding finite sequences of natural numbers by natural numbers. We shall not distinguish notationally between a finite sequence and the number which codes it.

Definition. If $X \subseteq \omega$, $A^X = \{n : 2n \in X\}$ and $R^X = \{\langle n,m\rangle : 2\langle n,m\rangle + 1 \in X\}$. X is a well-founded relation if $\langle A^X,R^X\rangle$ is a well-founded relation. For $e \in (A^X)'$, X_e is defined in the natural way. Height$(X) = \text{height}(\langle A^X,R^X\rangle)$, for X a well-founded relation.

Lemma A.1. $\{X : X \text{ is a well-founded relation}\}$ belongs to Γ.

Proof. This set is Π^1_1 and $\Pi^1_1 \subseteq \Gamma$ for every Spector pointclass Γ.

Lemma A.2. Let ρ be an ordinal. If X is a well-founded relation with height $< \rho$ and $Y \subseteq X$, then Y is a well-founded relation of height $< \rho$.

Proof. $R^X \supseteq R^Y$ and $(A^X)' \supseteq (A^Y)'$.

Definition. X is a prewellordering if $X = \{2n : n \in A^X\} \cup \{2\langle n,m\rangle + 1 : \langle n,m\rangle \in R^X\}$, and R^X is a prewellordering of A^X.

We now describe the idea for constructing Θ. Θ^α will always be either a prewellordering or all of ω. If X is not a well-founded relation of height $<$ some fixed ρ, $\Theta(X)$ will be ω. For well-founded relations X of height $< \rho$, $\Theta(X)$ will depend monotonically on height(X) and will depend only on height X.

To construct Θ, we define a sequence of approximations A^α_β to the Φ^β. For $\alpha \leq \rho$, the entire system $\langle A^{\alpha'}_\beta : \beta \leq \alpha' \leq \alpha\rangle$ will be, in an appropriate sense, uniformly Δ in well-founded relations of height α $(\Delta = \Gamma \cap \breve{\Gamma})$. Certain numbers, describing the transition from the A^α_β to the $A^{\alpha+1}_\beta$ will be designated as codes for α. If X is a well-founded relation of height $\alpha < \rho$, $\Theta(X)$ will be the prewellordering whose $\langle A,R\rangle$ satisfies

A is the set of all codes for ordinals $\leq \alpha$;

R is the obvious prewellordering of these codes .

The A^α_γ will be defined for $\gamma \leq \alpha \leq \rho$ and will satisfy:

(1) $A^\alpha_\gamma \subseteq \Phi(A^\alpha_{<\gamma}) \cup A^\alpha_{<\gamma}$

(2) $A^\alpha_\gamma \supseteq A^\alpha_{<\gamma}$

(3) $A^\alpha_\alpha = A^\alpha_{<\alpha}$

where $A^\alpha_{<\gamma} = \bigcup_{\gamma'<\gamma} A^\alpha_{\gamma'}$.

Let us begin by setting A^0_0 = the empty set.

Suppose now that A^α_γ is defined for each $\gamma \leq \alpha$ so that (1)-(3) are satisfied.

Definition. If there exists a pair $\langle \beta, n\rangle$ such that $n \notin A^\alpha_\beta$ but $n \in \Phi(A^\alpha_{<\beta})$, let $\langle \beta(\alpha), n(\alpha)\rangle$ be the least such pair, where we minimize, in order,

(a) $\varphi(A^\alpha_{<\beta}, n)$

(b) β

(c) n.

Lemma A.3. If $\alpha < o(\Phi)$, then $\langle\beta(\alpha), n(\alpha)\rangle$ is defined.

Proof. If $\langle\beta(\alpha), n(\alpha)\rangle$ is undefined, then $A^{\alpha}_{\beta} \supseteq \Phi(A^{\alpha}_{<\beta})$ for every $\beta \leq \alpha$. Thus (1) and (2) above imply that $A^{\alpha}_{\beta} = \Phi(A^{\alpha}_{<\beta}) \cup A^{\alpha}_{<\beta}$ for every $\beta \leq \alpha$. This means that $A^{\alpha}_{\beta} = \Phi^{\beta}$ for every $\beta \leq \alpha$. But then (3) implies that $\Phi^{<\alpha} = \Phi^{\infty}$ and so $\alpha \geq o(\Phi)$.

If $\langle\beta(\alpha), n(\alpha)\rangle$ is undefined, then we let $\rho = \alpha$. If $\langle\beta(\alpha), n(\alpha)\rangle$ is defined, set

$$A^{\alpha+1}_{\gamma} = \begin{cases} A^{\alpha}_{\gamma} & \text{if } \gamma < \beta(\alpha) ; \\ A^{\alpha}_{\beta(\alpha)} \cup \{n(\alpha)\} & \text{if } \gamma \geq \beta(\alpha) . \end{cases}$$

Let λ be a limit ordinal, and assume that the A^{α}_{γ} have been defined and satisfy (1)-(3) for all $\gamma \leq \alpha < \lambda$. Assume also that, for every $\gamma \leq \alpha' < \alpha < \lambda$,

(i) $A^{\alpha}_{\gamma} \neq A^{\alpha'}_{\gamma} \rightarrow \exists\alpha^{*}(\alpha \leq \alpha^{*} < \alpha' \ \&\ \beta(\alpha^{*}) \leq \gamma)$

(ii) $A^{\alpha}_{\gamma} \not\subseteq A^{\alpha'}_{\gamma} \rightarrow \exists\alpha^{*}(\alpha \leq \alpha^{*} < \alpha' \ \&\ \beta(\alpha^{*}) < \gamma)$.

Note that these conditions hold for $\alpha' = \alpha + 1$.

Let $\nu(\lambda)$ be the least $\gamma \leq \lambda$ such that, for arbitrarily large $\alpha < \lambda$, $\beta(\alpha) \leq \gamma$.

If $\gamma < \nu(\lambda)$, let $\alpha < \lambda$ be large enough so that $\alpha \leq \alpha' < \lambda$ implies $\beta(\alpha') > \gamma$. Let $A^{\lambda}_{\gamma} = A^{\alpha}_{\gamma}$. Condition (i) implies that if $\alpha < \alpha' < \lambda$ then $A^{\alpha'}_{\gamma} = A^{\lambda}_{\gamma}$.

Suppose there are arbitrarily large $\alpha < \lambda$ such that $\beta(\alpha) < \nu(\lambda)$. Set $A^{\lambda}_{\gamma} = A^{\lambda}_{<\nu(\lambda)}$ for all $\gamma \geq \nu(\lambda)$.

Assume finally that there is an $\alpha < \lambda$ such that $\alpha \leq \alpha' < \lambda$ implies $\beta(\alpha) \geq \nu(\lambda)$. Set $A^{\lambda}_{\gamma} = \bigcup_{\alpha\leq\alpha'<\lambda} A^{\alpha'}_{\nu(\lambda)}$ for all $\gamma \geq \nu(\lambda)$.

Note that (i) and (ii) hold for all $\gamma \leq \alpha < \alpha' \leq \lambda$. It is easy to check also that (1)-(3) hold for $\alpha = \lambda$.

Definition. If $\gamma \leq \alpha \leq \rho$, $\langle\beta, n\rangle$ is γ-minimal at α if

(a) $\beta < \gamma$;

(b) $n \notin A^{\alpha}_{\beta}$;

(c) Either

(i) $\beta' < \gamma \rightarrow A^{\alpha}_{\beta'} \supseteq \Phi(A^{\alpha}_{<\beta'})$ or

(ii) $\langle\beta, n\rangle$ is the least pair satisfying (a) and (b) such that $n \in \Phi(A^{\alpha}_{<\beta})$, where we minimize as in the definition of $\langle\beta(\alpha), n(\alpha)\rangle$ above.

Lemma A.4. Suppose $\alpha \leq \alpha' < \rho$ and $\beta(\alpha') < \gamma$ and that α' is the least ordinal with these properties. Then $\langle\beta(\alpha'),n(\alpha')\rangle$ is the unique pair γ-minimal at α and satisfies (ii).

Proof. We know that $\gamma' < \gamma$ implies that $A^{\alpha'}_{\gamma'} = A^{\alpha}_{\gamma'}$. The lemma follows directly from the definitions of $\langle\beta(\alpha'),n(\alpha')\rangle$ and of γ-minimality at α, since $\beta(\alpha') < \gamma$.

Note that γ-minimal pairs always exist unless $\gamma = 0$ or $A^{\alpha}_{0} = \omega$.

Definition. If $\alpha < \rho$, a precode for α is a $\langle\langle\beta_1,\ldots,\beta_k\rangle\rangle,\langle n_1,\ldots,n_k\rangle\rangle$ such that $k \geq 1$, $\langle\beta_1,n_1\rangle = \langle\beta(\alpha),n(\alpha)\rangle$, and $1 \leq i < k$ implies that $\langle\beta_{i+1},n_{i+1}\rangle$ is β_i-minimal at α, and $\gamma < \beta_k$ implies that $A^{\alpha}_{\gamma} \supseteq \Phi(A^{\alpha}_{<\gamma})$.

Lemma A.5. If $\alpha < \rho$ there is a precode for α.

Proof. Let $\langle\beta_1,n_1\rangle = \langle\beta(\alpha),n(\alpha)\rangle$. If $\langle\beta_i,n_i\rangle$ has been chosen, $\beta_i \neq 0$, and $A^{\alpha}_{0} \neq \omega$, let $\langle\beta_{i+1},n_{i+1}\rangle$ be β_i-minimal at α. Since the β_i are strictly decreasing, we finally reach a β_k such that $\beta_k = 0$ or $A^{\alpha}_{0} = \omega$. Then $\gamma < \beta_k$ implies that $A^{\alpha}_{\gamma} = \omega \supseteq \Phi(A^{\alpha}_{<\gamma})$.

Lemma A.6. If $\langle\langle\beta_1,\ldots,\beta_k\rangle\rangle,\langle n_1,\ldots,n_k\rangle\rangle$ is a precode for $\alpha < \rho$, then, for every $\alpha' > \alpha$, there is an i such that $n_i \in A^{\alpha'}_{\beta_i}$.

Proof. Let β be the least value of $\beta(\alpha^*)$ attained for $\alpha' > \alpha^* \geq \alpha$. By the last clause of the definition of precode, $\beta \geq \beta_k$. If $\beta_{i+1} \leq \beta < \beta_i$, then the β_i-minimality of $\langle\beta_{i+1},n_{i+1}\rangle$ and Lemma A.4 imply that $\beta = \beta_{i+1}$ and that there is an α^*, $\alpha' > \alpha^* > \alpha$ such that $\langle\beta(\alpha^*),n(\alpha^*)\rangle = \langle\beta_{i+1},n_{i+1}\rangle$. It follows that $n_{i+1} \in A^{\alpha^*+1}_{\beta_{i+1}}$. Since $\beta(\alpha'') \geq \beta_{i+1}$ for $\alpha' > \alpha'' > \alpha^*$, we we have $n_i \in A^{\alpha'}_{\beta_{i+1}}$. It remains to consider the case $\beta \geq \beta_1$. Since $\langle\beta_1,n_1\rangle = \langle\beta(\alpha),n(\alpha)\rangle$, we have $n_1 \in A^{\alpha+1}_{\beta_1}$. Since $\beta \geq \beta_1$, $n_1 \in A^{\alpha'}_{\beta_1}$.

In our representation of finite sequences by natural numbers, let us assume that the number 0 does not represent any finite sequence.

Definition. If $\alpha < \rho$, a code for α is a number representing $\langle\langle c_1,\ldots,c_k\rangle,\langle n_1,\ldots,n_k\rangle\rangle$ such that, for some $\langle\beta_1,\ldots,\beta_k\rangle$, $\langle\langle\beta_1,\ldots,\beta_k\rangle,\langle n_1,\ldots,n_k\rangle\rangle$ is a precode for α and, for all i, either c_i is a code for β_i and $\beta_i < \alpha$ or else $i = 1$, $c_i = 0$ and $\beta_i = \alpha$.

Lemma A.7. For every $\alpha < \rho$, there is a code for α.

Lemma A.7 follows easily from Lemma A.5. The following is our main technical lemma.

Lemma A.8. For every $\alpha < \rho$, there are no number c and ordinal α', $\alpha < \alpha' < \rho$, such that c is a code for both α and α'.

Proof. Let α be the least ordinal for which the lemma is false. Let $\alpha' > \alpha$ and let c be a code for both α and α'. Let $c = \langle\langle c_1,\ldots,c_k\rangle, \langle n_1,\ldots,n_k\rangle\rangle$. Let $\langle \beta_1,\ldots,\beta_k\rangle$ and $\langle \beta_1',\ldots,\beta_k'\rangle$ witness that c is a code for α and α' respectively. By the minimality of α, $\beta_i = \beta_i'$ unless $i = 1$, $c_1 = 0$, $\beta_i = \alpha$, and $\beta_i' = \alpha'$. By the definition of precode, no n_i belongs to $A^{\alpha'}_{\beta_i'}$. By Lemma A.6, there is an i such that $n_i \in A^{\alpha'}_{\beta_i}$. We have a direct contradiction unless this $i = 1$ and $\beta_1 = \alpha$. But then, since $A^{\alpha'}_{\alpha} \subseteq A^{\alpha'}_{\alpha'}$, we have the contradiction that $n_1 \in A^{\alpha'}_{\alpha'} = A^{\alpha'}_{\beta_1'}$.

We are now ready to define the operation Θ. If X is not a well-founded relation of height $< \rho$, then $\Theta(X) = \omega$. If X is a well-founded relation of height $\alpha < \rho$, then $2n \in \Theta(X)$ if and only if n is a code for some $\alpha' \leq \alpha$, and $2n + 1 \in \Theta(X)$ if and only if n represents $\langle m_1, m_2\rangle$, m_2 is a code for $\alpha_2 \leq \alpha$, and m_1 is a code for $\alpha_1 < \alpha_2$. By Lemma A.8, $\Theta(X)$ is a prewellordering of height $\alpha + 1$.

Lemma A.9. Θ is monotone. $o(\Theta) = \rho + 1 > o(\Phi)$. $\Theta^\infty = \omega$.

Proof. The first assertion follows from Lemma A.2 and the definition. The second assertion follows from the definition and Lemma A.3. The third assertion follows from the definition.

The theorem will be proved when we show that Θ is $\breve{\Gamma}$. We first show that the system $\langle A^{\alpha'}_{\gamma} : \gamma \leq \alpha' \leq \alpha\rangle$ is uniformly Δ in well-founded relations of height $\alpha \leq \rho$.

Lemma A.10. There are Γ relations P_1, Q_1 and S_1 and $\breve{\Gamma}$ relations P_2, Q_2 and S_2, such that, if X is a well-founded relation of height $\alpha \leq \rho$, then for all e_1, e_2, n, γ and α',

(1) $n \in A^{\alpha}_{\alpha} \leftrightarrow P_1(X,n) \leftrightarrow P_2(X,n)$;

(2) $n \in A^{\alpha}_{\gamma} \leftrightarrow Q_1(X,e_1,n) \leftrightarrow Q_2(X,e_1,n)$ if $e_1 \in (A_X)'$ and $\gamma = \text{height}(X_{e_1})$;

(3) $n \in A^{\alpha'}_{\gamma} \leftrightarrow S_1(X,e_1,e_2,n) \leftrightarrow S_2(X,e_1,e_2,n)$ if $e_1, e_2 \in (A_X)'$, $\gamma = \text{height}(X_{e_1})$, and $\alpha' = \text{height}(X_{e_2})$.

Proof. We define the relations inductively and use the recursion theorem. Since this is a routine application of the recursion theorem, we content ourselves with indicating the main points.

A^0_0 is trivial. For λ a limit ordinal, the A^{λ}_{γ} are defined from the system $\langle A^{\alpha}_{\gamma} : \gamma \leq \alpha < \lambda \rangle$ using quantification over numbers and ordinals $< \alpha$. Such quantification becomes, in our situation, number quantification. To decide whether a well-founded relation has limit height involves is Δ^1_1. The $A^{\alpha+1}_{\gamma}$ are defined trivially from the A^{α}_{γ} and $\langle \beta(\alpha), n(\alpha) \rangle$. $\langle \beta(\alpha), n(\alpha) \rangle$ is defined from the A^{α}_{γ} by minimization with respect to the norm φ, which can be done in either a Γ or $\breve{\Gamma}$ fashion.

Lemma A.11. The property "X is a well-founded relation of height $< \rho$," is Γ.

Proof. By Lemma A.1, we need only show that there is a Γ property U such that, if X is a well-founded relation, then

$$\text{height}(X) < \rho \leftrightarrow U(X) .$$

For $\text{height}(Y) \leq \rho$, $\text{height}(Y) < \rho \leftrightarrow \exists \beta \leq \text{height}(Y) \exists n(n \in \Phi(A^{\text{height}(Y)}_{\beta}) - A^{\text{height}(Y)}_{<\beta})$. By Lemma A.10, this is equivalent (for $\text{height}(Y) \leq \rho$) to a Γ condition V on Y. Now let

$$U(X) \leftrightarrow [V(X) \ \& \ \forall e \in (A^X)' V(X_e)] .$$

Lemma A.12. There are a $\breve{\Gamma}$ relations C and C^* such that, for X a well-founded relation of height $< \rho$,

$$C(X,c) \leftrightarrow c \text{ is a code for } \text{height}(X) ;$$
$$C^*(X,e,c) \leftrightarrow c \text{ is a code for } \text{height}(X_e) ,$$
$$\text{if } e \in (A_X)' .$$

Proof. Note that, for such X, the properties "$\langle \beta, n \rangle = \langle \beta(\text{height}(X)), n(\text{height}(X)) \rangle$" and "$\langle \beta, n \rangle$ is γ-minimal at height(X)" are, when turned into relations between X and numbers, $\breve{\Gamma}$. It follows that the property of being a precode for height(X) is similarly $\breve{\Gamma}$, as is the property of being a precode for $\text{height}(X_e)$. The lemma follows by an easy application of the recursion theorem.

Lemma A.13. Θ is $\breve{\Gamma}$.

Proof. $n \in \Theta(X) \leftrightarrow X$ is not a well-founded relation of height $< \rho$ or one of the following:

(a) $\exists c(n = 2c \ \& \ C(X,c))$
(b) $\exists c \exists e(n = 2c \ \& \ e \in (A_X)' \ \& \ C^*(X,e,c))$
(c) $\exists c_1 \exists c_2, \exists e(n = 2\langle c_1, c_2 \rangle + 1 \ \& \ e \in (A_X)' \ \& \ C^*(X,e,c_1) \ \& \ C(X,c))$

(d) $\exists c_1 \exists c_2 \exists e_1 \exists e_2 (n = 2\langle c_1, c_2\rangle + 1 \;\&\; e_1 \in (A_X)' \;\&\; e_2 \in (A_X)' \;\&\; C^*(X, e_1, c_1)$ $\&\; C^*(X, e_2, c_2) \;\&\; \langle e_1, e_2\rangle \in R_X)$.

§3. Nice Γ monotone induction definitions. In this section we give a construction which will help us solve two kinds of problems. To state the first problem, we give the following definition.

Definition. Let Γ be a Spector pointclass. Ψ is a universal $\breve{\Gamma}$-monotone operation if Ψ is a $\breve{\Gamma}$ monotone operation, $o(\Psi) = o(\breve{\Gamma}\text{ mon})$, and, if $\langle \Phi_e : e \in \omega\rangle$ is an enumeration of $\breve{\Gamma}$ operations given by ω-parameterization, there is a recursive function g such that, if Φ_e is monotone, then

$$\forall n(n \in \Phi_e^\infty \longleftrightarrow g(e,n) \in \Psi^\infty).$$

Our first problem is to show that universal $\breve{\Gamma}$ monotone operations exist. For $\breve{\Gamma} = \Pi^1_2$, this is trivial, but we wish to prove it in general.

Our second problem is as follows. Suppose $\eta < o(\breve{\Gamma}\text{ mon})$. To prove various results about $o(\breve{\Gamma})$, we wish to find a $\breve{\Gamma}$ monotone Θ such that $o(\Theta) > \eta$ and such that the stages Θ^α are just like those of the Θ constructed in the proof of Theorem A, in particular, $\alpha < o(\Theta)$ implies that $\Theta^{<\alpha}$ is a prewellordering of length α.

Our plan is to give a construction which solves the second problem, but to give it in sufficient generality to solve the first problem as well.

Definition. If X is a prewellordering, Y is an end extension of X if Y is a prewellordering and, for some $e \in A_Y$, $X = Y_e$.

Note that, for the Θ defined in the proof of Theorem A, $\alpha < \alpha' \le \rho$ implies that $\Theta^{<\alpha'}$ is an end extension of Θ^α.

Theorem B. Let Γ be a Spector pointclass and let $\langle \Phi_e : e \in \omega\rangle$ be an enumeration of $\breve{\Gamma}$ operations as given by ω-parameterization. There is a recursive function f such that

(a) for each e and e_0, $\Phi_{f(e,e_0)}$ is a monotone operation;

(b) for each e and e_0 and each $\alpha < o(\Phi_{f(e,e_0)})$, $\Phi^{<\alpha}_{f(e,e_0)}$ is a prewellordering of height α and, for $\alpha' < \alpha$, is an end extension of $\Phi^{<\alpha'}_{f(e,e_0)}$;

(c) for each e and e_0, if X and Y are either both not wellfounded relations or both wellfounded relations and of the sme height, then $\Phi_{f(e,e_0)}(X) = \Phi_{f(e,e_0)}(Y)$;

(d) for each e and e_0 such that Φ_e is monotone:

(i) $e_0 \in \Phi_e^{\eta} - \Phi_e^{<\eta}$ implies that $o(\Phi_{f(e,e_0)}) > \eta$ and $\Phi_{f(e,e_0)}^{\infty} = \omega$;

(ii) $e_0 \notin \Phi_e^{\infty}$ implies that $\Phi_{f(e,e_0)}^{\infty}$ is a prewellordering.

Proof. Let φ be a norm on $\{\langle X,e,n\rangle : n \notin \Phi_e(X)\}$ as given by the fact that Γ is a Spector pointclass. Let $\Phi = \Phi_e$. We suppress the argument e of φ to simplify notation. We construct a Θ which will be $\Phi_{f(e,e_0)}$. The reader can easily verify that our construction is sufficiently uniform that f exists. (Recall that we assume that the ω-parameterization satisfies the s-m-n theorem.)

As in the proof of Theorem A, we define a sequence of approximations A_γ^α, $\gamma \leq \alpha \leq \rho$, to the Φ^γ. As before, let $A_{<\gamma}^\alpha = \bigcup_{\gamma'<\gamma} A_{\gamma'}^\alpha$. The A_γ^α will have the following properties:

(1) $A_\gamma^\alpha \supseteq A_{<\gamma}^\alpha$

(2) $A_\alpha^\alpha = \omega$

(3) $\alpha' < \alpha \rightarrow A_\gamma^{\alpha'} \supseteq A_\gamma^\alpha$

(4) If Φ is monotone, $A_\gamma^\alpha \supseteq \Phi(A_{<\gamma}^\alpha)$.

Suppose the $A_\gamma^{\alpha'}$ are defined for $\gamma \leq \alpha' \leq \alpha$ and satisfy (1)-(4).

Definition. If there is a $\langle\beta,n\rangle$ such that

(a) $\beta \leq \alpha$

(b) $e_0 \in A_{<\alpha}^\alpha \rightarrow \beta < \alpha$

(c) $n \in A_\beta^\alpha - A_{<\beta}^\alpha$

(d) $n \notin \Phi(A_{<\beta}^\alpha)$

then let $\langle\beta(\alpha),n(\alpha)\rangle$ be the least such pair, where we minimize first with respect to $\varphi(A_{<\beta}^\alpha,n)$, then with respect to β, and then with respect to n.

Lemma B.1. If Φ is monotone and $\langle\beta(\alpha),n(\alpha)\rangle$ is undefined, then $e_0 \in \Phi^\alpha$.

Proof. Assume Φ is monotone. If $e_0 \in A_{<\alpha}^\alpha$ and $\langle\beta(\alpha),n(\alpha)\rangle$ is undefined, then (1) and (4) imply that $A_\gamma^\alpha = \Phi(A_{<\gamma}^\alpha) \cup (A_{<\gamma}^\alpha)$ for all $\gamma < \alpha$. Thus $A_\gamma^\alpha = \Phi^\gamma$ for all $\gamma < \alpha$, and so $e_0 \in \Phi^{<\alpha}$. If $e_0 \notin A_{<\alpha}^\alpha$ and $\langle\beta(\alpha),n(\alpha)\rangle$ is undefined, a similar argument shows $e_0 \in \Phi^\alpha$ (in fact, that $\Phi^\alpha = \omega$).

If $\langle\beta(\alpha),n(\alpha)\rangle$ is undefined, then $\alpha = \rho$. If $\langle\beta(\alpha),n(\alpha)\rangle$ is defined, set

$$A_\gamma^{\alpha+1} = \begin{cases} A_\gamma^\alpha & \text{if } \gamma \leq \alpha \text{ and } \gamma \neq \beta(\alpha); \\ A_\gamma - \{n(\alpha)\} & \text{if } \gamma = \beta(\alpha); \\ \omega & \text{if } \gamma = \alpha + 1 . \end{cases}$$

It is easy to see that (1)-(4) still hold.

Suppose λ is a limit ordinal and the A_γ^α are defined for $\gamma \leq \alpha < \lambda$ and satisfy (1)-(4). Set

$$A_\gamma^\lambda = \begin{cases} \bigcap_{\alpha<\lambda} A_\gamma^\alpha & \text{if } \gamma < \lambda \\ \omega & \text{if } \gamma = \lambda . \end{cases}$$

Clearly (1)-(4) remain true.

<u>Definition.</u> $\langle\beta,n\rangle$ is α-<u>minimal</u> if $\beta < \alpha$, $n \in A_\beta^\alpha - A_{<\beta}^\alpha$, and either (1) also $n \notin \Phi(A_{<\beta}^\alpha)$ and $\langle\beta,n\rangle$ is the least (in our usual sense) pair with these properties or (2) for every $\gamma < \alpha$, $A_\gamma^\alpha \subseteq \Phi(A_{<\gamma}^\alpha) \cup A_{<\gamma}^\alpha$.

<u>Definition.</u> If $\alpha < \rho$ and $e_0 \in A_{<\alpha}^\alpha$, a <u>precode for</u> α is $\langle\beta(\alpha),n(\alpha)\rangle$. If $\alpha < \rho$ and $e_0 \notin A_{<\alpha}^\alpha$, a precode for α is any $\langle m,\beta,n\rangle$, where $m \in \Phi(A_{<\alpha}^\alpha) - A_{<\alpha}^\alpha$ and either $\langle\beta,n\rangle$ is α-minimal or $\beta = \alpha$ and $(\forall\gamma < \alpha)(A_\gamma^\alpha \subseteq \Phi(A_{<\gamma}^\alpha) \cup A_{<\gamma}^\alpha)$.

<u>Lemma B.2.</u> If $\alpha < \rho$ and there is no precode for α, then $A_{<\alpha}^\alpha$ is a fixed point of Φ with $e_0 \notin A_{<\alpha}^\alpha$.

<u>Proof.</u> If $\alpha < \rho$ and $e_0 \in A_{<\alpha}^\alpha$, then obviously $\langle\beta(\alpha),n(\alpha)\rangle$ is a precode for α. Suppose $e_0 \notin A_{<\alpha}^\alpha$. If $\Phi(A_{<\alpha}^\alpha) \subseteq A_{<\alpha}^\alpha$, then $A_{<\alpha}^\alpha$ is a fixed point of Φ. If no α-minimal pair exists, then $(\forall\gamma < \alpha)(A_\gamma^\alpha \subseteq \Phi(A_{<\gamma}^\alpha) \cup A_{<\gamma}^\alpha$ and, in fact, $\alpha = 0$. Thus either $A_{<\alpha}^\alpha$ is a fixed point of Φ or some $\langle m,\beta,n\rangle$, with $\langle\beta,n\rangle$ α-minimal, is a precode for α or some $\langle m,\alpha,n\rangle$ is a precode for α.

<u>Corollary.</u> Let ρ^* be the least ordinal such that $\rho^* = \rho$ or there is no precode for ρ^*. If Φ is monotone and $e_0 \in \Phi^\infty$, then $\rho^* = \rho$.

<u>Proof.</u> If Φ is monotone and $\rho^* < \rho$, then $A_{<\rho^*}^{\rho^*}$ is a fixed point of Φ not containing e_0. Thus $e_0 \notin \Phi^\infty$.

<u>Definition.</u> A <u>code for</u> α is a $\langle c,n\rangle$ or an $\langle m,c,n\rangle$ such that, for some β, $\langle\beta,n\rangle$ or $\langle m,\beta,n\rangle$ respectively is a precode for α, and either $\beta < \alpha$ and c is a code for β or else $\beta = \alpha$ and $c = 0$.

Lemma B.3. If $\alpha < \rho^*$, there is a code for α.

Proof. This follows by induction from the definition.

Lemma B.4. For every $\alpha < \rho^*$ there is no α', $\alpha < \alpha' < \rho^*$, such that some number is a code for both α and α'.

Proof. Let α be a minimal counterexample. Let $\alpha' > \alpha$ be such that some number is a code for both α and α'.

Case 1. Some $\langle c,n\rangle$ codes both α and α'. Let $\langle \beta,n\rangle$ and $\langle \beta',n\rangle$ be the precodes for α and α' respectively which witness that $\langle c,n\rangle$ codes these two ordinals. Since $\beta < \alpha$, the minimality of α implies that $\beta = \beta'$. By the definition of $A_\beta^{\alpha+1}$, $n \notin A_\beta^{\alpha+1}$. By property (3) of the sequence of approximations, $A_\beta^{\alpha+1} \supseteq A_\beta^{\alpha'}$. Thus $n \notin A_\beta^{\alpha'}$ and so $\langle \beta,n\rangle \neq \langle \beta(\alpha'),n(\alpha')\rangle$. This is a contradiction.

Case 2. Some $\langle m,c,n\rangle$ codes both α and α'. Let β and β' be such that $\langle m,\beta,n\rangle$ and $\langle m,\beta',n\rangle$ are precodes for α and α' respectively and c codes both β and β' or else $c = 0$, $\beta = \alpha$ and $\beta' = \alpha'$.

If $A_\gamma^\alpha = A_\gamma^{\alpha'}$ for every $\gamma < \alpha$, then $A_{<\alpha}^\alpha = A_{<\alpha}^{\alpha'}$. Since $m \in \Phi(A_{<\alpha}^\alpha)$ it follows that $m \in A_\alpha^{\alpha^*}$ for every $\alpha^* \leq \alpha'$. Thus $m \in A_{<\alpha'}^{\alpha'}$, a contradiction.

If $A_\gamma^\alpha \neq A_\gamma^{\alpha'}$ for some $\gamma < \alpha$, we must have $c \neq 0$, $\beta = \beta'$, and, for the least $\alpha^* > \alpha$ such that $A_\gamma^{\alpha^*} \neq A_\gamma^{\alpha^*+1}$ for some $\gamma < \alpha$, $\gamma = \beta$ and $\langle \beta,n\rangle = \langle \beta(\alpha),n(\alpha)\rangle$. Thus $n \notin A_\beta^{\alpha^*+1}$. It follows that $n \notin A_\beta^{\alpha'}$. But this contradicts the fact that $\langle \beta,n\rangle$ is α'-minimal.

The definition of Θ and the rest of the proof are similar to the corresponding parts of the proof of Theorem A. We omit them, except for a remark or two.

Note that if Φ is monotone and $e_0 \in \Phi^\eta - \Phi^{<\eta}$, then $\rho^* = \rho \geq \eta$ by Lemma B.1 and the corollary to Lemma B.2. Thus $o(\Theta) = \rho + 1$ and $\Theta^\infty = \omega$ as in the proof of Theorem A.

If Φ is monotone and $e_0 \notin \Phi^\infty$, then $\rho = \mathrm{On}$ by Lemma B.1. Thus $\rho^* < \rho$ and $\Theta^{\rho^*} = \Theta^\infty$. Hence Θ^∞ is a prewellordering.

If we let $f(e,e_0)$ be an index for Θ, f can be chosen to be recursive, and is as required.

Corollary. If Γ is a Spector pointclass, there is a universal $\breve{\Gamma}$ monotone operation.

Proof. Let f be as in the statement of the Theorem. Let us think of Ψ as a map from $P(\omega \times \omega \times \omega)$ into itself. We define

$$\Psi(X) = \{\langle e,e_0,n\rangle : n \in \Phi_{f(e,e_0)}(\{m : \langle e,e_0,m\rangle \in X\})\} .$$

Clearly $o(\Psi) = \sup_{\langle e,e_0\rangle} o(\Phi_{f(e,e_0)})$. If Φ_e is monotone, and $\eta < o(\Phi_e)$, let $e_0 \in \Phi_e^{\eta} - \Phi_e^{<\eta}$. $o(\Phi_{f(e,e_0)}) > \eta$. We have then that $\sup_{e_0 \in \Phi_e^{\infty}} o(\Phi_{f(e,e_0)}) \geq o(\Phi_e)$. Hence $o(\Psi) \geq o(\Phi_e)$ as required.

Now let $g(e,e_0) = \langle e,e_0,1\rangle$. Since 0 does not represent an ordered pair, $2\cdot 0 + 1 = 1$ is not a member of any prewellordering. Hence $e_0 \notin \Phi_e^{\infty} \to \Phi_{f(e,e_0)}^{\infty}$ is a prewellordering $\to 1 \notin \Phi_{f(e,e_0)}^{\infty} \to \langle e,e_0,1\rangle \notin \Psi^{\infty} \to g(e,e_0) \notin \Psi^{\infty}$. If $e_0 \in \Phi_e^{\infty}$, then $\Phi_{f(e,e_0)}^{\infty} = \omega \to 1 \in \Phi_{f(e,e_0)}^{\infty} \to \langle e,e_0,1\rangle \in \Psi^{\infty} \to g(e,e_0) \in \Psi^{\infty}$.

§4. A closure property of $o(\breve{\Gamma}\text{ mon})$. In this section we prove a strengthened version of Theorem A, allowing a stage of a $\breve{\Gamma}$ monotone inductive definition to appear as a parameter in the Γ operation.

Definition. Let $B \subseteq \omega$. $\Gamma(B) = \{G_B : G \in \Gamma\}$, where $G_B = \{Y : \langle B,Y\rangle \in G\}$. Note that $\Gamma(B)$ is a Spector class if Γ is, and $(\Gamma(B))^{\smile} = \breve{\Gamma}(B)$.

Theorem C. Let Γ be a Spector pointclass. If $B = \Phi^{<\eta}$, with Φ $\breve{\Gamma}$ montone and $\eta < o(\Phi)$, then

$$o(\Gamma(B)) < o(\breve{\Gamma}\text{ mon}) .$$

Proof. Let Γ be a Spector pointclass, let Φ be a $\breve{\Gamma}$ monotone operation, and let $\eta < o(\Phi)$.

We first show that we can replace $\Phi^{<\eta}$ with any prewellordering of length η. The simplest way to do this is to use the proof of Theorem B. Let $e_0 \in \Phi^{\eta} - \Phi^{<\eta}$. Let ρ and $\langle A_\beta^\alpha : \beta \leq \alpha \leq \rho\rangle$ be as in the proof of Theorem B. Since $\langle\beta(\rho),n(\rho)\rangle$ is undefined, $A_\beta^\rho = \Phi^\beta$ for every $\beta < \rho$. Since $\eta \leq \rho$, $A_{<\eta}^\rho = \Phi^{<\eta}$, and so the latter is Δ in any wellfounded relation of height ρ. Thus we have shown that B is Δ in any wellfounded relation of height at least ρ, where $\rho < o(\breve{\Gamma}\text{ mon})$.

Applying Theorem B, let Θ be $\breve{\Gamma}$ montone such that, for X a wellfounded relation, $\Theta(X)$ depends only on height(X), such that $\Theta^{<\alpha}$ is a prewellordering of height α for $\alpha \leq$ some ordinal ρ, such that $\Theta^\rho = \omega$, such that B is Δ is Θ^α for some $\alpha < \rho$, and such that, if X, $\tilde{X}$ are wellfounded relations with height$(X) <$ height$(\tilde{X}) < \rho$, then $\Theta(\tilde{X})$ is an end extension of $\Theta(X)$.

Let $\Phi : P(\omega) \times P(\omega) \to P(\omega)$ be Γ and such that, for each X, $o(\Phi_X) = o(\Gamma(X))$, where $\Phi_X(Y) = \Phi(X,Y)$.

Since the proof of Theorem A was uniform, let $\Phi^*(X,Y)$ be $\breve{\Gamma}$ and such that each Φ_X^* is monotone and satisfies $o(\Phi_X^*) \geq o(\Phi_X)$ and $\Phi_X^{*\infty} = \omega$.

We define an operation Ψ. It is convenient to think of Ψ as an operation sending each sequence $\langle X, Y_0, Y_1, \ldots \rangle$ of sets of natural numbers into another such sequence $\langle X', Y'_0, Y'_1, \ldots \rangle$.

We define Ψ as follows:

(1) If $\Theta(X)$ is not a wellfounded relation, then $\Psi(X, Y_0, \ldots) = \omega$.

(2) Suppose $\Theta(X)$ is a wellfounded relation. If $m \in A^{\Theta(X)}$ let

$$Y'_m = \Phi^*((\Theta(X))_m, Y_m)$$

and let $m \in X' \leftrightarrow Y_m = \omega$.

<u>Lemma C.1.</u> Ψ is monotone.

<u>Proof.</u> Let $X \subseteq \tilde{X}$ and let $Y_m \subseteq \tilde{Y}_m$ for each m. If $\Theta(X)$ is not a wellfounded relation, then $\Theta(Y)$ is not a wellfounded relation.

It suffices, then, to prove the lemma under the assumption that both $\Theta(X)$ and $\Theta(\tilde{X})$ are wellfounded relations. Either $\Theta(X) = \Theta(\tilde{X})$ or else $\Theta(\tilde{X})$ is an end extension of the prewellordering $\Theta(X)$. In either case $(\Theta(X))_m = (\Theta(\tilde{X}))_m$ for all $m \in A^{\Theta(X)}$. For such m, $m \in A^{\Theta(\tilde{X})}$ and

$$\Phi^*((\Theta(X))_m, Y_m) \subseteq \Phi^*((\Theta(\tilde{X}))_m, \tilde{Y}_m)$$

since Φ^* is monotone in Y. For such m, if $m \in X'$ then $m \in X$ or $Y_m = \omega$; hence $m \in \tilde{X}$ or $\tilde{Y}_m = \omega$ and so $m \in \tilde{X}'$.

<u>Lemma C.2.</u> For any $\gamma < \rho$, $o(\Psi) > o(\Phi^*_{\Theta^{<\gamma}})$.

<u>Proof.</u> For each β, let $\Psi^\beta = \langle X^\beta, Y_0^\beta, \ldots \rangle$. Suppose inductively that there is a β such that $X^{<\beta} = \Theta^{<\gamma}$ and $Y_m^{<\beta} = \omega$ for each $m \in A^{\Theta^{<\gamma}}$ and is empty otherwise. It is easily seen that $X^\beta = X^{<\beta}$ and $Y_m^\beta = Y_m^{<\beta}$ unless $m \in X^\beta - X^{<\beta}$. For such m, $Y_m^\beta = \Phi^*(X^{<\beta}, Y_m) \cup Y_m$. These conditions continue to hold until a β' is reached when each of these $Y_m^{<\beta'} = \omega$. When that happens, $X^{\beta'} = \Theta^\gamma$ and each $Y^{\beta'} = Y^{<\beta'}$. This completes the successor stage of our induction. The limit stage is clear.

Since, for β, β' as in our argument, $\beta' - \beta = o(\Phi^*_{\Theta^{<\gamma}})$, we have proved the lemma.

Since Ψ is clearly $\check{\Gamma}$, the theorem is proved.

§5. Admissibility properties. Theorem C shows that $o(\Pi^1_2 \text{ mon})$ is admissible. For any set B, $o(\Sigma^1_2(B) \text{ mon})$ is stable. Theorem C thus shows that $o(\Pi^1_2 \text{ mon})$ is a limit of stables, thus stable, and thus admissible.

We do not know whether $o(\breve{\Gamma} \text{ mon})$ is admissible for every Spector pointclass Γ. The following theorem, however, gives a positive answer for classes like Σ^1_2 (e.g., for Γ the result of applying the game quantifier to Σ^0_n, $n \geq 3$).

Theorem D. Let $\breve{\Gamma}$ be a Spector pointclass. Suppose that, for every $X \subseteq \omega$ and every $\breve{\Gamma}(X)$ monotone Φ, $\Phi^\infty \in \breve{\Gamma}(X)$. $o(\breve{\Gamma} \text{ mon})$ is admissible - indeed, is non-projectible.

Proof. Let $\lambda = o(\breve{\Gamma} \text{ mon})$. We show that λ is a limit or ordinals stable in λ. Let $\beta < \lambda$. By Theorem B, let X be a prewellordering of height β such that $X = \Phi^{<\eta}$ for some $\breve{\Gamma}$ monotone Φ. Since $\lambda_X = o(\breve{\Gamma}(X) \text{ mon}) \geq \lambda$, it is enough to show some $\delta > \beta$ is smaller than λ and is stable in λ_X. Let δ_X be the height of a prewellordering of a universal $\Gamma(X)$ set, as given by the fact that $\Gamma(X)$ is a Spector class. $\delta_X = o(\Gamma(X) \text{ mon}) < o(\Gamma(X))$, which is smaller than λ by Theorem C.

Lemma D.1. δ_X is stable in λ_X.

Proof. Let $\rho < \delta_X$ and suppose that $\exists \gamma < \lambda_X P(\rho,\gamma)$, where P is Δ_0. Let E be a $\Delta(X)$ wellordering of order type ρ. Obviously $P(\rho,\gamma) \leftrightarrow \tilde{P}(\gamma)$, where $\tilde{P}$ is Δ_0 in E. Let $\gamma < \lambda_X$ be minimal such that $\tilde{P}(\gamma)$. By the relativization of Theorem B, let Θ be $\breve{\Gamma}(X)$ monotone such that, for $\eta \leq \gamma + 1$, $\Theta^{<\eta}$ is a prewellordering of height η and such that, if Y and Y' are wellfounded relations of the same height, $\Theta(Y) = \Theta(Y')$.

Define $\tilde{\Theta}(Y)$ as follows:

(a) If Y is not a wellfounded relation, let $\tilde{\Theta}(Y) = \omega$.

(b) If Y is a wellfounded relation of height $\leq \gamma$, let $\tilde{\Theta}(Y) = \Theta(Y)$.

(c) If Y is a wellfounded relation of height $> \gamma$, let $e_0 \in A'(Y)$ be such that $\text{height}(Y_{e_0}) = \gamma$. Let $\tilde{\Theta}(Y) = \Theta(Y_{e_0})$.

$\tilde{\Theta}$ is clearly $\breve{\Gamma}(X)$ monotone. $\Theta^\infty = \Theta^\gamma$, which is a prewellordering of height $\gamma + 1$. Thus some prewellordering of height $\gamma + 1$ is $\breve{\Gamma}(X)$, by hypothesis.

It suffices, then to show that every $\breve{\Gamma}(X)$ prewellordering has height $< \delta_X$. This is a standard result about Spector classes, but we don't know of a proof in print. For simplicity, we forget about the X, and prove the result for any Spector class Γ.

Let Y be a $\breve{\Gamma}$ prewellordering of height exactly δ. (If Y' has height $> \delta$, let $Y = Y'_e$ for some appropriate e). Let φ be a prewellordering of a universal Γ set W of height δ, as given by Γ being a Spector pointclass.

By the recursion theorem, define $\langle e,n\rangle \in V \in \breve{\Gamma}$ just in case $e \in A^Y$ and for every m such that $m \in W$ and $\varphi(m) < \varphi(n)$, there is an $e' \in A^Y$ such that $\langle e',e\rangle \in R^Y$ and $\langle e',m\rangle \in V$. (Here $\varphi(n) = \infty$ if $n \notin W$.) We have

$$n \in W \longleftrightarrow \exists e \langle e,n\rangle \in V .$$

This yields the contradiction that $W \in \breve{\Gamma}$.

§6. <u>The game quantifier</u>. If Γ is a pointclass then $\Game(\Gamma)$ is the set of all sets of the form

$\{Y \in \mathcal{Y} :$ I has a winning strategy for the game defined by $\mathcal{B}$ and $Y\}$

where $\mathcal{B} \subseteq \mathcal{Y} \times P(\omega)$, $\mathcal{B} \in \Gamma$, and the game is given by I and II alternately choosing 0 or 1, thus producing a characteristic function of an $X \subseteq \omega$, and with the winning condition for I that $\langle Y,X\rangle \in \mathcal{B}$.

The following theorem shows that Theorem D applies to many interesting classes, e.g. $\Game(\Sigma^0_n)$ for $n \geq 3$.

<u>Theorem E</u>. Suppose Γ is closed under union, intersection, recursive preimages and existential number quantification. Suppose $\Sigma^0_3 \subseteq \Gamma$. Suppose the game defined by $\mathcal{B}$ and Y is determined for every Y and every $\mathcal{B} \in \Gamma$. Suppose $\Game(\Gamma)$ has the prewellordering property. Suppose Φ is $\Game(\breve{\Gamma})$ monotone. (Note that $\Game(\breve{\Gamma}) = (\Game(\Gamma))^{\smile}$.) We have then $\Phi^\infty \in \Game(\breve{\Gamma})$.

<u>Proof</u>. Let φ be a norm on $\{\langle X,m\rangle : m \notin \Phi(X)\}$ as given by the prewellordering property. Let $n \in \omega$. We describe a game G(n). It will be clear from the hypotheses that

$$\{n : \text{I has a winning strategy for } G(n)\} \in \Game(\breve{\Gamma}) .$$

We shall prove the theorem by showing that this set is precisely Φ^∞.

G(n) is played as follows:

II produces a relation R in ω and, for each $b \in \omega$, subsets $B_{\lhd b}$ and B_b of ω (by giving their characteristic functions). I similarly chooses, for each $c \in \omega$, subsets $C_{<c}$ and C_c of ω.

(*) II loses unless $B_b \supseteq B_{\lhd b}$, $B_0 = B_{\lhd 0}$, $n \notin B_0$, $B_{\lhd b} = \bigcup \{B_{b'} : R(b',b)\}$, and R is a linear ordering of ω with 0 maximal.

As the listing of R and the B_b, $B_{\lhd b}$, C_c, and $C_{\lhd c}$ continues, I may make a series of challenges to II. Each challenge, once made, may be cancelled, and no new challenge may be made until the existing one is cancelled.

(1) If II lists $m \in B_b$, I may <u>challenge</u> $\langle b,m \rangle$. While listing continues, the two players play the $\check{\Gamma}$ game corresponding to $m \in \Phi(B_{\lhd b})$.

(2) If II lists $m \in B_b - B_{\lhd b}$ and $m' \notin B_{b'}$, I may <u>challenge</u> $\langle b,m,b',m' \rangle$. In this case the players play the game corresponding to $\varphi(B_{\lhd b},m) \leq \varphi(B_{\lhd b'},m')$.

If a challenge is cancelled, play of the corresponding game stops.

I wins $G(n)$ if he wins because of (*), if he makes and never cancels some challenge and wins the corresponding game, or if there is a b such that

$$\exists c(B_{\lhd b} \supseteq C_{\lhd c} \ \&\ B_b \not\supseteq C_c)$$

and there is no R-least such B.

<u>Lemma E.1</u>. If $n \notin \Phi^{\infty}$, II has a winning strategy for $G(n)$.

<u>Proof</u>. II plays a wellordering R of order type $o(\Phi) + \omega + 1$, and, for each b, plays $B_{\lhd b} = \Phi^{<|b|}$ and $B_b = \Phi^{|b|}$, where $|b|$ is the order type of the initial segment of b with respect to R. Whenever I makes a challenge, II plays a winning strategy for the corresponding game.

<u>Lemma E.2</u>. If $n \in \Phi^{\infty}$, I has a winning strategy for $G(n)$.

<u>Proof</u>. For each ordinal α we define inductively $A^{\alpha}_{<\beta}$ and A^{α}_{β} for each ordinal β. Let $A^{\alpha}_{<\beta} = \bigcup_{\beta'<\beta} A^{\alpha}_{\beta'}$. Let

$$A^{\alpha}_{\beta} = A^{\alpha}_{<\beta} \cup \{m : \varphi(A^{\alpha}_{<\beta},m) \not< \alpha\} .$$

Note that $A^{\alpha}_{\beta} \supseteq \Phi^{\beta}$ for each α and β, and so $n \in \Phi^{o(\Phi)} \subseteq A^{\alpha}_{o(\Phi)}$ for each α.

Suppose inductively that there are only countably many α such that $\exists \beta' < \beta(A^{\alpha+1}_{\beta'} \neq A^{\alpha}_{\beta'})$. Let the set of such α be $M(\beta)$. If $A^{\alpha+1}_{\beta} \neq A^{\alpha}_{\beta}$, then either $\alpha \in M(\beta)$ or else $A^{\alpha+1}_{\beta} \subsetneqq A^{\alpha}_{\beta}$. Thus A^{α}_{β} is monotone decreasing between successive members of $M(\beta)$. Hence $M(\beta + 1)$ is countable. It follows that

$$\mathcal{D} = \{A^{\alpha}_{<\beta} : \alpha \in On \ \&\ \beta \leq o(\Phi)\}$$

is countable.

Using $\mathcal{D}$, we give a system of challenges (based on joint work of Solovay and the author). Let $\{\langle b_i,m_i,b'_i,m'_i,D_i,D'_i \rangle : i \in \omega\}$ be all sextuples with

b_i, m_i, b'_i, and $m_i \in \omega$ and D_i and $D'_i \in \mathcal{D}$. I considers each i in turn, and in two steps:

Step 1. I waits to see whether $m_i \in B_{b_i}$. If $m_i \in B_{b_i}$ or if $m_i \notin \Phi(D_i)$ he goes on to Step 2. Otherwise he challenges $\langle b_i, m_i \rangle$ and plays his winning strategy for showing $m_i \in \Phi(D_i)$. If it is ever seen that $B_{\lhd b_i} \neq D_i$, the challenge is cancelled.

Step 2. I waits to see whether $m_i \in B_{b_i} - B_{\lhd b_i}$ and $m'_i \notin B_{b_i}$. If not, or if $\varphi(D'_i, m') < \varphi(D_i, m)$, he goes on to $i+1$. Otherwise he challenges $\langle b_i, m_i, b'_i, m'_i \rangle$ and plays his winning strategy for showing $\varphi(D_i, m) \leq \varphi(D'_i, m')$. If it is ever seen that $D_i \neq B_{\lhd b_i}$ or $D'_i \neq B_{\lhd b'_i}$, then the challenge is cancelled.

Note that if II does not lose because of some challenge, then Step 1 assures that

$$B_{\lhd b_i} \in \mathcal{D} \to B_{b_i} \supseteq \Phi(B_{\lhd b_i}) .$$

Under the same conditions, Step 2 assures that there is a single ordinal $\bar{\alpha}$ such that, for all $B_{\lhd b} \in \mathcal{D}$,

$$B_b = B_{\lhd b} \cup \{m : \varphi(B_{\lhd b}, m) \nless \bar{\alpha}\} .$$

An easy induction shows that, if b belongs to the wellordered initial segment of R and II does not lose because of (*) or challenges, then

$$B_b = A^{\bar{\alpha}}_{|b|} \supseteq \Phi^{|b|}$$

as long as $|b| \leq o(\Phi)$. Thus II must lose if R is a wellordering.

We still must describe the C_c and $C_{\lhd c}$. Let the C_c be an enumeration of all Φ^β for $\beta \leq o(\Phi)$ and let $C_{\lhd c}$ be $\Phi^{<\beta}$ whenever $C_c = \Phi^\beta$.

If II does not lose and b belongs to the wellordered initial segment of R, then $B_{\lhd b} \in \mathcal{D}$, so $B_b \supseteq \Phi(B_{\lhd b}) \supseteq C_c$ whenever $B_{\lhd b} \supseteq C_{\lhd c}$.

Thus, if II does not lose there must be a b_0 not in the wellordered initial segment of R such that

$$R(b, b_0) \ \& \ B_{\lhd b} \supseteq C_{\lhd c} \to B_b \supseteq C_c \supseteq \Phi(C_{\lhd c}) .$$

To complete the proof, we show that, if II does not lose every b not in the wellordered initial segment of R satisfies

$$B_b \supseteq \Phi^\infty .$$

Assume that all such b satisfy $B_b \supseteq \Phi^{<\beta}$. We show that all such b satisfy $B_b \supseteq \Phi^{\beta}$. It is enough to show this for all b not in the well-ordered initial segment and such that $R(b,b_0)$. For such b, there is a b' with $R(b',b)$ such that $B_{b'} \supseteq \Phi^{<\beta}$. Hence $B_{<b} \supseteq B_{b'} \supseteq \Phi^{<\beta} = C_{<c}$ for some c. Thus $B_b \supseteq C_c = \Phi^{\beta}$.

Corollary. If Γ is as in the statement of the theorem and $\mathfrak{D}(\Gamma)$ is a Spector pointclass, then $o(\mathfrak{D}(\breve{\Gamma})\text{ mon})$ is non-projectible.

Proof. $\Gamma(X)$ satisfies the hypotheses of the theorem for every X, since Γ does. The Corollary follows from the theorem and Theorem D.

References

Y. N. Moschovakis [1980], Descriptive Set Theory, North Holland (1980).

TREES AND DEGREES

Piergiorgio Odifreddi*

Department of Mathematics
University of California
Los Angeles, California 90024

Istituto Matematico
Università di Torino
Torino, Italy 10100

§1. The finite extensions method and its limitations. In their classical paper [1954] Kleene and Post undertook the study of degree theory, and proved the very first results by using the Baire category method. In this section we will briefly review the method and see why something more powerful is needed in order to prove certain kinds of results. In the remaining sections we will introduce many tools which have been very useful in the development of degree theory, and which ultimately make use of the notion of tree. More specifically, in Section 2 we work with the simplest kind of trees: total recursive ones. Some limitations are introduced in Section 3, where we require the trees to have certain uniformities. Section 4 is devoted to a vast area of results obtained by the use of uniform trees: the construction of initial segments of the degrees, of which we only give some basic examples. In Section 5 we relax the requirement that our trees be recursive and study recursively pointed trees, while in Section 6 recursiveness is kept but we no longer insist on our trees being total. Finally, in Section 7 we build trees of trees for every notion of tree dealt with in the paper and see how the proofs of some results consist in choosing some branches of these trees in appropriate ways.

With the exception of results in this section, no proof is given unless the method of trees is genuinely used in it. We refer to the original papers or to our forthcoming book [198?] for a complete treatment.

We hope that the reader who is familiar with the basic notions concerning recursive functions will find our development of the methodology of trees self-contained. This will not stop us from quoting other results which are related to our exposition and which may be of interest to the more experienced reader.

---

*The author was partially supported by a grant of CNR, Italy. He wishes to thank Dick Epstein and Carl Jockusch, Jr. for their valuable teaching, suggestions, corrections and help.

We also use standard notations, as in Epstein [1979] or Rogers [1967]. In particular $\{e\}^{\sigma}$ is the e-th recursive partial function using as oracle the string σ of 0's and 1's, $\{e\}^{A}$ is the same with oracle A, W_e is the domain of $\{e\}$ etc. We will also confuse A and its characteristic function, and if a degree contains a set with a certain property we will say that the degree itself has that property.

The very first non-trivial result of degree theory is:

Proposition 1.1. There are two incomparable degrees (Kleene-Post). We want A and B, each of them not recursive in the other. The requirements to be satisfied are, for each e:

$$A \not\simeq \{e\}^{B} \quad \text{and} \quad B \not\simeq \{e\}^{A} .$$

We build A and B simultaneously by initial segments σ_n and τ_n. Let $\sigma_0 = \tau_0 = \emptyset$. Suppose we already have σ_e and τ_e and are trying to satisfy the above requirement for e. Choose x outside the domain of σ_e (i.e. A(x) is still undetermined) and see if there is some $\tau \supseteq \tau_e$ s.t. $\{e\}^{\tau}(x)\downarrow$. If there is, take any such τ and define A(x) as whichever of 0, 1 is different from $\{e\}^{\tau}(x)$ (i.e. take $\sigma \supseteq \sigma_e$ s.t. $\sigma(x)$ is as described). Otherwise, $\{e\}^{B}(x)$ will be undefined and we can let $\sigma = \sigma_e$, $\tau = \tau_e$. Then repeat the same process, with A and B interchanged (using now σ and τ in place of σ_e and τ_e). Finally, let σ_{e+1} and τ_{e+1} be the two strings found in the last step. Let $A = \bigcup \sigma_e$ and $B = \bigcup \tau_e$. □

We can think of the subsets of ω as a topological space ${}^{\omega}2$, with the topology induced by the basic open sets $O_{\sigma} = \{X : X \supseteq \sigma\}$. On ${}^{\omega}2 \times {}^{\omega}2$ we put the product topology. What the proof above tells us is that e.g. the requirement $R_e = \{(A,B) : A \not\simeq \{e\}^{B} \wedge B \not\simeq \{e\}^{A}\}$ is dense in ${}^{\omega}2 \times {}^{\omega}2$, in the sense that given (σ,τ) we can always find (σ',τ') such that $\sigma \subseteq \sigma'$, $\tau \subseteq \tau'$ and $O_{(\sigma',\tau')} \subseteq R_e$. Of course the $O_{(\sigma',\tau')}$'s are open and the pair (A,B) of 1.1 is in a countable intersection of open dense sets. In the usual terminology, the set of pairs (A,B) with A and B incomparable is comeager. We can think of the proof of 1.1 as being the construction of one set $A \oplus B$ with two incomparable degrees below it. Only notational changes (or appeal to the homeomorphism ${}^{\omega}2 \cong {}^{\omega}2 \times {}^{\omega}2$) are needed to build a set X s.t. $Od(X) = \{z : 2z \in X\}$ and $Ev(X) = \{z : 2z + 1 \in X\}$, (Od and Ev stand for odd and even parts) are incomparable. Again, since $Od(X) \leq_T X$ and $Ev(X) \leq_T X$, we have a comeager set of sets of non minimal degree. It follows at once that the class of sets of minimal degree is meager and, by the Baire category theorem, it cannot be comeager. Hence sets of minimal degree - if they exist - cannot be built by the same method as before, which we call the finite extension method (for obvious reasons). These observations answer Sacks [1963], page 159 and Rogers [1967], page 276.

By using the finite extension method to build degrees with certain properties, it is clear that we simultaneously are showing that degrees without those properties cannot be built by the same method. We present two more examples.

Definition. A non-zero degree $\underset{\sim}{a}$ is called hyperimmune-free if for every $A \in \underset{\sim}{a}$ and every $f \leq_T A$, f is majorized by a recursive function.

The name comes from the fact that $\underset{\sim}{a}$ is hyperimmune-free iff it does not contain any hyperimmune set (Martin-Miller [1968]). We use the definition above because it does not involve the concept of hyperimmunity (a hyperimmune-free definition!) and it is more suitable to our treatment in Section 2. A useful fact to keep in mind is that if $\underset{\sim}{a}$ is hyperimmune-free then it consists of only one tt-degree (Jockusch [1969a]) and so do all the degrees below it (since by definition if $\underset{\sim}{a}$ is hyperimmune-free, so is every degree below it).

Proposition 1.2. There is a hyperimmune degree (Post).

We define the elements of A in increasing order, with the condition that if a_n is the n-th such element and $\{n\}(n)\downarrow$, then $\{n\}(n) < a_n$. Actually, this is already the construction! And if $g(n) = a_n$, g is recursive in A and not majorized by any (total) recursive function. □

Definition. A degree $\underset{\sim}{a}$ is called completely autoreducible if it contains only autoreducible sets, i.e. sets A such that (for all x) the question "$x \in A$?" can be answered recursively in A itself but without making any use of the oracle A on the input x.

Autoreducible sets are very easy to exhibit: for any A, $A \oplus A$ is such because $2x$ and $2x + 1$ give the same information (so if a computation uses one of them, there is an equivalent computation using the other). Thus every degree contains autoreducible sets.

Proposition 1.3. There is a non-autoreducible degree (Trahtenbrot).

It's enough to make A s.t. for every e there is an n such that $A(n) \not\simeq \{e\}^{A-\{n\}}(n)$. We do this at stage $e + 1$, given σ_e ($\sigma_0 = \emptyset$). Choose n outside the domain of σ_e and see if there is $\sigma \supseteq \sigma_e$ with $\sigma(n) = 0$ and $\{e\}^{\sigma}(n)\downarrow$. If not, let σ_{e+1} be any $\sigma \supseteq \sigma_e$ with $\sigma(n) = 0$. Otherwise, let σ_{e+1} be such a σ except on n, where we let $\sigma_{e+1}(n)$ be the element of $\{0,1\}$ different from $\{e\}^{\sigma}(n)$. □

We conclude this section with a simple proof (due to Jockusch) of a result we will need, in an unexpected way, in Section 3.

Proposition 1.4. There exists a recursive partition of $[\omega]^2$ (the set of unordered pairs of natural numbers) into two classes, without recursive homogeneous sets (i.e. for no infinite recursive set, any two elements of it are in the same fixed class of the partition) (Specker).

By diagonalization we can easily build an infinite, coinfinite set $B \leq_T 0'$ s.t. if W_e is infinite then $B \cap W_e \neq \emptyset$ and $\overline{B} \cap W_e \neq \emptyset$. Then by the limit lemma, $B(n) = \lim_{s\to\infty} g(n,s)$ for some recursive (0-1 valued) g. Define

$$f(\{n,s\}) = g(n,s) \quad \text{if} \quad n < s .$$

This gives a recursive partition of $[\omega]^2$ into two classes. Let $A = \{a_0 < a_1 < \dots\}$ be infinite and homogeneous for this partition. By homogeneity

$$g(a_n,a_{n+1}) = g(a_n,a_{n+2}) = \cdots = B(a_n)$$

and, moreover, $B(a_n)$ does not depend on a_n. Hence $A \subseteq B$ or $A \subseteq \overline{B}$ and, by the choice of B, A cannot be r.e. □

Incidentally, this shows that the recursive analogue of Ramsey's theorem is false. See Jockusch [1972] for an analysis of the possible complexity of homogeneous sets for recursive partitions of $[\omega]^n$.

§2. The trees method. From the previous section we have plenty of reasons to be unhappy with the finite extension method. Since the core of it is to construct sets $A \in \bigcap_{n\in\omega} T_n$ where $T_n = \{X : X \supseteq \sigma_n\}$ for some increasing sequence $\{\sigma_n\}_{n\in\omega}$ of strings of 0's and 1's (and hence for some decreasing sequence $\{T_n\}_{n\in\omega}$ of sets of strings), our first though is to allow more general kinds of sets of strings.

Definition. A tree is a function T from strings of 0's and 1's to strings of 0's and 1's, with the following properties:

- $T(\sigma)\downarrow$ and $\tau \subseteq \sigma \Rightarrow T(\tau)\downarrow$ and $T(\tau) \subseteq T(\sigma)$
- if one of $T(\sigma * 0)$, $T(\sigma * 1)$ is defined, then both are defined and they are incompatible.

What really matters in a tree is its range, but to think of it as a function is a useful tool: e.g. we can simply talk of $T(\sigma * 0)$, $T(\sigma * 1)$

instead of the two smallest incompatible extensions of $T(\sigma)$ on the tree. Similarly, to state that both or neither of $T(\sigma * 0)$, $T(\sigma * 1)$ are defined is no loss of generality since if we are only interested in the paths extending $T(\sigma * 0)$, we may define $T(\sigma * 1)$ as well and then let the tree grow only above $T(\sigma * 0)$.

We refer to properties of the function T as properties of the tree determined by it. So we say that a tree is recursive if T is (perhaps in this case it would be better to speak of r.e. trees, but this is now in common usage). A total tree (i.e. a tree whose function is total) really determines a perfect closed subset of ${}^{\omega}2$ (in the topology of Section 1). We say that A is on T, or that A is a branch of T, if for infinitely many σ's we have $T(\sigma) \subseteq A$. We say that σ is on T if it is on the range of it.

T' is a subtree of T ($T' \subseteq T$) if every σ on T' is also on T: the concept of subtree is for trees what the concept of extension is for strings. T' is the full subtree of T above σ if it consists of every string on T extending σ. E.g. the full subtree of T above $T(\sigma)$ is defined: $T'(\tau) = T(\sigma * \tau)$.

We say that A is constructed by the method of trees if for some decreasing sequence $\{T_n\}_{n\in\omega}$ of trees, $A \in \bigcap_{n\in\omega} T_n$. Of course the arguments using the finite extensions method can be recast in the language of trees, and this immediately shows where the limitation of that method lies: only full trees are used. Perhaps the simplest genuine application of the method of trees is the construction of hyperimmune-free degrees. As in the rest of the paper, we first single out the ingredients of the proof.

In this and the next two sections, Q and T stand for recursive total trees.

Lemma 2.1. Given T and e, there is $Q \subseteq T$ s.t. for every A on Q, $A \neq W_e$.

Since $T(0)$, $T(1)$ are incomparable, one of them must disagree with W_e. Take Q = full subtree of T above it. □

The next lemma says that our trees may always be chosen of a very special kind.

Lemma 2.2. Given T and e, there is a $Q \subseteq T$ s.t. one of the following holds:

(1) for every A on Q, $\{e\}^A$ is not total

(2) for every A on Q, $\{e\}^A$ is total and

$$\forall n \forall \sigma \, (|\sigma| = n \Rightarrow \{e\}^{Q(\sigma)}(n)\downarrow) .$$

($|\sigma|$ stands for the length of σ). See if

$$\exists\sigma \in T\ \exists x\ \forall\tau \supseteq \sigma\ (\tau \in T \Rightarrow \{e\}^{\tau}(x)\uparrow)\ .$$

If yes, take such a σ and Q = full subtree of T above σ. Then case 1 holds. Otherwise, define Q as follows. $Q(\emptyset)$ = least $\tau \in T$ s.t. $\{e\}^{\tau}(0)\downarrow$. Given $Q(\sigma)$ take $\tau \supseteq Q(\sigma)$ on T s.t. $\{e\}^{\tau}(\mathrm{lh}\ \sigma + 1)\downarrow$, and let $Q(\sigma * i) = \tau * i$ for $i = 0,1$ (to have two incomparable extensions). □

Theorem 2.3. There exists a hyperimmune-free degree (Martin-Miller).

The requirements on A are:

R_{2e} : $A \neq W_e$

R_{2e+1} : $\{e\}^A$ total $\Rightarrow$ for some recursive f, $(\forall n)\{e\}^A(n) \leq f(n)$.

We define a sequence $\{T_n\}_{n\in\omega}$ of total recursive trees s.t. $T_{n+1} \subseteq T_n$ and such that whenever A is a branch of T_{n+1}, A satisfies R_n.

T_0 = the full binary tree.

T_{2e+1} is the Q of Lemma 2.1 for $T = T_{2e}$.

T_{2e+2} is the Q of Lemma 2.2 for $T = T_{2e+1}$.

To see that every branch of Q satisfies R_{2e+1}, let A be on it and $\{e\}^A$ be total. If $f(n) = \max_{|\sigma|=n} \{e\}^{T_{2e+2}(\sigma)}$, then $\{e\}^A(n) \leq f(n)$. □

Note that the degree constructed is actually below $\underset{\sim}{0}''$, since the construction (of the lemmas) uses at most two-quantifier questions. This bound is optimal since Martin-Miller [1968] have proved that any non-zero degree comparable with $\underset{\sim}{0}'$ is hyperimmune. A relativization of the above proof gives: given $\underset{\sim}{b}$ hyperimmune-free, there is $\underset{\sim}{a}$ hyperimmune-free s.t. $\underset{\sim}{b} < \underset{\sim}{a} < \underset{\sim}{b}''$. Jockusch [1969b] has generalized the results above to degrees bi-immune-free.

The same ideas of the proof of 2.3 give the following: the hyperimmune-free degrees are a basis for Π^0_1 classes of sets (Jockusch-Soare [1972a,b]). A Π^0_1 class is simply the set of branches of a recursive tree with infinite recursive range. This is of interest because typical examples of Π^0_1 classes are the classes of complete (resp. consistent) extensions of any formalizable theory, e.g. Peano arithmetic. From the previous basis theorem follows the existence of, say, complete extensions of Peano arithmetic which blow up every hypersimple set.

Originally, the first application of the method of trees was Spector's construction of a minimal degree, which is by far the most important result of this section. We know from 2.1 how to deal with diagonalization. Now we must learn how to deal with the requirements:

$$\{e\}^A \text{ total} \Rightarrow \{e\}^A \text{ recursive or } A \leq_T \{e\}^A .$$

<u>Definition</u>. σ, τ <u>e-split</u> if for some x both $\{e\}^\tau(x)$ and $\{e\}^\sigma(x)$ are defined, and they are different (we say σ, τ e-split on x, and we call them an e-splitting).

T is an <u>e-splitting tree</u> if for every σ, $T(\sigma * 0)$ and $T(\sigma * 1)$ e-split.

The interest of this notion comes from the next lemma:

<u>Lemma</u> 2.4. <u>Given</u> T, e <u>and</u> A <u>on</u> T <u>if</u> $\{e\}^A$ <u>is total then</u>

(1) <u>if there is no e-splitting on</u> T, $\{e\}^A$ <u>is recursive</u>

(2) <u>if</u> T <u>is e-splitting, then</u> $A \leq_T \{e\}^A$.

(<u>Spector</u>).

(1) Given x, $\{e\}^A(x)\downarrow$ because it's total. So for some $\sigma \subseteq A$ on T, $\{e\}^\sigma(x)\downarrow$. Since there is no e-splitting on T, it is enough to search for any $\sigma \in T$ s.t. $\{e\}^\sigma(x)\downarrow$ to have the right value of $\{e\}^A(x)$, and this is a recursive procedure. (2) We generate initial segments of A. If $T(\sigma) \subseteq A$, either $T(\sigma * 0) \subseteq A$ or $T(\sigma * 1) \subseteq A$. To decide which, note that for some x, both $\{e\}^{T(\sigma * i)}(x)$ $(i = 0,1)$ are defined and they differ. Choose i s.t. $\{e\}^{T(\sigma * i)}(x) \simeq \{e\}^A(x)$. Then $T(\sigma * i) \subseteq A$, and the procedure is recursive in $\{e\}^A$. □

At this point we could immediately get, given T and e, a subtree of T with one of the hypotheses of 2.4 fulfilled. By adding as a previous step an application of 2.2 we can actually get more:

<u>Lemma</u> 2.5. <u>Given</u> T <u>and</u> e, <u>there is</u> $Q \subseteq T$ <u>s.t. one of the following holds</u>:

(1) <u>for every</u> A <u>on</u> Q, $\{e\}^A$ <u>is not total</u>

(2) <u>for every</u> A <u>on</u> Q, $\{e\}^A$ <u>is recursive</u>

(3) <u>for every</u> A <u>on</u> Q, $A \leq_T \{e\}^A$

(<u>Spector</u>).

First apply 2.2 to get $T' \subseteq T$ s.t. either for every A on it $\{e\}^A$ is nto total, or for every A on it $\{e\}^A$ is total. If the first case arises, let $Q = T'$ (1 is thus satisfied). Otherwise ask if there are e-splittings above every σ, i.e. if

$$\forall \sigma \in T' \; \exists x \exists \tau_1, \tau_2 \in T' \; (\tau_1, \tau_2 \supseteq \sigma \wedge \{e\}^{\tau_1}(x) \neq \{e\}^{\tau_2}(x))$$

where both the computations converge. If not, take σ with no e-splittings above it and let Q = the full subtree of T' above σ ((2) is satisfied by 2.4.1, since for every A on Q $\{e\}^A$ is total). Otherwise, define Q as follows. $Q(\emptyset) = T'(\emptyset)$. Given $Q(\sigma)$, search for x and $\tau_1, \tau_2 \supseteq Q(\sigma)$ on T' which are e-splitting on x, and let $Q(\sigma * i) = \tau_i$. Then Q is e-splitting and (3) is satisfied by 2.4.2. □

For later use we note that in case T is not recursive, Lemma 2.5 is still correct with the addition: $Q \leq_T T$, $\{e\}^A \leq_T Q$ in part 2 and $A \leq_T \{e\}^A \oplus Q$ in part 3.

Theorem 2.6. There exists a minimal degree (Spector).

Let $A \in \bigcap_{n\in\omega} T_n$ where:

T_0 = full binary tree
T_{2e+1} is the Q of Lemma 2.1 for $T = T_{2e}$
T_{2e+2} is the Q of Lemma 2.4 for $T = T_{2e+1}$. □

The slight complication of Lemma 2.4 has an immediate pay-off:

Theorem 2.7. There exists a set which is simultaneously of minimal degree and of minimal tt-degree.

Let A be as in 2.6. Since we used Lemma 2.2 in 2.5, the degree of A is actually a tt-degree (see the observations before 1.2). □

We will see at the end of this section that there is no implication between the concepts of hyperimmune-free degree and of minimal degree. The relativization of 2.6 says: given $\underset{\sim}{b}$ there is $\underset{\sim}{a}$ minimal over $\underset{\sim}{b}$, i.e. no degree is strictly between $\underset{\sim}{b}$ and $\underset{\sim}{a}$. In general it is not true that there is such an $\underset{\sim}{a}$ with the stronger property that any degree less than $\underset{\sim}{a}$ is less than or equal to $\underset{\sim}{b}$: if $\underset{\sim}{b} = \underset{\sim}{0}'$ then any $\underset{\sim}{a}$ above it is of the form $\underset{\sim}{c}'$ for some $\underset{\sim}{c}$ and hence is the l.u.b. of two lesser degrees (see Epstein [1979], page 140-141), and they cannot be both below $\underset{\sim}{b}$ if $\underset{\sim}{a} > \underset{\sim}{b}$.

Open problem: if b is minimal, does there always exist $\underset{\sim}{a} > \underset{\sim}{b}$ s.t. the only degrees below $\underset{\sim}{a}$ are $\underset{\sim}{b}$ and $\underset{\sim}{0}$?
We will see in Section 4 that for some $\underset{\sim}{b}$ this happens.

Theorem 2.8. If $n \geq 1$, there is a minimal degree which is below $\underset{\sim}{0}^{n+1}$ but not below $\underset{\sim}{0}^n$ (Manaster).

Since the construction of 2.5 only asks two-quantifier questions, it is recursive in $\underset{\sim}{0}''$. If we do the same construction of 2.6, only diagonalizing against the e-th Σ_n^0 set at step $2e+1$ (in a way similar to 2.1) we get a set of minimal degree which is not recursive in $\underset{\sim}{0}^n$ and it is recursive in $\max\{\underset{\sim}{0}'', \underset{\sim}{0}^{n+1}\} = \underset{\sim}{0}^{n+1}$ (because the diagonalization only requires one-quantifier questions on Σ_n^0 sets). □

A consequence of (the proof of) the last result is that, given any $n \geq 1$, there is a set $A \in \Delta_{n+2}^0 - \Sigma_{n+1}^0$ of minimal degree. Its complement is a set $\Delta_{n+2}^0 - \Pi_{n+1}^0$ of minimal degree. In particular, <u>the fact that A is of minimal degree does not impose limits on its arithmetical complexity.</u> In fact, we can diagonalize against any countable set of sets, having for example non-analytical sets of minimal degree.

No set $A \in \Sigma_1^0$ has minimal degree (Muchnik). In 6.1 we will exhibit $A \in \Delta_2^0$ of minimal degree. A finite injury argument using 1-trees (see §3) produces, for any $n \geq 3$, $A \in \Sigma_3^0 - \Delta_3^0$ of minimal degree (communication of Jockusch).

<u>Open problem</u>: is there a set $A \in \Sigma_2^0 - \Delta_2^0$ of minimal degree?

We can draw additional information from the constructions above. Let e.g. $\underset{\sim}{a}$ be the minimal degree below $\underset{\sim}{0}''$ given by 2.6: then $\underset{\sim}{a}'' = \underset{\sim}{0}''$. One direction is always true. For the other we simply use the fact (Epstein [1979], page 157) that $\{e : \{e\}^A \text{ is total}\} \equiv_T A''$: we can decide whether $\{e\}^A$ is total or not simply by looking at the stage T_{2e+2} of the construction and see if case 1 of Lemma 2.5 holds or not. This is possible to do recursively in the construction, hence recursively in $\underset{\sim}{0}''$.

But for $\underset{\sim}{a}$ minimal we don't always have $\underset{\sim}{a}'' = \underset{\sim}{0}''$. As a matter of fact, we will see in Section 7 that $\underset{\sim}{a}''$ for $\underset{\sim}{a}$ minimal can be anything it could be (i.e. every degree $\geq \underset{\sim}{0}''$). Moreover, a similar result holds for $\underset{\sim}{a}'$ as well (Cooper [1973]). We may nonetheless suspect that the minimality of $\underset{\sim}{a}$ implies the lowness of its jump, in some sense. Our first thought is: a minimal degree $\underset{\sim}{a}$ realizes the minimal possible jumps. This amounts to saying: $\underset{\sim}{a}^n = \underset{\sim}{a} \cup \underset{\sim}{0}^n$. The next lemma will give us counterexamples to this.

<u>Lemma</u> 2.9. <u>Given</u> T, B <u>and</u> e <u>there is</u> $Q \subseteq T$ <u>s.t. for every</u> A <u>on</u> Q, $A' \neq \{e\}^{A \oplus B}$ (<u>Sasso</u>).

The idea is: we only need to ruin the equality on one witness, and to ask if A goes always left at even levels of T can be phrased as a question about A' (a finite amount of information about A can decide it negatively but not positively), uniformly in A. Precisely, let a be s.t. $\{a\}^\sigma(x) \simeq 0$ iff σ branches right at some even level of T (i.e. $\sigma = T(\tau * 0)$ for some τ of even length). Then $a \in A'$ iff $\{a\}^A(a)\downarrow$

iff A branches right at some even level of T .

See if for some $\tau \in T$, $\{e\}^{\tau\oplus B}(a) \simeq 0$. If yes we take Q = full subtree of T above $\tau * 00$. If A is on Q then $A \supseteq \tau$, so $\{e\}^{A\oplus B}(a) \simeq 0$. But either $\mathrm{lh}\tau + 1$ or $\mathrm{lh}\tau + 2$ is even, and since $A \supseteq \tau * 00$ then A branches right at some even level of T, so $a \in A'$.

Otherwise it's enough to take Q so as to insure that $a \notin A'$ if A is on Q. Take Q = subtree of T that branches left at every even level:

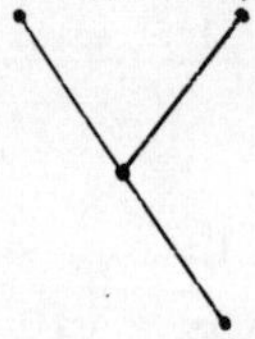

□

Theorem 2.10. Not every minimal degree realizes the minimal jump (Sasso).

Insert in the usual construction of a minimal degree, steps to insure $A' \not\simeq \{e\}^{A\oplus B}$ for each e (using $B \in \underset{\sim}{0}'$ in 2.9). □

Jockusch and Posner [1978] have proved that the next best possible result is true: if $\underset{\sim}{a}$ is minimal then $\underset{\sim}{a}'' = (\underset{\sim}{a} \cup \underset{\sim}{0}')'$. The proof consists of building a function recursive in $\underset{\sim}{0}'$ which majorizes almost everywhere every function of minimal degree. A suggestive consequence of this particular fact is: every minimal degree is "hyperimmune-relative-to-0'"-free. The unrelativized version of that is false (2.13.c). Note that $\underset{\sim}{a} \cup \underset{\sim}{0}'' \leq (\underset{\sim}{a} \cup \underset{\sim}{0}')'$. The above result cannot be improved to read: if $\underset{\sim}{a}$ is minimal then $\underset{\sim}{a}'' = \underset{\sim}{a} \cup \underset{\sim}{0}''$. A counterexample is obtained by using 2.9 with $B \in \underset{\sim}{0}''$ (which gives $\underset{\sim}{a}' \not\leq \underset{\sim}{a} \cup \underset{\sim}{0}''$ and hence $\underset{\sim}{a}'' \not\leq \underset{\sim}{a} \cup \underset{\sim}{0}''$).

We turn to our last example of Section 1. Here the relevant concept is:

Definition. T is doubly e-splitting if for every σ there are x_0, x_1 different and s.t. for $i = 0,1$, $\{e\}^{T(\sigma*0)}(x_i) \not\simeq \{e\}^{T(\sigma*1)}(x_i)$ (both computations convergent).

Lemma 2.11. If T is doubly e-splitting, A is on T and $\{e\}^A$ is a characteristic function then $\{e\}^A$ is autoreducible (Jockusch-Paterson).

We are given an oracle for $\{e\}^A$. What we want to do is to calculate $\{e\}^A(n)$ using any value of the oracle except $\{e\}^A(n)$ itself. We first use the oracle to find large enough initial segments of A, exactly as we did in 2.4.2. The point here is that we have double e-splittings, so we can always avoid the use of $\{e\}^A(n)$. Then having big enough segments of A, we can calculate $\{e\}^A(n)$. Note here the back and forth argument: from $\{e\}^A$ we go to A and then come back to $\{e\}^A$ again. □

Theorem 2.12. There exists a completely autoreducible degree (Jockusch-Paterson).

The only thing we are missing is the analogue of 2.5. We prove that when there we built an e-splitting tree, we could have made it doubly e-splitting. The hypothesis is that on T we can always find e-splittings above any given string. Given $Q(\sigma)$ we have this picture:

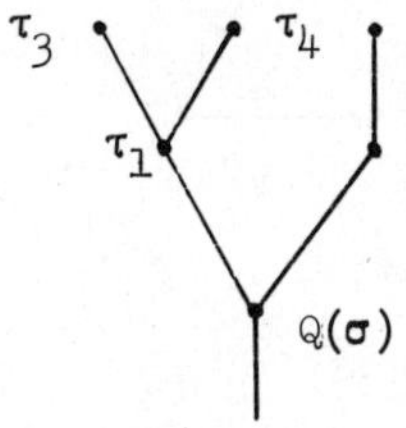

Here τ_1, τ_2 are e-splitting extensions of $Q(\sigma)$, say on x; τ_3, τ_4 are e-splitting extensions of τ_1, say or y (necessarily different from x) and τ_5 is an extension of τ_2 s.t. $\{e\}^{\tau_5}(y)\downarrow$ (which exists as in 2.5 since case 1 there fails). Since τ_3 and τ_4 e-split on y, τ_5 and τ_i e-split on y for i = 3 or i = 4. And by definition τ_5 and τ_i e-split also on x. Let $Q(\sigma * 0) = \tau_5$ and $Q(\sigma * 1) = \tau_i$. □

Jockusch and Paterson have proved that the r.e. non recursive degrees and the degrees $\geq \underset{\sim}{0}'$ are not completely autoreducible. Ladner [1973] has however proved the existence of non recursive r.e. degrees in which every r.e. set is autoreducible.

We look now at the implications among the three concepts introduced.

Theorem 2.13. (a) There is a degree which is minimal, hyperimmune-free and completely autoreducible.

(b) There is a minimal hyperimmune-free degree not completely autoreducible.

(c) There is a minimal completely autoreducible degree not hyperimmune-free.

(d) There is a hyperimmune-free completely autoreducible degree not minimal.

(a) The proof of 2.12 actually gives this, since of course a doubly e-splitting tree is also e-splitting.

(b) See 3.7.

(c) See 6.4 and recall that no nontrivial degree below $\mathbf{0}'$ is hyperimmune-free.

(d) See 4.7. □

§3. Uniform trees or admissible triples. The proof of the existence of minimal degrees we gave in Section 2 is a simplification (due to Shoenfield [1966]) of Spector's original one. The trees used by Spector [1956] were of this form:

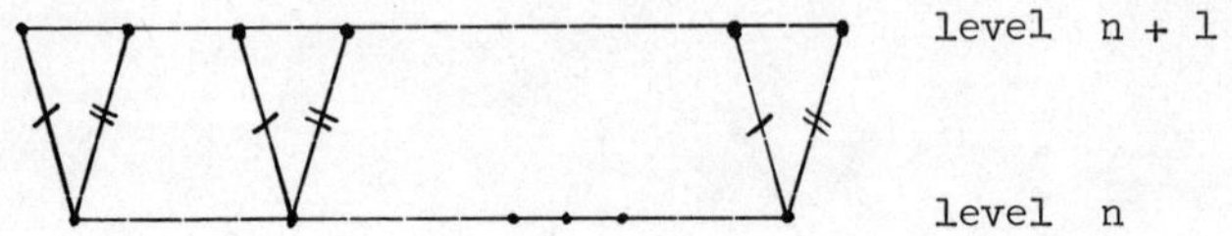

I.e. at every level the immediate strings after each node were not dependent on the node itself. By $T(\sigma * i) - T(\sigma)$ we mean the string τ_i s.t. $T(\sigma * i) = T(\sigma) * \tau_i$.

Definition. T is a uniform tree (or a Spector tree) if for every σ and i $(i = 0,1)$, $T(\sigma * i) - T(\sigma)$ only depends on $|\sigma|$ and for every σ, $|T(\sigma)|$ only depends on $|\sigma|$.

A useful way of representing the situation for uniform trees is in terms of admissible triples: there are three functions g (strictly increasing) and f_L, f_R (left and right functions) s.t. f_L, f_R take values in $\{0,1\}$ and are incompatible in every interval $[g(n), g(n+1))$. Formally:

(1) $(\forall n)(g(n) < g(n+1))$

(2) $\forall n \exists x(g(n) \leq x < g(n+1) \wedge f_L(x) \neq f_R(x))$.

Here $g(n) = |T(\sigma)|$ if $|\sigma| = n$. The picture is:

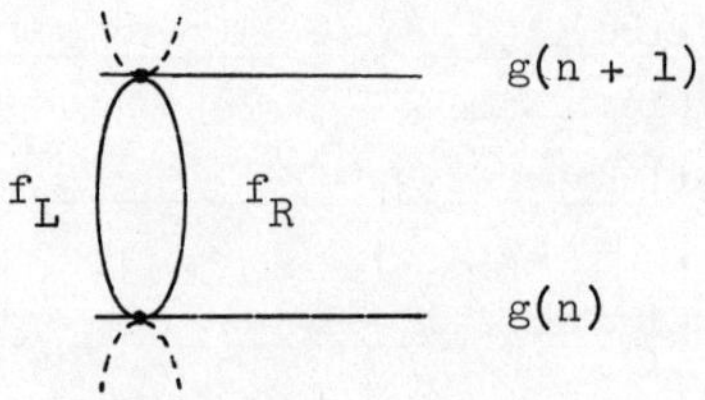

A branch of the tree is simply a path which at every node follows one of f_L, f_R up to the next node. A tree is recursive if g, f_L, f_R are. The next lemma proves that hyperimmune-free degrees can be constructed by uniform trees:

Lemma 3.1. *Given* T *uniform and given* e, *there is* $Q \subseteq T$ *uniform s.t. one of the following holds:*

(1) *for every* A *on* Q, $\{e\}^A$ *is not total*

(2) *for every* A *on* Q, $\{e\}^A$ *is total and*

$$\forall n \forall \sigma \quad (|\sigma| = n \Rightarrow \{e\}^{Q(\sigma)}(n)\downarrow) .$$

See if

$$\exists \sigma \in T \, \exists x \forall \tau \supseteq \sigma \; (\tau \in T \Rightarrow \{e\}^{\tau}(x)\uparrow) .$$

If yes take Q = the full subtree of T above σ. Otherwise, given $Q(\sigma_i)$ for $1 \leq i \leq 2^n$ (σ_i strings of length n) take:

$$\tau_1 \quad \text{s.t.} \quad \{e\}^{Q(\sigma_1)*\tau_1}(n+1)\downarrow$$

$$\tau_2 \quad \text{s.t.} \quad \{e\}^{Q(\sigma_2)*\tau_1*\tau_2}(n+1)\downarrow$$

....

Let $\tau = \tau_1 * \tau_2 * \ldots * \tau_{2^n}$: for each i, $\{e\}^{Q(\sigma_i)*\tau}(n+1)\downarrow$. Then let $Q(\tau_i * 0) = Q(\tau_i) * \tau * 0$ and $Q(\tau_i * 1) = Q(\tau_i) * \tau * 1$. □

Similarly, the next lemma (inserted in the proof of 2.5) proves that minimal degrees can be constructed by uniform trees:

Lemma 3.2. *Given* T *uniform and given* e, *if*

(a) *every* $\sigma \in T$ *has e-splitting extensions on* T

(b) $\forall \sigma \in T \; \forall x \, \exists \tau \in T \; (\tau \supseteq \sigma \wedge \{e\}^{\tau}(x)\downarrow)$

then T *has an e-splitting uniform subtree* Q (*Spector*).

We define Q inductively. Let $Q(\sigma_i)$ for $1 \leq i \leq 2^n$ be given (σ_i strings of length n).

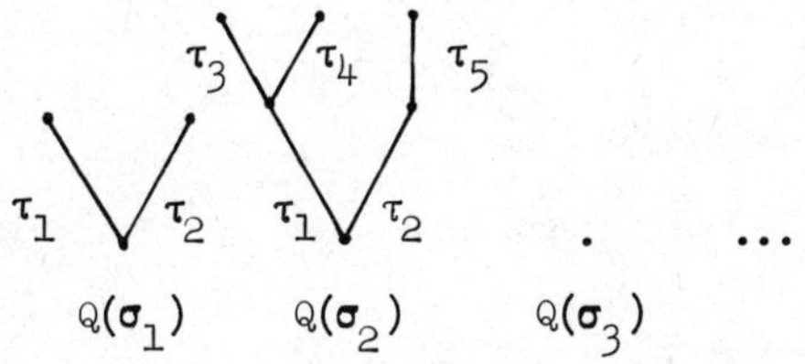

By (a) there are τ_1, τ_2 s.t. $Q(\sigma_1) * \tau_1$, $Q(\sigma_1) * \tau_2$ e-split. We reproduce τ_1, τ_2 above $Q(\sigma_2)$. By (a) again there are τ_3, τ_4 s.t. $Q(\sigma_2) * \tau_1 * \tau_3$,

$Q(\sigma_2) * \tau_1 * \tau_4$ e-split, say on x. By (b) there is τ_5 s.t. $\{e\}^{Q(\sigma_2)*\tau_2*\tau_5}(x)\downarrow$, so for $i = 3$ or $i = 4$ $Q(\sigma_2) * \tau_1 * \tau_i$, $Q(\sigma_2) * \tau_2 * \tau_5$ e-split on x. We reproduce $\tau_1 * \tau_i$, $\tau_2 * \tau_5$ above $Q(\sigma_3)$ etc. This way we get two big strings τ, τ' s.t. for each i $Q(\sigma_i) * \tau$, $Q(\sigma_i) * \tau'$ e-split. By extending one of them we may actually get two such strings of the same length. □

We noted in Section 1 that minimal degrees cannot be obtained by finite extensions. Spector tried, without success, to build them by coinfinite recursive extensions: this method is the next natural step after the finite extensions one. It uses pairs (R,S) of disjoint recursive sets s.t. $R \cup S$ is coinfinite (to leave enough room for the other requirements to be satisfied). A is on (R,S) if $R \subseteq A$ and $S \subseteq \overline{A}$. A construction consists of the definition of $\{(R_n,S_n)\}_{n\in\omega}$ as above and s.t. $R_n \subseteq R_{n+1}$, $S_n \subseteq S_{n+1}$. In the end we will have $\bigcup_{n\in\omega} R_n = \overline{\bigcup_{n\in\omega} S_n}$, so that exactly one set is determined.

Spector predicted that a construction of a minimal degree via coinfinite recursive extensions would be a great simplification of his own construction via recursive trees. Sacks [1963], page 162 conjectured the impossibility of such a construction. Lachlan [1971] proved both of them wrong: minimal degrees can be constructed by coinfinite recursive extensions, and the construction is much more difficult than Spector's original one. This is perhaps not surprising, since we are trying to obtain the same result by means of less powerful methods, so that more ingenuity is required.

Definition. T is a 1-tree if it is uniform and $T(\sigma * 0)$, $T(\sigma * 1)$ are adjacent, i.e. they differ on only one argument.

In terms of admissible triples, condition 2 becomes:

$$(2') \quad \forall n\, \exists!\, x\, (g(n) \leq x < g(n+1) \wedge f_L(x) \neq f_R(x)) \, .$$

In this case we call g, f_L, f_R an isolating triple. It's enough to build minimal degrees by 1-trees to solve Spector's problem, since given an isolating triple the corresponding pair (R,S) is defined as:

if $f_L(x) \neq f_R(x)$ then $x \in \overline{R \cup S}$

if $f_L(x) = f_R(x) = 1$ then $x \in R$

if $f_L(x) = f_R(x) = 0$ then $x \in S$.

Note that Lemma 3.1 holds for 1-trees, since by definition $Q(\tau_i * 0)$ and $Q(\tau_i * 1)$ in its proof only differ on their last component. The analogue of 3.2 is now the following:

Lemma 3.3. Given T 1-tree and given e, if

(a) every $\sigma \in T$ has e-splitting extensions on T

(b) $\forall \sigma \in T\, \forall x\, \exists \tau \in T\, (\tau \supseteq \sigma \wedge \{e\}^{\tau}(x)\downarrow)$

(c) T does not have 1-subtrees without e-splittings

then T has an e-splitting 1-subtree Q (Lachlan).

Since (b) holds, we may suppose that

$$\forall n \forall \sigma \ (|\sigma| = n \Rightarrow \{e\}^{T(\sigma)}(n)\downarrow)$$

(otherwise we first apply 3.1). As in 3.2, we proceed by induction, showing the first two steps of the construction of the level n of Q.

Given σ, by (a) there are τ, τ' extensions of it on T which are e-splitting, say on x. We may suppose that τ, τ' have the same length (otherwise extend the shorter) and that for all strings μ of that length, $\{e\}^{\mu}(x)\downarrow$ (by the initial observation). Since T is a 1-tree, there is a sequence $\tau_0 \ldots \tau_i$ of strings on it of the same length, each adjacent to the following in the list and s.t. $\tau_0 = \tau$, $\tau_i = \tau'$. Since for all $j \leq i$ $\{e\}^{\tau_j}(x)\downarrow$ and $\{e\}^{\tau_0}(x) \neq \{e\}^{\tau_i}(x)$, two of these strings e-split on x. Hence each $\sigma \in T$ has e-splitting adjacent extensions.

We attack now the crucial case of two strings σ_1, σ_2 (after which we can proceed by induction).

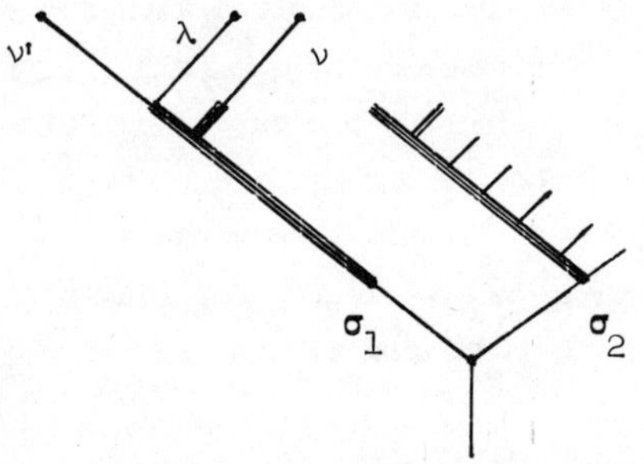

We first build a 1-subtree of T above σ_2 this way: we take a pair of e-splitting adjacent extensions of it, and this is the first level of the 1-subtree; then we consider the leftmost branch, take a pair of e-splitting adjacent extensions of it and reproduce it on the rightmost one, and this is the second level; we go on by considering the leftmost branch of each level, finding an e-splitting adjacent extension and reproducing it on every node of the same level.

Then we take this subtree and reproduce it brutally above σ_1. By (c) there must be an e-splitting on this 1-subtree. By methods we know we can choose an e-splitting ν, ν' s.t. $|\nu| = |\nu'|$ and

- ν' is on the leftmost branch of the 1-subtree (given any e-splitting, say on x, it's enough to wait until a big enough segment ν' on the leftmost branch is s.t. $\{e\}^{\nu'}(x)\downarrow$: then one of the two original branches and ν' are e-splitting)
- ν goes right on the tree as late as possible (i.e. the common part of ν, ν' is maximal)
- if ν, ν' e-split on x and $\mu \in T$, then

$$|\mu| = |\nu| = |\nu'| \Rightarrow \{e\}^{\mu}(x)\downarrow$$

Now we take λ as ν, only λ goes right one level after ν does: λ, ν are obviously adjacent and $\{e\}^{\lambda}(x) = \{e\}^{\nu'}(x)$ by the choice of ν (the common part of ν, ν' is maximal). Then $\{e\}^{\lambda}(x) \not\simeq \{e\}^{\nu}(x)$ and λ, ν are an adjacent e-splitting above σ_1. And by definition they are also an (adjacent) e-splitting above σ_2 (since they extend the boldface e-splitting in the picture). □

<u>Theorem</u> 3.4. <u>It is possible to build a minimal degree below</u> $\underset{\sim}{0}''$ <u>by 1-trees (Lachlan).</u>

Using 3.3 we can prove the analogue of 2.5. One little point needs attention: when (a) and (b) have been verified, using two-quantifier questions, we can't brutally go on and ask if T has a 1-subtree without e-splittings (if we want to remain below $\underset{\sim}{0}''$), since the question $(\exists i)(\{i\}$ is total $\wedge \ldots)$ will be Σ^0_3. But we can ask if the inductive process of building the e-splitting 1-subtree of the proof of 3.3 does terminate or not. If yes then we can take the e-splitting 1-subtree. If not then we know that there is a 1-subtree of T without e-splitting, and we search for (an index of) it. When we find it, we let it be our Q (and for all A on it, $\{e\}^A$ is recursive). □

The use of 1-trees gives more information:

<u>Theorem</u> 3.5. <u>If</u> A <u>has minimal degree and it is constructed by</u> 1-<u>trees, it also has minimal</u> m-<u>degree (Lachlan)</u>.

It's enough to prove that if T is an e-splitting 1-tree, A is on T and $\{e\}^A$ is an m-reduction then $A \leq_m \{e\}^A$. To say that $\{e\}^A$ is an m-reduction means that for some recursive h, $B = \{e\}^A = h^{-1}(A) : x \in B \Leftrightarrow h(x) \in A$. We want to prove $A \leq_m B$. Let $a \in B \wedge b \in \overline{B}$ (if they do not exist, then B is recursive). We will define f s.t. $x \in A \Leftrightarrow f(x) \in B$. Consider T as an isolating triple:

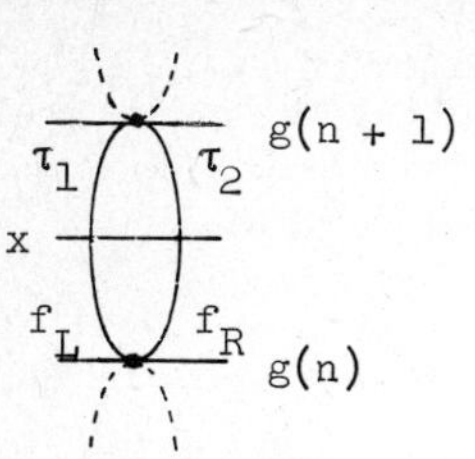

if $f_L(x) = f_R(x) = 1$ then $x \in A$ since it is on every branch of T. Let $f(x) = a$.

if $f_L(x) = f_R(x) = 0$ then $f(x) = b$.

If $f_L(x) = f_R(x)$ then $x \in$ range h. Indeed, since T is e-splitting, for some z is $\{e\}^{\sigma * \tau_1}(z) \not\simeq \{e\}^{\sigma * \tau_2}(z)$ where σ is any string on T of length g(n) and τ_1, τ_2 are f_L, f_R between g(n) and g(n + 1). Now $\sigma * \tau_1$, $\sigma * \tau_2$ differ only on x since T is a 1-tree, and by hypothesis $\{e\}^{\sigma * \tau_i} = h^{-1}(\sigma * \tau_i)$: hence $x \in$ range h (otherwise the inverse images would be the same). Then let w be s.t. $x = h(w)$ and define $f(x) = w$, so that

$$x \in A \Leftrightarrow h(w) \in A \Leftrightarrow w \in B \Leftrightarrow f(x) \in B \ . \qquad \square$$

Perhaps the next application is the most important one for the method of 1-trees:

<u>Theorem 3.6. Given</u> $\underset{\sim}{a}$ <u>there are</u> $\underset{\sim}{m}_1, \underset{\sim}{m}_2, \underset{\sim}{m}_3, \underset{\sim}{m}_4$ <u>minimal degrees s.t.</u> $\underset{\sim}{a} = (\underset{\sim}{m}_1 \cup \underset{\sim}{m}_2) \cap (\underset{\sim}{m}_3 \cup \underset{\sim}{m}_4)$:

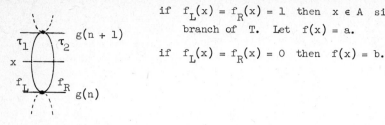

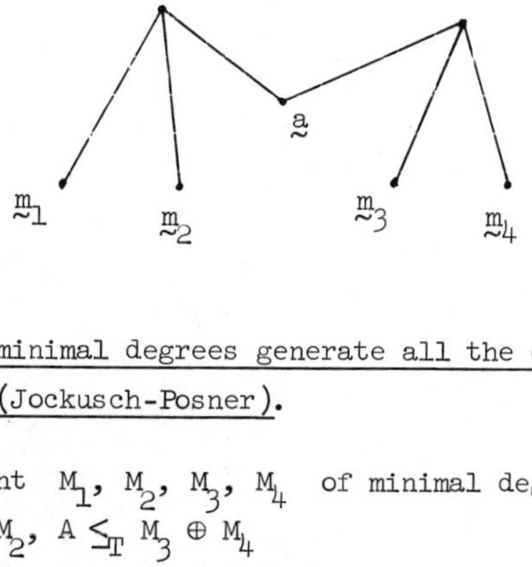

<u>In particular, the minimal degrees generate all the degrees and hence are an automorphism basis (Jockusch-Posner).</u>

Given A we want M_1, M_2, M_3, M_4 of minimal degree s.t.

(1) $A \leq_T M_1 \oplus M_2$, $A \leq_T M_3 \oplus M_4$

(2) $C \leq_T M_1 \oplus M_2, M_3 \oplus M_4 \Rightarrow C \leq_T A$.

We give the strategies for the separate requirements. To have e.g. $A \leq_T M_1 \oplus M_2$ we give in advance the reduction procedure. Let P be a recursive partition of $[\omega]^2$ without recursive homogeneous sets (1.4). We want $M_1 \cap M_2$ infinite and s.t. if $M_1 \cap M_2 = \{a_0 < a_1 < \dots\}$ then $n \in A \Leftrightarrow \{a_{2n}, a_{2n+1}\} \in P$. To build M_1, M_2 of minimal degrees we use coinfinite recursive extensions with these inductive provisos: if (R_i, S_i) are conditions for M_i,

- $R_1 \cap R_2$ is finite and of even cardinality
- if $n \in R_1 \cap R_2 = \{a_0 < \dots < a_{2k-1}\}$ then for $n < k$, $n \in A \Leftrightarrow \{a_{2n}, a_{2n+1}\} \in P$
- $R_1 \cup S_1 = R_2 \cup S_2$.

We only show how to extend them so as to have the inductive hypothesis preserved. Suppose we do one step of the construction of M_1 minimal, i.e. we find $R_1' \supseteq R_1$, $S_1' \supseteq S_1$, $R_1' \cap S_1' = \emptyset$ and $R_1' \cup S_1'$ coinfinite. Then we impose

$$R_2' = R_2$$

$$S_2' = S_2 \cup (R_1' \cup S_1' - R_2)$$

so that $R_1 \cap R_2 = R_1' \cap R_2'$ and we don't have new elements to worry about.

We also want $M_1 \cap M_2$ infinite, so from time to time we will have to put something into it. Let

$$x \in Q \Leftrightarrow x \in R_1 \cup S_1 = R_2 \cup S_2 \wedge x > \max(R_1 \cap R_2)$$

Q is recursive and infinite, hence it's not homogeneous for P. Given k, for some $a,b \in Q$ is $k \in A \Leftrightarrow \{a,b\} \in P$. Let

$$R_i' = R_i \cup \{a,b\} \qquad (i = 1,2)$$

$$S_i' = S_i \ .$$

Finally, to satisfy $\{e\}^{M_1 \oplus M_2} \simeq \{i\}^{M_3 \oplus M_4}$ total $\Rightarrow$ recursive in A, we use the usual procedure: if there is a way, consistent with the previous codes, to extend (at a certain stage) a given condition and make the antecedent false, we do it. Otherwise every extension which is consistent gives the same value when defined, hence the function is recursive in A (because of: $n \in A \Leftrightarrow \{a_{2n}, a_{2n+1}\} \in P$). □

Our last application of 1-trees gives:

Theorem 3.7. There is a minimal hyperimmune-free degree not completely autoreducible (Jockusch-Paterson).

Insert in the construction of 3.4 steps like in 1.3 to make the set not autoreducible. The use of 1-trees in this construction plays a role which is the dual of the role played by doubly e-splitting trees in 2.12. They tell us that there is only one way to recover certain values, so that we can force them to do what we want. □

§4. Initial segments of the degrees. In the two previous sections we considered the problem of constructing a minimal degree in many different ways.

In this section we go in a different direction, touching on more general initial segments.

<u>Definition.</u> A set of degrees is an <u>initial segment</u> if it is closed downward.

The set $\{\underset{\sim}{0}, \underset{\sim}{a}\}$ with $\underset{\sim}{a}$ minimal is the simplest non trivial initial segment (the two-element chain). The next step is the construction of two degrees $\underset{\sim}{a}_0, \underset{\sim}{a}_1$ s.t. $\underset{\sim}{0} < \underset{\sim}{a}_1 < \underset{\sim}{a}_2$ and no other degrees except $\underset{\sim}{0}, \underset{\sim}{a}_1$ are below $\underset{\sim}{a}_2$ (the three-element chain).

The idea is to construct a set A in such a way that a subset of A recursive in A, e.g. Od(A), does the job of the intermediate degree. The requirements are:

- Od(A) not recursive
- $A \not\leq_T \mathrm{Od}(A)$
- if $\{e\}^A$ is total then $\{e\}^A$ is recursive, or $\{e\}^A \equiv_T \mathrm{Od}(A)$ or $\{e\}^A \equiv_T A$.

These are just generalizations of the conditions for A being minimal, and in many respects so is the construction. Perhaps the most crucial parts are the idea of forcing Od(A) to be minimal and Lemma 4.2 dealing with diagonalization.

<u>Lemma 4.1.</u> <u>Given</u> T <u>and</u> e, <u>if for some</u> σ $T(\sigma * 0), T(\sigma * 1)$ <u>disagree on their odd parts, then there is</u> $Q \subseteq T$ <u>s.t. for every</u> A <u>on</u> Q, $\mathrm{Od}(A) \neq W_e$.

Since $T(\sigma * 0), T(\sigma * 1)$ disagree on their odd parts, the odd part of one of them, $T(\sigma * i)$, disagrees with W_e. Take Q as the full subtree of T above $T(\sigma * i)$. □

<u>Lemma 4.2.</u> <u>Given</u> T <u>uniform and given</u> e, <u>if for some</u> σ $T(\sigma * 0)$, $T(\sigma * 1)$ <u>agree on their odd parts, then there is</u> $Q \subseteq T$ <u>s.t. for every</u> A <u>on</u> Q, $A \not\simeq \{e\}^{\mathrm{Od}(A)}$ (<u>Titgemeyer</u>).

Take x s.t. $T(\sigma * 0)(x) \neq T(\sigma * 1)(x)$: such an x exists because $T(\sigma * 0), T(\sigma * 1)$ are incompatible and it's not on the odd part of them by hypothesis. See if

$$\exists \tau \supseteq T(\sigma * 0)(\tau \in T \wedge \{e\}^{\mathrm{Od}(\tau)}(x)\downarrow) .$$

If not, let Q = full subtree of T above $T(\sigma * 0)$: if A is on Q then $\{e\}^{\mathrm{Od}(A)}$ is not total, and hence differs from A. Otherwise take such a τ, and let τ' be such that $\tau' - T(\sigma * 1) = \tau - T(\sigma * 0)$:

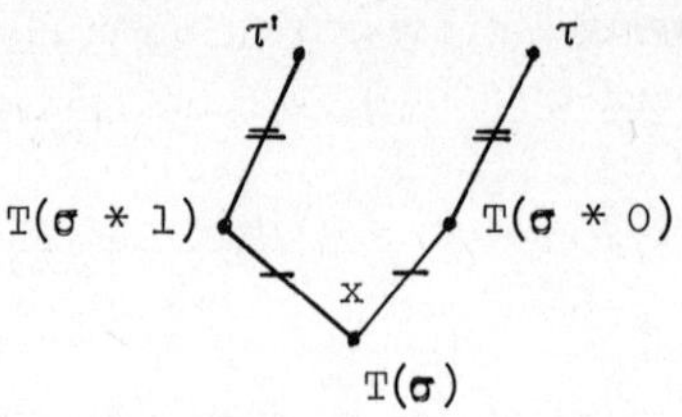

Since T is uniform, $\tau' \in T$. Now τ and τ' have the same odd parts (because they extend in the same way $T(\sigma * 0)$ and $T(\sigma * 1)$, which have the aame odd parts). Then $\{e\}^{Od(\tau)}(x) \simeq \{e\}^{Od(\tau')}(x)$ but $\tau(x) \neq \tau'(x)$ (by the choice of x at the very beginning). Hence for example $\tau(x) \not\simeq \{e\}^{Od(\tau)}(x)$ and we can let Q = the full subtree of T above τ. □

From these two lemmas we know that we can diagonalize if we use uniform trees s.t. $T(\sigma * 0)$, $T(\sigma * 1)$ agree on the odds (i.e. on their odd parts) for infinitely many σ's, and disagree on the odds for infinitely many σ's. Of course Lemma 2.4 is still valid, so we have conditions to insure $\{e\}^A$ recursive (A is on T with no e-splittings) or $A \leq_T \{e\}^A$ (A is on T e-splitting). With a similar proof we get a similar lemma that takes care of the remaining case:

Lemma 4.3. Given T, e and A on T if $\{e\}^A$ is total then:

(1) if there is no e-splitting on T which agrees on the odds, $\{e\}^A \leq_T Od(A)$

(2) if whenever $T(\sigma * 0)$, $T(\sigma * 1)$ disagree on the odds they e-split, $Od(A) \leq_T \{e\}^A$.

(Titgemeyer, Hugill).

(1) Since $\{e\}^A(x)\downarrow$, to get the right value it's enough to search for σ on T s.t. $\{e\}^\sigma(x)\downarrow$ and $Od(\sigma) \subseteq Od(A)$, and this is recursive in $Od(A)$.

(2) Given $T(\sigma)$ s.t. $Od(T(\sigma)) \subseteq Od(A)$, either $T(\sigma * 0)$, $T(\sigma * 1)$ agree on the odds and then both work, or they disagree on the odds and then they e-split, say on x. Then if $\{e\}^{T(\sigma * i)}(x) \simeq \{e\}^A(x)$, $Od(T(\sigma * i)) \subseteq Od(A)$. □

We now have two possibilities to get an analogue of 2.5.

Lemma 4.4. Given T and e, there is $Q \subseteq T$ s.t. one of the following holds:

(1) for every A on Q, $\{e\}^A$ is not total

(2) for every A on Q, $\{e\}^A$ is recursive in $Od(A)$

(3) for every A on Q, $A \leq_T \{e\}^A$

(Titgemeyer).

Similar to 2.5. First apply 2.2 and get $T' \subseteq T$ s.t. either for every A on T' $\{e\}^A$ is not total (and then let $Q = T'$) or for every A on T', $\{e\}^A$ is total. Then ask if there is $\sigma \in T'$ with no e-splitting above it agreeing on the odds. If yes, take Q = the full subtree of T' above σ. ((2) is satisfied by 4.3.1). Otherwise we can always find e-splittings (agreeing on the odds, but this doesn't matter) and hence let Q = the e-splitting subtree of T'. □

<u>Lemma</u> 4.5. <u>Given T and e, there is $Q \subseteq T$ s.t. one of the following holds:</u>

(1) <u>for every A on Q, $\{e\}^A$ is not total</u>

(2) <u>for every A on Q, $\{e\}^A$ is recursive</u>

(3) <u>for every A on Q, $\{e\}^A \equiv_T Od(A)$</u>

(4) <u>for every A on Q, $A \leq_T \{e\}^A$</u>

<u>(Hugill)</u>.

As before, but when we are in the case in which for some σ on T' there is no e-splitting above it agreeing on the odds, we could only claim $\{e\}^A \leq_T Od(A)$ there. Here we go on and ask if above σ there are always e-splittings disagreeing on the odds. If yes then we build a subtree of T' satisfying both (1) and (2) of 4.3, i.e. alternating branches agreeing on the odds (which are not e-splitting because above σ) and e-splitting (which have to disagree on the odds). So (3) is satisfied. Otherwise, we can find a string (above σ) with no e-splittings at all above it. (Since an e-splitting has to either agree or disagree on the odds) and then we take the full subtree above it, and (2) is satisfied. □

<u>Theorem</u> 4.6. <u>The three-element chain is embeddable as initial segment of the degrees (Titgemeyer).</u>

The basic ingredients of the proof are in the lemmas above. What we have to do is to define $\{T_n\}_{n \in \omega}$, a sequence of decreasing recursive trees which are uniform and alternate branches agreeing on the odds and branches disagreeing on the odds. This last part is not difficult. The ideas to make the trees uniform (when dealing with splittings) are in 3.1 and 3.2. We preferred to prove 4.4 and 4.5 without uniformities, so as to let the new ideas emerge. (See Epstein [1979] for a complete proof.) A last word is needed: if we choose to follow 4.5 in the construction, that is enough. If instead we follow 4.4 then we have less work to do, but we need to add additional steps to insure that Od(A) has minimal degree. □

<u>Theorem</u> 4.7. <u>There are completely autoreducible hyperimmune-free degrees</u>

which are not minimal (Jockusch-Paterson).

Use the ideas of 2.12 to get, in the construction above, double splittings. Then the degrees obtained are completely autoreducible and hyperimmune-free, but the top one is not minimal. □

Simple modifications of the procedure to get 4.6 give: every finite chain is embeddable as initial segment (Titgemeyer). We will extend this later in this section. A different kind of modification gives instead:

Theorem 4.8. The diamond is embeddable as an initial segment of the degrees (Sacks).

We want two distinct degrees $\underset{\sim}{a}$, $\underset{\sim}{b}$ minimal and such that the only degrees below $\underset{\sim}{a} \cup \underset{\sim}{b}$ are $\underset{\sim}{0}$, $\underset{\sim}{a}$, $\underset{\sim}{b}$. What we do is to build A s.t. Od(A) and Ev(A) do the job of $\underset{\sim}{a}$ and $\underset{\sim}{b}$. The only modification in the proof of 4.6 is that we have to treat Od(A) and Ev(A) symmetrically, and whenever there we had branches or splittings disagreeing on the odds we now want them to actually agree on the evens. We made a trivial observation in 4.5, when we noticed that an e-splitting has to either agree or disagree on the odds. The pertinent observation is now that - on trees on which all branches either agree on the odds or agree on the evens - whenever there are e-splittings we may find them agreeing on the odds or on the evens (we go by adjacent paths as in the first part of 3.3, where now adjacent means agreeing on the odds or the evens). □

Note that Sacks [1963], page 67 has proved that if a degree is r.e. in some degree less than itself then it is the l.u.b. of the degrees less than itself. Theorem 4.8 proves that the converse of this is not true, since if a degree is r.e. in some degree less than itself then it is not minimal over it.

The ideas involved in 4.6 can be pushed further to prove that every finite distributive lattice is embeddable as an initial segment and, as a corollary, that the theory of degrees is undecidable. A key fact in the proof of the general result is that if a finite lattice is distributive then it is isomorphic to a sublattice of the power set of some finite set. It is then enough, given a finite distributive lattice, to find a partition of ω into infinite coinfinite recursive sets whose lattice structure under inclusion mirrors the given lattice (like $\emptyset$, Od, Ev, ω did for the diamond) and build the top degree of the initial segment. What is needed is some kind of representation theorem for the lattices, but no new recursion-theoretic ideas are involved. Incidentally, this is not the way Lachlan [1968] originally proved the theorem. This approach is due to Yates and has been generalized by Epstein [1979] to deal with the case of countable bottomed distributive lattices. We only go one step in this direction.

Theorem 4.9. Every recursive linear ordering is embeddable as initial segment of the degrees (Hugill).

We may restrict ourselves to infinite orderings $<_L$ of ω in which 0 is the least element and 1 is the greatest (if $<_L$ was not topped before, top it). Choose a set $\{X_i\}_{i\in\omega}$ of disjoint uniformly recursive sets s.t.

$$i \leq_L j \Rightarrow X_i \subseteq X_j$$

$$i <_L j \Rightarrow X_j - X_i \text{ infinite}$$

$$X_0 = \emptyset, \quad X_1 = \omega$$

We construct A s.t. if

$$x \in X_i(A) \Leftrightarrow \text{the } x\text{-th element (in order of magnitude) of } X_i \text{ is in } A$$

then $\{X_i(A)\}_{i\in\omega}$ is an initial segment isomorphic to L. Note that $X_0(A) = \emptyset$, $X_1(A) = A$ and $X_i(A) \leq_T A$. E.g. in 4.6 we had only $X_2 = Od$. The idea is simply to approximate L, i.e. at each stage consider the order induced by L on $\{0,\ldots,n\}$ and insure conditions on A that make $\{X_i(A)\}_{i\leq n}$ isomorphic to it. Lemmas 4.1, 4.2 and 4.3 can be rewritten with reference to any X_i instead of the odds (we already used this in 4.8). It is then enough to build uniform trees s.t. whenever, at a certain point, we have $i <_L j$, those trees have infinitely many branches agreeing on X_i but disagreeing on X_j, and infinitely many disagreeing on X_i. Note that we have to use the approach of Lemma 4.5 now, because the one of 4.4 is not enough unless the order is well-founded. Otherwise, $\{e\}^A$ could be pushed below any $X_i(A)$ $(i \neq 0)$ without being eventually recursive. The reason why dealing with only finitely many conditions at each stage is sufficient, is that every function recursive in A has infinitely many indices and for all but finitely many such our trees will have the desired properties. □

The extension of this method to embed all the countable bottomed distributive lattices is not trivial: if the lattice is not recursive, the representation of it via recursive subsets cannot be chosen ahead of time and has to be built along the way (this is true even for non-recursive linear orderings).

Theorem 4.10. There is a degree with no minimal predecessors (Martin, Hugill).

Embed the ordering $1 + \omega^*$, where ω^* is the reverse ordering of the integers. None of the non-zero degrees in this initial segment is minimal, and hence all the non-zero degrees in it lack a minimal predecessor. □

With reference to the kind of methodological questions analyzed in Section 1, it must be noted that Martin's proof of 4.10 uses the finite extension method. His result implies that the upper closure of the set of minimal degrees is meager, and hence has a non-empty complement. Thus the finite extensions method, although severely limited in power, can nonetheless give interesting results.

Theorem 4.11. (1) There is a tt-degree with no minimal predecessors.
(2) There is a m-degree with no minimal predecessors.

(1) Make the top degree in 4.10 hyperimmune-free.
(2) Use 1-trees in 4.10. □

Theorem 4.12. There is an ascending sequence of degrees with an upper bound below which there is no minimal upper bound.

Use 4.9 to embed $\omega + \omega^*$. □

We will see in Section 5 that every ascending sequence of degrees has a minimal upper bound.

The best possible result with respect to countable initial segments of the degrees has been obtained by Lachlan-Lebeuf [1976]: any countable bottomed upper-semilattice is isomorphic to an initial segment. Basic steps toward this were Lerman [1969], [1971].

§5. Recursively pointed trees. In this section our trees are not recursive anymore. We introduce a method, due to Sacks [1971], to force the paths of a tree to have degrees in the cone determined by a given degree $\underset{\sim}{a}$.

Definition. T is recursively pointed if whenever A is on T, $T \leq_T A$. I.e. T is recursive in any of its branches.

Of course recursive trees are recursively pointed.

Lemma 5.1. If T is recursively pointed, T has branches of any degree above the degree of T (Sacks).

Let $T \leq_T A$: we find a branch B of T with $B \equiv_T A$. Let $B = \bigcup_{\sigma \subseteq A} T(\sigma)$. Then:

- $B \leq_T T \oplus A$ by definition, and $T \leq_T A$ by hypothesis: $B \leq_T A$.

- $A \leq_T T \oplus B$, but B is a branch of T and $T \leq_T B$: $A \leq_T B$. □

Lemma 5.2. *If T is recursively pointed and $T \leq_T A$ then there is $Q \subseteq T$ recursively pointed s.t. $Q \equiv_T A$ (Sacks).*

We define Q by induction. Given $Q(\sigma)$ (on T) we look at the next level of T above $Q(\sigma)$, and let A decide whether to go right or left:

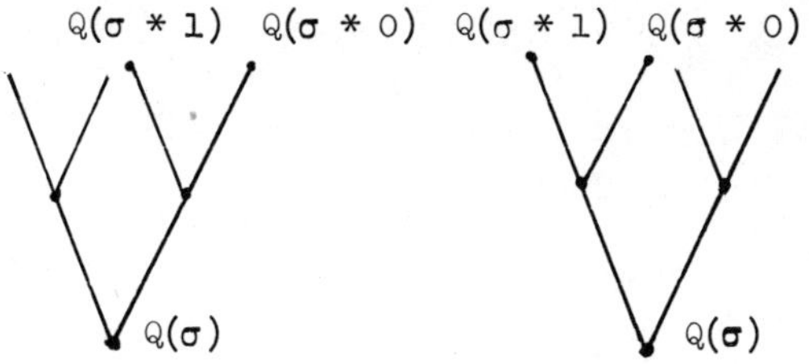

$lh\ \sigma \in A$ $\qquad$ $lh\ \sigma \notin A$

Then $Q \leq_T T \oplus A$ and, since $T \leq_T A$, $Q \leq_T A$. Note also that from any path B (of Q) and T itself we can recover A, so $A \leq_T B$ (since $T \leq_T B$ by pointedness). And to have $A \leq_T Q$ it's enough to choose as B a path recursive in Q (e.g. the leftmost branch).

Finally Q is pointed since given any $B \in Q$, by the uniformity used in the construction of Q we can actually recover Q itself from B and T. Again $T \leq_T B$ and hence $Q \leq_T B$. □

Lemma 5.3. *If T is recursively pointed and $Q \subseteq T$, $Q \leq_T T$ then Q is recursively pointed and $Q \equiv_T T$ (Sacks).*

To see that Q is pointed, let $A \in Q \subseteq T$: $Q \leq_T T \leq_T A$. If A is the leftmost branch of Q then $T \leq_T A$ by pointedness of T and $A \leq_T Q$ by definition, so $T \leq_T Q$. □

Theorem 5.4. *Every countable set of degrees has a minimal upper bound (Sacks).*

Let $\{\underset{\sim}{b}_n\}_{n\in\omega}$ be a countable set of degrees. Define

$$\underset{\sim}{a}_0 = \underset{\sim}{b}_0 \qquad \underset{\sim}{a}_{n+1} = \underset{\sim}{a}_n \cup \underset{\sim}{b}_{n+1} .$$

Then $\{\underset{\sim}{a}_n\}_{n\in\omega}$ is a countable ascending sequence of degrees and any minimal upper bound of it is also one for $\{\underset{\sim}{b}_n\}_{n\in\omega}$. We now reproduce 2.6 but with the following changes: we arrange T_n to be a recursively pointed tree with the required property and having degree $\underset{\sim}{a}_n$. This is possible because e.g. if we have T_{2e+1} recursively pointed of degree $\underset{\sim}{a}_{2e+1}$ then, since

$\underset{\sim}{a}_{2e+1} \leq \underset{\sim}{a}_{2e+2}$, we take $T \subseteq T_{2e+1}$ recursively pointed and of degree $\underset{\sim}{a}_{2e+2}$ (by 5.2). Then we apply 2.5 (see the observation after it) to get $Q = T_{2e+2}$. By 5.3, T_{2e+2} is again recursively pointed and of degree $\underset{\sim}{a}_{2e+2}$.

Now let $A \in \bigcap_{n\in\omega} T_n$. Since $A \in T_n$ and T_n is recursively pointed of degree $\underset{\sim}{a}_n$, $\underset{\sim}{a}_n \leq \deg(A)$ and A is hence an upper bound of $\{a_n\}_{n\in\omega}$. Moreover, if $\{e\}^A$ is total by the choice of T_{2e+2} (via 2.5) we either have $\{e\}^A \leq_T T_{2e+2}$ (i.e. $A \leq_T \underset{\sim}{a}_{2e+1}$) or $A \leq_T \{e\}^A \oplus T_{2e+2}$. So if $\{e\}^A$ is itself an upper bound to $\{\underset{\sim}{a}_{2n}\}_{n\in\omega}$ then $A \leq_T \{e\}^A$ and A is a minimal upper bound. □

Theorem 5.5. Every countable ascending sequence of hyperimmune-free degrees has a hyperimmune-free minimal upper bound (Martin-Miller).

Same proof as above, starting from $\{\underset{\sim}{a}_n\}_{n\in\omega}$ directly. If $\{e\}^A$ is total, since $A \in T_{2e+2}$ we have by the usual arguments (see 2.3) that $\{e\}^A$ is majorized by a function g recursive in T_{2e+2}. But $T_{2e+2} \in \underset{\sim}{a}_{2e+2}$ which is hyperimmune-free, hence g is in turn majorized by a recursive function. □

Note that we don't have any reason to believe that we can extend 5.5 in the style of 5.4 simply by passing from a set $\{\underset{\sim}{b}_n\}_{n\in\omega}$ of hyperimmune-free degrees to the associated chain $\{\underset{\sim}{a}_n\}_{n\in\omega}$. What we lack is the fact that the hyperimmune-free degrees are closed under join. We will see in Section 7 that this is actually false, hence so is the analogue of 5.4, even for finite sets of degrees.

Definition. $\underset{\sim}{0}^{\omega}$ is the degree of $A^{(\omega)}$ for A recursive, where

$$\langle x,n\rangle \in A^{(\omega)} \Leftrightarrow x \in A^{(n)} .$$

($A^{(n)}$ is the n-th jump of A).

Theorem 5.6. $\{\underset{\sim}{0}^{(n)}\}_{n\in\omega}$ has a minimal upper bound less than $\underset{\sim}{0}^{(\omega)}$ (Sacks).

Note that the least constructive step in the construction of 5.4 was the application of Lemma 2.5, which requires a two-quantifier question on the tree from which we start. So the whole construction, in the case when $\underset{\sim}{b}_n = \underset{\sim}{a}_n = \underset{\sim}{0}^{(n)}$, is such that for each n the step from T_n to T_{n+1} is recursive in $\underset{\sim}{0}^{(n+2)}$, uniformly in n. Hence the construction is recursive in $\underset{\sim}{0}^{(\omega)}$, and so is A. To have $\deg(A) <_T \underset{\sim}{0}^{(\omega)}$ it's enough to prove that $\underset{\sim}{0}^{(\omega)}$ is not itself a minimal upper bound. But using 1.1 we can easily construct two minimal upper bounds below $\underset{\sim}{0}^{(\omega)}$ which are incomparable. □

We now show that the degree $\underset{\sim}{0}^{(\omega)}$ is actually definable in the theory of degrees with jump.

<u>Definition.</u> $\underset{\sim}{a}$ is the <u>n-least upper bound</u> (n-l.u.b.) of a set A of degrees if it is the least element of $\{\underset{\sim}{x}^{(n)} : (\forall \underset{\sim}{c} \in A)\underset{\sim}{c} \leq \underset{\sim}{x}\}$.

5.6 implies in particular that $\underset{\sim}{0}^{(\omega)}$ is not the l.u.b. = 0-l.u.b. of $\{\underset{\sim}{0}^{(n)}\}_{n\in\omega}$.

<u>Theorem</u> 5.7. (1) <u>there is no 1-l.u.b. of</u> $\{\underset{\sim}{0}^{(n)}\}_{n\in\omega}$
(2) $\underset{\sim}{0}^{\omega}$ <u>is the 2-l.u.b. of</u> $\{\underset{\sim}{0}^{(n)}\}_{n\in\omega}$
<u>(Sacks)</u>.

We first recall a simple computation due to Enderton and Putnam: if $(\forall n)(\underset{\sim}{0}^{(n)} \leq \underset{\sim}{a})$ then $\underset{\sim}{0}^{(\omega)} \leq \underset{\sim}{a}^{(2)}$. We get this result if, under the hypothesis, we prove $\underset{\sim}{0}^{(n)} \leq \underset{\sim}{a}^{(2)}$ uniformly in n. This comes from the observation that since $\underset{\sim}{0}^{(n+1)} = (\underset{\sim}{0}^{(n)})'$ then for some e

$$(\forall x)(x \in 0^{(n+1)} \Leftrightarrow \{e\}^{0^{(n)}}(x)\downarrow)$$

and if we express $0^{(n)}$ recursively in $A \in \underset{\sim}{a}$ (possible by hypothesis) then we can obtain such an e recursively in $\underset{\sim}{a}^{(2)}$ (both sides are recursive in $\underset{\sim}{a}'$, and the quantifier adds one more jump). To prove (2) it will then be enough to find A s.t. for all n $\underset{\sim}{0}^{(n)} \leq_T A$, and $A^{(2)} \leq_T \underset{\sim}{0}^{(\omega)}$ (so that $A^{(2)} \equiv_T 0^{(\omega)}$ and $\underset{\sim}{0}^{(\omega)}$ is actually attained as a 2-l.u.b.). If we build $A \in \bigcap_{n\in\omega} T_n$ where T_n is recursively pointed and of degree $\underset{\sim}{0}^{(n)}$, the first condition will be automatically true. For the second we appeal to the observations on the double jump made after 2.8 to claim that actually the degree of 5.6 is already as needed. If we prefer, we can build it more easily by using 2.2 instead of all of 2.5 (since we don't need it here to be a minimal upper bound).

To prove (1) we build two such degrees as in 5.6, but we request here that their jumps be incomparable. □

We will characterize in Section 7 the ascending sequences of degrees having l.u.b. as those which are eventually constant. Similarly it has been noted by Jockusch and Simpson [1975] that the ascending sequences of degrees having 1-l.u.b. are exactly those whose jumps are eventually constant

§6. <u>Partial recursive trees</u>. We now attack the problem of constructing a minimal degree below $\underset{\sim}{0}'$. Recall the basic ingredients of 2.6: we have to diagonalize against the r.e. sets and build trees which are either e-splitting or with no e-splitting. The first step requires only one-quantifier questions and is thus recursive in $\underset{\sim}{0}'$, while the second apparently requires two-quantifier questions. The idea of the new proof is to mimic as far as possible the second step by asking only one-quantifier questions. Suppose we are given

T and e, and we want to build $Q \subseteq T$. Instead of asking if for every $\sigma \in T$ there are e-splitting extensions, we pretend it is so and we go ahead to define Q: $Q(\emptyset) = T(\emptyset)$ and $Q(\sigma * 0)$, $Q(\sigma * 1)$ = the smallest e-splitting extensions of $Q(\sigma)$. Of course it may very well be that T is not as we pretended, and that for some σ $Q(\sigma)$ doesn't have e-splitting extensions on T. This makes $Q(\sigma * i)$ $(i = 0,1)$ undefined and Q partial. If at a later stage of the construction we ask if a given string σ we are interested in (namely a beginning of the set we are building) has incompatible extensions on Q (this is a one-quantifier question) and we will find out that it doesn't, then we will know that our pretending was fake and we will change our mind about Q - taking the full subtree of T above σ. That's the whole idea.

<u>Theorem</u> 6.1. <u>There exists a minimal degre below</u> $\underset{\sim}{0}'$ <u>(Sacks)</u>.

We will build A by initial segments σ_s, so that the trees here play an auxiliary role. We have an auxiliary function $g(s)$ which at stage s tells us how many trees we continue to believe are good. T_e^s is what we think T_e is at stage s.

Step 0: $\sigma_0 = \emptyset$, $g(0) = 1$, T_0^0 = the full binary tree, T_1^0 = 0-splitting subtree of T_0^0.

Step $s + 1$: We have σ_s and $T_0^s \supseteq T_1^s \supseteq \cdots \supseteq T_{g(s)}^s$. Let i_0 be the greatest $i \leq g(s)$ s.t. σ_s has a proper extension on T_i^s (such an i exists because $T_0^s = T_0^0$ = the full binary tree). If there is an extension, there are two incompatible ones (by definition of tree). Let σ_{s+1} = a proper extension of σ_s on $T_{i_0}^s$ incompatible with W_s (to diagonalize). Also $T_0^{s+1} = T_0^s, \ldots, T_{i_0}^{s+1} = T_{i_0}^s$ (since there is no reason to change them). If $i_0 = g(s)$ then we still believe that all our trees at step s were good, so we simply add one more tree to keep the construction going: $T_{i_0+1}^{s+1}$ = i_0-splitting subtree of $T_{i_0}^{s+1}$.

If $i_0 < g(s)$ then we discovered that $T_{i_0+1}^s$ is not i_0-splitting above σ_s (because σ_s has no proper extensions on it), and we let $T_{i_0+1}^{s+1}$ = the full subtree of $T_{i_0}^{s+1}$ above σ_{s+1}. All the other T_i^s for $i > i_0 + 1$ are dropped because we built them as subtrees of $T_{i_0+1}^s$, which was a wrong guess.

In both cases, $g(s + 1) = i_0 + 1$. Note that $\lim_{s\to\infty} T_e^s = T_e$ exists because T_e^s can change at most $2^e - 1$ times (T_0^s never changes; T_1^s can only change once; T_2^s can change once for each choice of T_1^s, hence 3 times, etc.). Also, by construction, T_{e+1} is either e-splitting or has no e-splittings above σ_s if $T_{e+1}^s = T_{e+1}$. Since $A \in \bigcap_{e\in\omega} T_e$, A has minimal degree. □

An interesting fact is that if we want to know T_e, or to find s such that

$T_e^s = T_e$, we still need $\underset{\sim}{0}''$ as oracle. But A itself was built recursively in $\underset{\sim}{0}'$. We are going to investigate now the jump behavior of minimal degrees below $\underset{\sim}{0}'$, and a little result is needed.

Lemma 6.2. Given T and e, there is $Q \subseteq T$ s.t. one of the following holds:

(1) for every A on Q, $e \in A'$.

(2) for every A on Q, $e \notin A'$.

Simply see if for some $\sigma \in T$, $\{e\}^{\sigma}(e)\downarrow$. If yes take Q = full subtree above σ; if not take Q = T. □

A similar lemma allows the proof of the low basis theorem: the low degrees are a basis for Π_1^0 classes of sets (Jockusch-Soare [1972b]).

Theorem 6.3. (1) There is a minimal degree $\underset{\sim}{a} < \underset{\sim}{0}'$ s.t. $\underset{\sim}{a}' = \underset{\sim}{0}'$ (Yates).
(2) There is a minimal degree $\underset{\sim}{a} < \underset{\sim}{0}'$ s.t. $\underset{\sim}{a}' \neq \underset{\sim}{0}'$ (Sasso, Cooper, Epstein).

(1) In 6.1 insert an additional step at stage s + 1, determining if $s \in A'$ or not once and for all. This, like diagonalization, is a one-quantifier question and hence keeps the construction recursive in $\underset{\sim}{0}'$.

(2) In 6.1 insert additional guesses relative to Lemma 2.9, taking as first guess the tree which always branches left at some even level (which amounts to implicitly guessing that for no τ is $\{e\}^{\tau\oplus 0'}(a) \simeq 0$, in the notation of 2.9. If at some later stage we find an initial segment of A for which this holds, we will change our tree). □

The result of Jockusch and Posner quoted after 2.10 becomes here: if $\underset{\sim}{a} < \underset{\sim}{0}'$ is minimal, then $\underset{\sim}{a}'' = \underset{\sim}{0}''$ (since $\underset{\sim}{a} \cup \underset{\sim}{0}' = \underset{\sim}{0}'$). The characterization of the possible jumps of minimal degrees is still open, but 6.3 proves that the result on double jumps is optimal.

Theorem 6.4. There exists a completely autoreducible degree below $\underset{\sim}{0}'$ (Jockusch-Paterson).

Same technique as in 6.1, but now the first guess is the doubly e-splitting subtree. □

Of course nothing like this holds for hyperimmune-free degrees. What

---

Added in proof. The sketch of proof of 6.3.(1) is oversimplified. See Epstein [1975] for a correct argument, using the full approximation method. Lerman has recently found an oracle proof of this result.

goes wrong in the use of the analogue of 2.2 for partial trees is that now we can only have $f(n) = \max\{\{e\}^{T(\sigma)}(n) : |\sigma| = n \wedge \{e\}^{T(\sigma)}(n)\downarrow\}$ and this is not recursive. Of course it is recursive in $\underset{\sim}{0}'$ and this is not accidental, because of the observations after 2.10.

The next result is a generalization of 6.1.

Theorem 6.5. Below any high degree there is a minimal degree (Cooper).

Jockusch [1977] has found a simple proof of this. Let $\underset{\sim}{b}$ be high, i.e. $\underset{\sim}{b} \leq \underset{\sim}{0}' \wedge \underset{\sim}{b}' = \underset{\sim}{0}''$. We look at the modifications needed in 6.1 and sketch the new ideas involved. The diagonalization is not a problem: since $\underset{\sim}{b}$ is high, the recursive sets are uniformly recursive in $\underset{\sim}{b}$, and the diagonalization against the s-th recursive set can be carried out recursively in $\underset{\sim}{b}$.

We also have to look for an i s.t. σ_s has a proper extension on T_i^s. This is a question recursive in $\underset{\sim}{0}'$ and can hence be recursively approximated by the limit lemma. The trouble is that when we only look at approximations of it, we could well think that σ_s has extension on T_i^s for every s even if this is in reality false for every s. But suppose we already have A, and look at the question: has some initial segment of A extensions on T_i^s? This is a one quantifier question on $A \oplus 0'$, hence is recursive in $(A \oplus 0')'$. If $\deg(A) \leq \underset{\sim}{b}$ by construction, this is also recursive in $(\underset{\sim}{b} \cup \underset{\sim}{0}')' \leq \underset{\sim}{0}'' \leq \underset{\sim}{b}'$ ($\underset{\sim}{b}$ is high) and hence can be approximated recursively in $\underset{\sim}{b}$. So what we really do at stage s is to simultaneously search for either an extension of σ_s on $T_{i_0}^s$ or for a later stage in which our approximation tells us that no such extension exists. This avoids one problem but introduces a new one: to pick up the approximation recursive in $\underset{\sim}{b}$, we need to know A in advance! But here the magic comes in and the recursion theorem tells us that in constructing A we can actually use A itself, hence this is not a problem. □

Another step toward the characterization of the class of degrees below $\underset{\sim}{0}'$ bounding minimal degrees is the following result of Yates: below any non recursive r.e. degree there is a minimal degree (see Epstein [1975]). Lerman has recently announced various embeddings of initial segments below $\underset{\sim}{0}'$, from which follows: there is a degree below $\underset{\sim}{0}'$ with no minimal predecessors.

We are left with a methodological question: are partial trees really necessary in the constructions below $\underset{\sim}{0}'$? The next theorem gives the answer yes, for both minimal and completely autoreducible degrees. It also says that, even in the constructions of Section 2, we don't have to diagonalize since the sets constructed to satisfy the other conditions are already non recursive.

Theorem 6.6. Posner's lemma. Let $A \in \bigcap_{e\in\omega} T_e$, where each T_e is a partial recursive tree either e-splitting or with no e-splittings. Then

(1) A is not recursive

(2) if $A \leq_T 0'$ then A is the only infinite branch of some T_e.

(1) Suppose A were recursive. Then for some e

$$\{e\}^{\sigma}(x) \simeq \sigma(x) \text{ iff } \sigma \not\subseteq A .$$

Let $T = T_e$. T is not e-splitting since e.g. if $T(i) \subseteq A$ then $\{e\}^{T(i)}(x)$ is always undefined, so $T(0)$, $T(1)$ don't e-split. T has e-splittings since e.g. $T(01) \subseteq A$. Then $T(1)$, $T(00)$ are both defined and $\not\subseteq A$, hence they e-split (since are incompatible). Contradiction.

(2) Since $A \leq_T 0'$, by the limit lemma $A(x) = \lim_{s\to\infty} A_s(x)$ with A_s uniformly recursive. Let e be s.t.

$$\{e\}^{\sigma}(x) \simeq \sigma(x) \text{ iff } (\exists s)(\sigma \subseteq A_s) .$$

First note that T_e must be e-splitting, since there are e-splittings on it. Otherwise, to compute $A(x)$ recursively it's enough to search for $\sigma \in T$ s.t. $\{e\}^{\sigma}(x)\downarrow$ (which certainly exist since A is on T_e) and by (1) we would have a contradiction. Since then T_e is e-splitting, in particular $T(\sigma * 0)$, $T(\sigma * 1)$ (if defined) e-split on some argument and $\{e\}^{T(\sigma*i)}$ must be defined on that argument. By definition this implies $T(\sigma * i) \subseteq A_{s_i}$ for some s_i $(i = 0,1)$. And of course if two strings are incompatible, they must be contained in different A_s, so $s_0 \neq s_1$. Since $\lim_{s\to\infty} A_s(x)$ exists for each x, given $T(\sigma * 0)$, $T(\sigma * 1)$ one of them can only have finitely many extensions on T. □

§7. Trees of trees. We have seen in Section 1 that the set of minimal degrees is meager. We can define on $^{\omega}2$ the product measure of the measure on $\{0,1\}$ defined by $\mu(\{0\}) = \mu(\{1\}) = \frac{1}{2}$: the set of minimal degrees has then measure 0. So it is true that, in some sense, there are few minimal degrees. Nevertheless, the minimal degrees are uncountably many and, without any use of the continuum hypothesis, we prove that there is a continuum of them.

Definition. A tree of trees T is a function from St $\times$ St into St (where St is the set of strings of 0's and 1's) which is a tree on each component.

We adopt the notation $T_\sigma(\tau)$ for $T(\sigma,\tau)$. We say that A is a branch of T (or A is on T) if A is on $\lambda\sigma \cdot T_\sigma(\emptyset)$ (as a tree on σ).

Theorem 7.1. There is a continuum of minimal degrees (Lacombe).

We build a total tree of trees, each branch of which is of minimal degree.

$T_{\emptyset}$ = full binary tree.

Given T_σ, let T = full subtree of T_σ above $T_\sigma(0)$ and $T_{\sigma * 0}$ = the Q of 2.1 or 2.5 (depending on whether $lh\ \sigma = 2e$ or $lh\ \sigma = 2e + 1$). Similarly for $T_{\sigma * 1}$ (using $T_\sigma(1)$). □

Similarly we can prove that there is a continuum of hyperimmune-free degrees and there is a continuum of completely autoreducible degrees. Actually, the first fact is already implicit in 7.1. By the methods of Section 5 we can get that each countable set of degrees has a continuum of minimal upper bounds. This has important consequences:

Theorem 7.2. A countable set of degrees has a least upper bound iff it has a least upper bound which is also the join of a finite subset of it (Spector).

Let $\{\underset{\sim}{b}_n\}_{n\in\omega}$ be a countable set of degrees, and let $\{\underset{\sim}{a}_n\}$ be as in 5.4. Either the sequence $\{\underset{\sim}{a}_n\}$ is eventually constant (and then its limit is the desired l.u.b.) or there is a continuum of minimal upper bounds (and then no l.u.b.). □

It follows from this that no strictly ascending sequence of degrees has l.u.b. Another corollary is:

Theorem 7.3. There are two degrees with no greatest lower bound, so the degrees are not a lattice (Kleene-Post).

Take any strictly ascending sequence of degrees and two minimal upper bounds for it: any greatest lower bound for them should be above every element of the chain (being so an upper bound for the chain) and be below both of them, contradiction. □

Note that the last two theorems can be proved by appealing to a proof of the existence of a continuum of minimal upper bounds for any chain that used 1-trees, and hence by the coinfinite extensions method. Spector's original proof of 7.2 is much simpler than this, and uses the coinfinite extensions method directly (see Epstein [1979], page 23).

The tree of trees of 7.1 is recursive in $\underset{\sim}{0}''$, and this has the following immediate consequence:

Theorem 7.4. If $\underset{\sim}{c} \geq \underset{\sim}{0}^{(2)}$ then there is a minimal degree $\underset{\sim}{a}$ s.t. $\underset{\sim}{a}^{(2)} = \underset{\sim}{a} \cup \underset{\sim}{0}^{(2)} = \underset{\sim}{c}$ (Cooper).

Let T be the tree of trees recursive in $\underset{\sim}{0}^{(2)}$ of 7.1. Given $C \in \underset{\sim}{c}$, let $A = \cup_{\sigma\subseteq C} T_\sigma(\emptyset)$. Then A has minimal degree and $A^{(2)} \equiv_T A \oplus 0^{(2)}$ by the observation after 2.8 (there we had $A^{(2)} \leq_T 0^{(2)}$ because the tree was recursive in $0^{(2)}$; here the relevant tree is $T_\sigma(\emptyset)$ for $T_\sigma(\emptyset) \subseteq A$, hence it's recursive in $A \oplus \underset{\sim}{0}^{(2)}$). Moreover:

- $A \leq_T C \oplus T \leq_T C \oplus 0^{(2)} \leq_T C$ (because $\underset{\sim}{0}^{(2)} \leq \underset{\sim}{c}$), hence $A \oplus 0^{(2)} \leq_T C$
- $C \leq_T A \oplus T \leq_T A \oplus 0^{(2)}$. □

This is only a weak version of the much stronger and difficult result of Cooper: <u>if $\underset{\sim}{c} \geq \underset{\sim}{0}'$ then $\underset{\sim}{a}' = \underset{\sim}{a} \cup \underset{\sim}{0}' = \underset{\sim}{c}$ for some minimal degree $\underset{\sim}{a}$</u>. The proof is similar to the one above, only it uses a tree of trees recursive in $0'$. The construction of this is by full-approximation method, see Epstein [1975].

<u>Theorem</u> 7.5. <u>Given $\underset{\sim}{0}^{(2)} \leq \underset{\sim}{c}_1 \leq \underset{\sim}{c}_2$ there is an initial segment $\underset{\sim}{0} < \underset{\sim}{a}_1 < \underset{\sim}{a}_2$ s.t. $\underset{\sim}{a}_i^{(2)} = \underset{\sim}{a}_i \cup \underset{\sim}{0}^{(2)} = \underset{\sim}{c}_i$ $(i = 1,2)$ (Simpson).</u>

The idea is to use a tree of trees recursive in $0^{(2)}$, all of whose branches are top degrees of a three-element initial segment. Then if $C_2 \in \underset{\sim}{c}_2$ and $A = \cup_{\sigma\subseteq C_2} T_\sigma(\emptyset)$ we have the result for $i = 2$. To have the result for $i = 1$ as well, we build the tree T with the additional uniformity:

$$Od(B) = Od(C) \text{ iff } Od\Big(\bigcup_{\sigma\subseteq B} T_\sigma(\emptyset)\Big) = Od\Big(\bigcup_{\sigma\subseteq C} T_\sigma(\emptyset)\Big) .$$

I.e. if we follow paths with the same odd parts, we get sets with the same odd parts. This does the job because we can certainly choose $C_2 \in \underset{\sim}{c}_2$ s.t. $Od(C) \in \underset{\sim}{c}_1$. To prove now $Od(A)^{(2)} \leq_T Od(A) \oplus 0^{(2)}$ we note that, since $Od(A) \leq_T A$, for some recursive g is $\{e\}^{Od(A)} \simeq \{g(e)\}^A$ and hence $\{e\}^{Od(A)}$ total iff $\{g(e)\}^A$ total. And we don't have to know all of A: any path with the same odd part as A will do the job. □

The result is of course a generalization of 7.4.

<u>Open problem</u>: Is the same true when double jumps are replaced by jumps? Note that the analogue of 7.5 holds for chains as well, with a similar proof. It actually holds for any finite distributive lattice, although we don't think that this kind of result is particularly interesting. The special case of chains has however an important application in Simpson's way of proving his own beautiful result: <u>the theory of degrees is recursively isomorphic to second order arithmetic</u>. Another, different, proof has been devised by Nerode-Shore [198?]: this is very elegant but requires more results on initial segments (namely the embedding of all countable distributive lattices).

Simpson's method is explained in detail in Epstein [1979]. The paper of Nerode and Shore is very readable, and the necessary embeddings are again in Epstein's book.

We give now an improvement of 3.6 in a particular case:

Theorem 7.6. For every $\underset{\sim}{c} \geq \underset{\sim}{0}^2$ there are two minimal degrees $\underset{\sim}{a}$ and $\underset{\sim}{b}$ s.t. $\underset{\sim}{a} \cup \underset{\sim}{b} = \underset{\sim}{c}$.

We build a 1-tree of 1-trees recursive in 0^2, each of whose branches is of minimal degree. Then we choose two paths in the following way. Let $C \in \underset{\sim}{c}$. We define A, B inductively. Suppose we have already decided which way to go through level n. Let x be the only element on which f_L, f_R differ in $[g(n), g(n+1))$. Then we want A to follow the branch that on x agrees with $C(n)$. I.e. for $g(n) \leq z < g(n+1)$ we let:

$$A(z) = \begin{cases} f_L(z) & \text{if } f_L(x) = C(n) \\ f_R(z) & \text{otherwise .} \end{cases}$$

B follows the other branch. Then $A, B \leq_T C \oplus 0^2 \leq_T C$ (since $\underset{\sim}{0}^2 \leq \underset{\sim}{c}$), hence $A \oplus B \leq_T C$. And $C \leq_T A \oplus B$ because if

$$h(0) = \mu x\ (A(x) \neq B(x))$$

$$h(n+1) = \mu x\ (x > h(n) \wedge A(x) \neq B(x))$$

then $C(n) = A[h(n)]$. □

Cooper has proved that there are two minimal degrees $\underset{\sim}{a}$ and $\underset{\sim}{b}$ s.t. $\underset{\sim}{a} \cup \underset{\sim}{b} = \underset{\sim}{0}'$. Note that 7.6 gives as usual the same result for hyperimmune-free degrees. This was proved by Martin and Miller, and has the following consequence:

Theorem 7.7. The hyperimmune-free degrees are not closed under join (Martin-Miller).

Let $\underset{\sim}{a} \cup \underset{\sim}{b} = \underset{\sim}{0}^2$ with $\underset{\sim}{a}$, $\underset{\sim}{b}$ hyperimmune-free. $\underset{\sim}{0}^2$ is not hyperimmune-free, since it is comparable with $\underset{\sim}{0}'$. □

So far we had trees of trees for any notion of tree introduced in the paper, except for partial trees. We end with this.

Theorem 7.8. There is a minimal degree below $\underset{\sim}{0}'$ incomparable with every r.e. degree different from $\underset{\sim}{0}$, $\underset{\sim}{0}'$ (Sasso).

Let T be a tree of trees recursive in $0'$, with all the branches of minimal degree. We want to choose a branch A of T with the property that $A \leq_T W_n \Rightarrow 0' \leq_T W_n$, so that the only r.e. sets above A are those of degree $0'$. This is enough since A has minimal degree, so the only r.e. sets below it are the recursive ones. The idea to satisfy the requirements:

$$A \simeq \{e\}^{W_n} \Rightarrow 0' \leq_T W_n$$

is to either diagonalize against the premise, or to use the following fact. If f is a function majorizing almost everywhere $g(x) = \mu s \ (K[x] = K_s[x])$ (where K is a set of degree $\underset{\sim}{0}'$ and K_s is a recursive approximation to it), then $0' \leq_T f$ (in fact, to compute $K(x)$ it's enough to compute $K_{f(x+1)}(x)$ - except for finitely many x - and this is recursive in f). We define initial segments of A (recursively in $0'$). Suppose $T(\sigma) \subseteq A$ (we write $T(\sigma)$ for $T_\sigma(\emptyset)$) and attack the requirement above for e, n. We want to decide if $T(\sigma * i) \subseteq A$. Let $a = \max\{lh\ T(\sigma * i) : i = 0,1\}$ and search (recursively in $0'$) for an s such that $K[a] = K_s[a]$. Then choose $T(\sigma * i)$ disagreeing with $\{e\}_s^{W_n[a]}$ on some element on which $T(\sigma * i)$ is defined (less than a, in particular). Since $T(\sigma * 0)$, $T(\sigma * 1)$ are incompatible, one works. Now we have two cases:

(1) there is disagreement on an argument on which $\{e\}_s^{W_n[a]}$ is defined. Then $A \not\simeq \{e\}^{W_n}$.

(2) case 1 fails. Then we can still have in the end $A \simeq \{e\}^{W_n}$, but now the function

$$f(x) = \mu s \ (\{e\}_s^{W_n[s]}[x] = \{e\}^{W_n}[x])$$

is s.t. for all $lh\ T(\sigma) \leq x < a$, $g(x) \leq f(x)$.

Of course to have this just on these x's is not enough, so a little arrangement in the construction is needed to consider every condition from a certain stage on, so to take care of every x except for a finite number of them. We leave this to the reader (if we still have one at this point). □

In the proof no use is actually made of any property of the r.e. sets, except for the fact that they are uniformly recursive in $\underset{\sim}{0}'$. Hence any set of degrees uniformly recursive in $\underset{\sim}{0}'$ can be substituted for the r.e. degrees in 7.8. In particular:

<u>Theorem</u> 7.9. <u>Given any degree $\underset{\sim}{a}$ s.t. $\underset{\sim}{0} < \underset{\sim}{a} < \underset{\sim}{0}'$, there is a minimal degree below $\underset{\sim}{0}'$ incomparable with it (Shoenfield).</u>

References

S. B. Cooper [1973], Minimal degrees and the jump operator, J. Symb. Logic 38 (1973), 249-271.

R. L. Epstein [1975], Minimal degrees of unsolvability and the full approximation method, Memoirs AMS 162 (1975).

R. L. Epstein [1979], Degrees of unsolvability: Structure and theory, Lecture Notes 759, Springer (1979).

D. F. Hugill [1969], Initial segments of Turing degrees, Proc. London Math. Soc. 19 (1969), 1-16.

C. Jockusch [1969a], Relationships between reducibilities, Trans. AMS 142, (1969), 229-237.

C. Jockusch [1969b], The degrees of bi-immune sets, Zeit. Math. Logic. Grund. Math. 15 (1969), 135-140.

C. Jockusch [1972], Ramsey's theorem and recursion theory, J. Symb. Logic 37 (1972), 268-280.

C. Jockusch [1977], Simple proofs of some theorems on high degrees, Can. J. Math. 29 (1977), 1072-1080.

C. Jockusch and M. S. Paterson [1976], Completely autoreducible degrees, Zeit. Math. Logic Grund. Math. 22 (1976), 571-575.

C. Jockusch and D. Posner [1978], Double jumps of minimal degrees, J. Symb. Logic 43 (1978), 715-724.

C. Jockusch and D. Posner [198?], Automorphism bases for the degrees of unsolvability, to appear.

C. Jockusch and S. Simpson [1975], A degree-theoretic definition of the ramified analytic hierarchy, Ann. of Math. Logic 10 (1975), 1-32.

C. Jockusch and R. Soare [1972a], Degrees of members of Π_1^0 classes, Pac. J. Math. 40 (1972), 605-616.

C. Jockusch and R. Soare [1972b], Π_1^0 classes and degrees of theories, Trans. AMS 173 (1972), 33-56.

S. C. Kleene and E. L. Post [1954], The upper-semilattice of degrees of recursive unsolvability, Ann. of Math. 59 (1954), 379-407.

A. H. Lachlan [1968], Distributive initial segments of the degrees of unsolvability, Zeit. Math. Logic. Grund. Math. 14 (1968), 457-472.

A. H. Lachlan [1971], Solution to a problem of Spector, Can. J. Math. 23 (1971), 247-256.

A. H. Lachlan and R. Lebeuf [1976], Countable initial segments of the degrees of unsolvability, J. Symb. Logic 41 (1976), 289-300.

R. Ladner [1973], A completely mitotic non recursive r.e. set, Trans. AMS 184 (1973), 479-507.

M. Lerman [1969], Some non distributive lattices as initial segments of the degrees of unsolvability, J. Symb. Logic 34 (1969), 85-98.

M. Lerman [1971], Initial segments of the degrees of unsolvability, Ann. of Math. 93 (1971), 365-389.

A. B. Manaster [1971], Some contrasts between degrees and the arithmetical hierarchy, J. Symb. Logic 36 (1971), 301-304.

D. A. Martin and W. Miller [1968], The degrees of hyperimmune sets, Zeit. Math. Logic Grund. Math. 14 (1968), 159-166.

A. Nerode and R. A. Shore [198?], Second order logic and first order theories of reducibility orderings, to appear.

P. G. Odifreddi [198?], Classical recursion theory, to appear.

D. Posner and R. Epstein [1978], Diagonalization in degree constructions, J. Symb. Logic 43 (1978), 280-283.

E. L. Post [1944], Recursively enumerable sets of positive integers and their decision problems, Bull. AMS 50 (1944), 284-316.

H. Rogers, Jr. [1967], The theory of recursive functions and effective computability, McGraw-Hill, 1967.

G. E. Sacks [1961], A minimal degree less than $0'$, Bull. AMS 67 (1961), 416-419.

G. E. Sacks [1963], Degrees of unsolvability, Ann. Math. Studies n.55, Princeton, 1963.

G. E. Sacks [1971], Forcing with perfect closed sets, Proc. Symp. Pure Math. 13 (1971), 331-355.

L. Sasso [1970], A cornucopia of minimal degrees, J. Symb. Logic. 35 (1970), 383-388.

L. Sasso [1974], A minimal degree not realizing the least possible jump, J. Symb. Logic 39 (1974), 571-573.

J. R. Shoenfield [1966], A theorem on minimal degrees, J. Symb. Logic 31 (1966), 539-544.

E. Specker [1971], Ramsey's theorem does not hold in recursive set theory, Logic Colloquium '69, North Holland 1971, 439-442.

C. Spector [1956], On degrees of recursive unsolvability, Ann. of Math. 65 (1956), 581-592.

D. Titgemeyer [1965], Untersuchungen über die struktur des Kleene-Postchen Halbverbandes der Grade der rekursiven Unlösbarkeit, Arch. Math. Logic Grund. 8 (1965), 45-62.

Trathenbrot [1970], On autoreducibility, Sov. Math. Dokl. 11 (1970), 814-817.

APPENDIX: PROGRESS REPORT ON THE VICTORIA DELFINO PROBLEMS

Since the Delfino problems were proposed in [1] two and a half years ago, one has been solved and there have been some interesting results related to three of the remaining four problems.

In connection with the first problem, Martin has established the conjectured lower bound for $\underset{\sim}{\delta}^1_5$ by proving (from AD + DC) that

$$\underset{\sim}{\delta}^1_5 \geq \aleph_{\omega^3+1} \; ;$$

moreover Martin showed (from AD) that the ultrapowers of $\underset{\sim}{\delta}^1_3 = \aleph_{\omega+1}$ under the three normal measures on $\underset{\sim}{\delta}^1_3$ are exactly $\underset{\sim}{\delta}^1_4 = \aleph_{\omega+2}$ (this was known to Kunen), $\aleph_{\omega\cdot 2+1}$ and $\aleph_{\omega^2+1}$ and that these three cardinals are measurable (and hence regular), so that (in particular), $\underset{\sim}{\delta}^1_5$ is not the first regular cardinal after $\underset{\sim}{\delta}^1_4$. We still have no upper bounds for $\underset{\sim}{\delta}^1_5$ from AD.

The second problem was solved by Moschovakis who showed (from AD + DC) that every coinductive pointset admits a scale. If we put

$$\Sigma^*_0 = \text{all Boolean combinations of inductive and coinductive sets}$$

and then define Σ^*_n by counting quantifiers over $\mathcal{R} = {}^\omega\omega$ in front of a Σ^*_0 matrix in the usual way, then the proof shows that every coinductive set admits a scale $\{\varphi_n\}_{n\in\omega}$, where each φ_n is a Σ^*_{n+1}-norm, uniformly in n.

Martin and Steel extended the method used by Moschovakis in this proof and showed that

$$ZF + DC + AD + V = L(R) \Rightarrow \text{Every } \underset{\sim}{\Sigma}^2_1 \text{ set admits a } \underset{\sim}{\Sigma}^2_1\text{-scale} \; ;$$

this combines with an earlier result of Kechris and Solovay to show that

$$ZF + DC + AD + V = L(R)$$
$$\Rightarrow \text{A pointset admits a scale if and only if it is } \underset{\sim}{\Sigma}^2_1 .$$

Martin then combined these ideas with the technique of the Third Periodicity Theorem ([2], 6E.1) and showed that under reasonable hypotheses of determinacy for games on $\mathcal{R}$, (AD_R), the scale property is preserved by the game quantifier $\mathfrak{D}^2$ on $\mathcal{R}$, where

$$(\mathfrak{D}^2\alpha)P(x,\alpha) \Leftrightarrow (\exists\alpha_0)(\forall\alpha_1)(\exists\alpha_2)(\forall\alpha_3) \ldots P(x, \langle\alpha_0,\alpha_1,\ldots\rangle) .$$

This result produces scales for sets that are not $\underset{\sim}{\Sigma}_1^2$ in $L(R)$ and leaves open the general question of the extent of scales in the presence of axioms stronger than AD.

In connection with the third problem, Kechris showed in [3] that if $T^3 = T^3(\overline{\varphi})$ is the tree associated with some Π_3^1-scale $\overline{\varphi}$ on a Π_3^1-complete set P and if

$$\tilde{L}[T^3] = \cup_{\alpha \in \eta} L[T^3, \alpha] ,$$

then

$$ZF + AD + DC + \underset{\sim}{\delta}_3^1 \to (\underset{\sim}{\delta}_3^1)^{\underset{\sim}{\delta}_3^1}$$

$$\Rightarrow \tilde{L}[T^3] \text{ is independent of the choice of } P, \overline{\varphi} .$$

This partial result emphasizes the importance of the question of <u>the strong partition property</u> for $\underset{\sim}{\delta}_3^1$ which is still open.

Finally, in connection with the fifth problem, it follows from unpublished results of Kechris and Solovay that

$$ZF + AD + DC + V = L(R)$$

$$\Rightarrow \text{Every function } f : D \to D \text{ on the degrees is representable .}$$

Although this has no direct bearing on a possible solution of the fifth problem, it underscores the generality of the question.

References

[1] APPENDIX: The Victoria Delfino problems, in Cabal Seminar 76-77, Lecture Notes in Mathematics #689, Springer 1978.

[2] Y. N. Moschovakis, Descriptive Set Theory, Studies in Logic, North Holland 1980.

[3] A. S. Kechris, Homogeneous trees and projective scales, this volume.

Vol. 729: Ergodic Theory. Proceedings, 1978. Edited by M. Denker and K. Jacobs. XII, 209 pages. 1979.

Vol. 730: Functional Differential Equations and Approximation of Fixed Points. Proceedings, 1978. Edited by H.-O. Peitgen and H.-O. Walther. XV, 503 pages. 1979.

Vol. 731: Y. Nakagami and M. Takesaki, Duality for Crossed Products of von Neumann Algebras. IX, 139 pages. 1979.

Vol. 732: Algebraic Geometry. Proceedings, 1978. Edited by K. Lønsted. IV, 658 pages. 1979.

Vol. 733: F. Bloom, Modern Differential Geometric Techniques in the Theory of Continuous Distributions of Dislocations. XII, 206 pages. 1979.

Vol. 734: Ring Theory, Waterloo, 1978. Proceedings, 1978. Edited by D. Handelman and J. Lawrence. XI, 352 pages. 1979.

Vol. 735: B. Aupetit, Propriétés Spectrales des Algèbres de Banach. XII, 192 pages. 1979.

Vol. 736: E. Behrends, M-Structure and the Banach-Stone Theorem. X, 217 pages. 1979.

Vol. 737: Volterra Equations. Proceedings 1978. Edited by S.-O. Londen and O. J. Staffans. VIII, 314 pages. 1979.

Vol. 738: P. E. Conner, Differentiable Periodic Maps. 2nd edition, IV, 181 pages. 1979.

Vol. 739: Analyse Harmonique sur les Groupes de Lie II. Proceedings, 1976–78. Edited by P. Eymard et al. VI, 646 pages. 1979.

Vol. 740: Séminaire d'Algèbre Paul Dubreil. Proceedings, 1977–78. Edited by M.-P. Malliavin. V, 456 pages. 1979.

Vol. 741: Algebraic Topology, Waterloo 1978. Proceedings. Edited by P. Hoffman and V. Snaith. XI, 655 pages. 1979.

Vol. 742: K. Clancey, Seminormal Operators. VII, 125 pages. 1979.

Vol. 743: Romanian-Finnish Seminar on Complex Analysis. Proceedings, 1976. Edited by C. Andreian Cazacu et al. XVI, 713 pages. 1979.

Vol. 744: I. Reiner and K. W. Roggenkamp, Integral Representations. VIII, 275 pages. 1979.

Vol. 745: D. K. Haley, Equational Compactness in Rings. III, 167 pages. 1979.

Vol. 746: P. Hoffman, τ-Rings and Wreath Product Representations. V, 148 pages. 1979.

Vol. 747: Complex Analysis, Joensuu 1978. Proceedings, 1978. Edited by I. Laine, O. Lehto and T. Sorvali. XV, 450 pages. 1979.

Vol. 748: Combinatorial Mathematics VI. Proceedings, 1978. Edited by A. F. Horadam and W. D. Wallis. IX, 206 pages. 1979.

Vol. 749: V. Girault and P.-A. Raviart, Finite Element Approximation of the Navier-Stokes Equations. VII, 200 pages. 1979.

Vol. 750: J. C. Jantzen, Moduln mit einem höchsten Gewicht. III, 195 Seiten. 1979.

Vol. 751: Number Theory, Carbondale 1979. Proceedings. Edited by M. B. Nathanson. V, 342 pages. 1979.

Vol. 752: M. Barr, *-Autonomous Categories. VI, 140 pages. 1979.

Vol. 753: Applications of Sheaves. Proceedings, 1977. Edited by M. Fourman, C. Mulvey and D. Scott. XIV, 779 pages. 1979.

Vol. 754: O. A. Laudal, Formal Moduli of Algebraic Structures. III, 161 pages. 1979.

Vol. 755: Global Analysis. Proceedings, 1978. Edited by M. Grmela and J. E. Marsden. VII, 377 pages. 1979.

Vol. 756: H. O. Cordes, Elliptic Pseudo-Differential Operators – An Abstract Theory. IX, 331 pages. 1979.

Vol. 757: Smoothing Techniques for Curve Estimation. Proceedings, 1979. Edited by Th. Gasser and M. Rosenblatt. V, 245 pages. 1979.

Vol. 758: C. Năstăsescu and F. Van Oystaeyen; Graded and Filtered Rings and Modules. X, 148 pages. 1979.

Vol. 759: R. L. Epstein, Degrees of Unsolvability: Structure and Theory. XIV, 216 pages. 1979.

Vol. 760: H.-O. Georgii, Canonical Gibbs Measures. VIII, 190 pages. 1979.

Vol. 761: K. Johannson, Homotopy Equivalences of 3-Manifolds with Boundaries. 2, 303 pages. 1979.

Vol. 762: D. H. Sattinger, Group Theoretic Methods in Bifurcation Theory. V, 241 pages. 1979.

Vol. 763: Algebraic Topology, Aarhus 1978. Proceedings, 1978. Edited by J. L. Dupont and H. Madsen. VI, 695 pages. 1979.

Vol. 764: B. Srinivasan, Representations of Finite Chevalley Groups. XI, 177 pages. 1979.

Vol. 765: Padé Approximation and its Applications. Proceedings, 1979. Edited by L. Wuytack. VI, 392 pages. 1979.

Vol. 766: T. tom Dieck, Transformation Groups and Representation Theory. VIII, 309 pages. 1979.

Vol. 767: M. Namba, Families of Meromorphic Functions on Compact Riemann Surfaces. XII, 284 pages. 1979.

Vol. 768: R. S. Doran and J. Wichmann, Approximate Identities and Factorization in Banach Modules. X, 305 pages. 1979.

Vol. 769: J. Flum, M. Ziegler, Topological Model Theory. X, 151 pages. 1980.

Vol. 770: Séminaire Bourbaki vol. 1978/79 Exposés 525–542. IV, 341 pages. 1980.

Vol. 771: Approximation Methods for Navier-Stokes Problems. Proceedings, 1979. Edited by R. Rautmann. XVI, 581 pages. 1980.

Vol. 772: J. P. Levine, Algebraic Structure of Knot Modules. XI, 104 pages. 1980.

Vol. 773: Numerical Analysis. Proceedings, 1979. Edited by G. A. Watson. X, 184 pages. 1980.

Vol. 774: R. Azencott, Y. Guivarc'h, R. F. Gundy, Ecole d'Eté de Probabilités de Saint-Flour VIII-1978. Edited by P. L. Hennequin. XIII, 334 pages. 1980.

Vol. 775: Geometric Methods in Mathematical Physics. Proceedings, 1979. Edited by G. Kaiser and J. E. Marsden. VII, 257 pages. 1980.

Vol. 776: B. Gross, Arithmetic on Elliptic Curves with Complex Multiplication. V, 95 pages. 1980.

Vol. 777: Séminaire sur les Singularités des Surfaces. Proceedings, 1976-1977. Edited by M. Demazure, H. Pinkham and B. Teissier. IX, 339 pages. 1980.

Vol. 778: SK1 von Schiefkörpern. Proceedings, 1976. Edited by P. Draxl and M. Kneser. II, 124 pages. 1980.

Vol. 779: Euclidean Harmonic Analysis. Proceedings, 1979. Edited by J. J. Benedetto. III, 177 pages. 1980.

Vol. 780: L. Schwartz, Semi-Martingales sur des Variétés, et Martingales Conformes sur des Variétés Analytiques Complexes. XV, 132 pages. 1980.

Vol. 781: Harmonic Analysis Iraklion 1978. Proceedings 1978. Edited by N. Petridis, S. K. Pichorides and N. Varopoulos. V, 213 pages. 1980.

Vol. 782: Bifurcation and Nonlinear Eigenvalue Problems. Proceedings, 1978. Edited by C. Bardos, J. M. Lasry and M. Schatzman. VIII, 296 pages. 1980.

Vol. 783: A. Dinghas, Wertverteilung meromorpher Funktionen in ein- und mehrfach zusammenhängenden Gebieten. Edited by R. Nevanlinna and C. Andreian Cazacu. XIII, 145 pages. 1980.

Vol. 784: Séminaire de Probabilités XIV. Proceedings, 1978/79. Edited by J. Azéma and M. Yor. VIII, 546 pages. 1980.

Vol. 785: W. M. Schmidt, Diophantine Approximation. X, 299 pages. 1980.

Vol. 786: I. J. Maddox, Infinite Matrices of Operators. V, 122 pages. 1980.